**Irritant Dermatitis**

# Current Problems in Dermatology

## Vol. 23

Series Editor     *G. Burg,* Zurich

Basel · Freiburg · Paris · London · New York ·
New Delhi · Bangkok · Singapore · Tokyo · Sydney

..........................

# Irritant Dermatitis

## New Clinical and Experimental Aspects

Volume Editors

*P. Elsner,* Zurich
*H.I. Maibach,* San Francisco, Calif.

123 figures, 22 in color, and 60 tables, 1995

Basel · Freiburg · Paris · London · New York ·
New Delhi · Bangkok · Singapore · Tokyo · Sydney

.............................

# Current Problems in Dermatology

Library of Congress Cataloging-in-Publication Data
Irritant dermatitis: new clinical and experimental aspects /
volume editors, P. Elsner, H.I. Maibach.
(Current problems in dermatology; vol. 23)
Includes bibliographical references and index.  (alk. paper)
1. Contact dermatitis.   I. Elsner, Peter, 1955–.   II. Maibach, Howard I.   III. Series.
[DNLM:  1. Dermatitis, Irritant.   2. Dermatitis, Occupational.    3. Irritants.
W1 CU804L v. 23 1995 / WR 175 I715 1995]
RL244.I77   1995   616.5–dc20
ISBN 3–8055–6083–4

© Copyright 1995 by S. Karger AG, P.O. Box, CH–4009 Basel (Switzerland)
Printed in Switzerland on acid-free paper by Thür AG Offsetdruck, Pratteln
ISBN 3–8055–6083–4

........................

# Contents

## Pathogenesis

## Bioengineering Techniques

## Prevention and Therapy

## Product Testing and Regulatory Aspects

............................
# Preface

Although irritant contact dermatitis (ICD) is more frequent than allergic contact dermatitis, until now research on ICD had been relatively neglected, partly because experimental technology was insufficient, and perhaps because ICD seemed more trivial than the more intellectually appealing problem of allergic sensitization.

This perception has recently changed dramatically. With the advent of bioengineering technology, subtle changes of epidermal barrier perturbation became measurable. New electron-microscopic and biochemical techniques provided exciting insights into the secrets of barrier lipids. Most recently, immunological research has shown the importance of immunological events in ICD – formerly considered a 'non-immunological condition'.

This book provides an overview on many aspects of ICD: clinical features, epidemiology, ultrastructure and immunology, prevention and therapy, and the important problems of in vivo and in vitro predictive testing increasingly required by regulating agencies. We hope that it will be useful to a wide readership interested in ICD: clinical dermatologists and allergologists, occupational physicians, toxicologists, pharmacologists, cosmetologists and bioengineers.

We thank all the contributors for their stimulating cooperation and Ms. V. Steinemann for her skillful secretarial assistance.

*Peter Elsner,* MD
*Howard I. Maibach,* MD

Elsner P, Maibach HI (eds): Irritant Dermatitis. New Clinical and Experimental Aspects.
Curr Probl Dermatol. Basel, Karger, 1995, vol 23, pp 1–8

..........................

# Mechanisms of Skin Irritation

*Enzo Berardesca, Fernanda Distante*

Department of Dermatology, University of Pavia, IRCCS Policlinico S. Matteo,
Pavia, Italy

The physical, biochemical or cellular mechanisms which determine the outbreak of irritant dermatitis are far from being completely understood. Irritant contact dermatitis (ICD) may be characterized by several clinical aspects [1], corresponding to different pathological skin responses and behaviors, and the term irritant dermatitis should be converted into irritant dermatitis syndrome. Indeed, the more common clinical aspects of ICD syndrome are not just one but at least seven and include (table 1): acute ICD (when exposure is sufficient and the molecule is potent enough); delayed acute ICD (inflammation occurs 12–24 h after a single exposure and occurs only with certain very specific chemicals like anthralin); cumulative ICD (where the frequency of exposure is too high in relation to the skin recovery time).

Other specific clinical conditions of ICD are traumatic ICD induced by repeated trauma and pustular and acneiform irritant dermatitis which develops after exposure to some metals, tars, oils and greases. Furthermore, other recently described aspects of irritation include nonvisible clinical conditions such as the nonerythematous irritation characterized only by changes in the function of the stratum corneum and subjective irritation in which the affected patients report itching and stinging sensations without signs of erythema after exposure to chemicals like lactic acid. It is clear that different mechanisms may lie behind all these different conditions making it extremely difficult to hypothesize common pathways for all irritant reactions.

Furthermore, other factors, beside the basic mechanisms, may influence the outbreak of dermatitis. These can be divided into: (1) external (characteristics of the molecule, exposure, association with other irritants, environmental condi-

| Table 1. Types of irritation in man | Acute |
|---|---|
| | Acute delayed |
| | Cumulative |
| | Traumatic |
| | Pustular and acneiform |
| | Nonerythematous |
| | Subjective |

tions) [2], and (2) internal or endogenous and mostly related to age [3], race [4] (blacks react differently to whites), sex [5] (changes during the menstrual cycle), site [6] and pre-existing dermatitis [7].

## Histology

Many chemicals are capable of inducing ICD in man and a number are capable of doing so after a single exposure when applied long enough at a sufficiently high concentration: surfactants cause removal of surface lipids and water-holding substances, solvents are responsible for damage to cell membranes while other compounds like anthralin induce direct cytotoxic damage of the keratinocytes [8].

When we look at the histology, it can often be difficult to differentiate between allergic and ICD; ICD itself differs during acute and chronic phases.

Willis et al. [9] evaluated and classified histological changes according to visual scoring and concentration of the irritant tested. Forty-eight-hour patch application of 80% nonanoic acid gave moderate visual reactions and induced eosinophilic degeneration of keratinocytes with nuclear degeneration. On electron-microscopic examination, the cytoplasm of these cells was composed of dense aggregates of osmiophilic keratin filaments associated with prominent intercellular desmosomes. Lipid vacuoles and membrane-bound vesicles were present in varying number among the filaments. The cells in the stratum spinosum and the basal cells were mostly unaffected, except for the presence of lipid droplets. There was also minimal spongiosis with few infiltrating mononuclear cells. 100% propylene glycol gave all negative reactions. Histologically, at light microscopy only a striking basketwave pattern of the stratum corneum with slight spongiosis was evident. Reactions to 0.5% benzalkonium chloride were mild with spongiosis and occasional foci of necrotic damage in the stratum spinosum.

Ultrastructurally, the keratinocytes were characterized by shrunken pyknotic nuclei, disrupted organelles and membrane vacuolization. 0.8% croton oil produced mild-to-severe reactions. The main characteristic was considerable spongiosis, namely in worse reactions, which led to the formation of intercytoplasmic vesicles in the upper dermis. Five percent sodium lauryl sulfate (SLS) induced mild-to-moderate reactions characterized by alterations in keratinocyte morphology and parakeratosis. This may be due to increased epidermal turnover, as SLS has been shown to stimulate epidermal mitosis. Dithranol also gave spongiosis but associated with degeneration and ballooning of keratinocytes due to interference with mitochondrial function. On electron microscopy, mitochondria showed disrupted membranes. The damage also reached keratinocytes and basal cells which showed lipid droplets in the cytoplasm.

## Inflammation

Keratinocytes have a role in antigen presentation and secretion of cytokines. Willis et al. [10] showed upregulation of ICAM-1 with correlation with expression of lymphocyte function antigen-1-positive leukocytes in irritant reactions, indicating that ICAM-1 induction is not restricted to diseases characterized by antigen presentation. Also, expression of CD 36 has been reported to change in relation to the irritant tested: for instance, in the stratum granulosum it is increased by PG and SLS and decreased by dithranol and NAA. HLA-DR is expressed in allergic contact dermatitis (ACD) but not in ICD. Interestingly, CD 36 is expressed by irritants that induce hyperproliferation like PG and SLS while it is decreased with irritants like dithranol that have an antiproliferative effect [10, 11].

Other cytokines in a number of studies have been reported to increase during irritant reactions: TNF-α, IL-6-which are increased up to 10 times and IL-1β; granulocyte/macrophage colony-stimulating factor, IL-2 which are increased up to 3 times. IL-1 has been reported to increase in ACD but not in ICD [12, 13].

Since ICAM-1 expression, which is the predominant feature of ICD, is upregulated by TNF-α and CD 36 is only partially expressed in the stratum granulosum, but not in the stratum spinosum and basal layer, it seems that TNF-α may be the major mediator of inflammation in irritant reactions whereas interferon-γ is expressed in ACD. The possibility of ICAM-1 expression induced by the irritant itself without any involvement of TNF should also be taken into account.

Therefore, the pathogenetic pathway in the acute phases of ICD, common to chemically unrelated irritants, starts with penetration into the barrier, mild damage to keratinocytes, and release of mediators of inflammation with T-cell activation. In this manner, once the level of activation has been initiated via epidermal cells, continuous T-cell accumulation and activation independent of the exogenous antigen can be maintained. This could explain the similarity of infiltrates in ACD and ICD and the common involvement of cytokines and mediators of inflammation.

However, we know that inflammation is only one of the several aspects of the ICD syndrome. Indeed, there are subacute irritant reactions, or nonvisible conditions of irritation that can be developed after chronic subliminal exposure to an irritant. These reactions are characterized by minimal or even absent inflammation. In these cases, the skin response may be modulated in a different way, since the barrier function and not the keratinocyte is the major target of the insulting stimulus. Indeed, stratum corneum is a very effective barrier to many potentially irritant substances. It constantly renews itself and is quick to repair damage. Though some insults may have a persistent effect, the constant state of adaptation results in a new steady state [14]. In the case of mild chronic irritation which is usually cumulative, nonerythematous and exacerbated by environmental conditions, the role of the stratum corneum barrier is fundamental.

Damage of the lipid barrier is associated with loss of cohesion of corneocytes and desquamation with increase of transepidermal water loss (TEWL). This is one of the triggering stimuli that promote lipid synthesis, keratinocyte proliferation and transient hyperkeratosis which leads to a new steady state of the cutaneous barrier.

Damage of the barrier with removal of intercellular lipids increases TEWL which stimulates lipid synthesis and promotes barrier restoration [15]. Grubauer et al. [16] showed that water flux is responsible for the homeostasis of the barrier. Indeed, damage of the barrier with a solvent and subsequent occlusion with blockade of water evaporation stops intercellular lipid synthesis and therefore barrier recovery. However, this explains the dynamics of the barrier, but not the hyperproliferative response induced by some irritants.

It is well documented that irritants increase the epidermal turnover. Fisher and Maibach [17] have shown that topical application of 5% BC or SLS 1% induce an increase in the mitotic activity of basal keratinocytes after 48 h. Lower concentrations had no effect on the mitotic activity and no correlation between hyperproliferation and the degree of inflammation induced has been shown.

Also, after chronic exposure with induction of subclinical dermatitis an increased epidermal turnover has been shown [18]. The mechanisms may involve both keratinocytes and stratum corneum structures. Keratinocytes may be responsible in three different ways all related to a direct effect of the irritant: (1) the production of cytokines such as IL-6, IL-8 and 12-HETE which have been shown to induce epidermal hyperplasia; (2) the reduction of cAMP due to a disruption of cell membranes and therefore an increase in cell division [19], and (3) stimulation of ornithine decarboxylase (ODC). ODC is an intracellular enzyme which converts ornithine to putrescine. ODC is stimulated by stripping and it has been demonstrated to increase DNA synthesis and proliferation [20].

On the other hand, the stratum corneum also plays a role in regulating epidermal turnover [21]. Acetone treatment or stripping of the stratum corneum

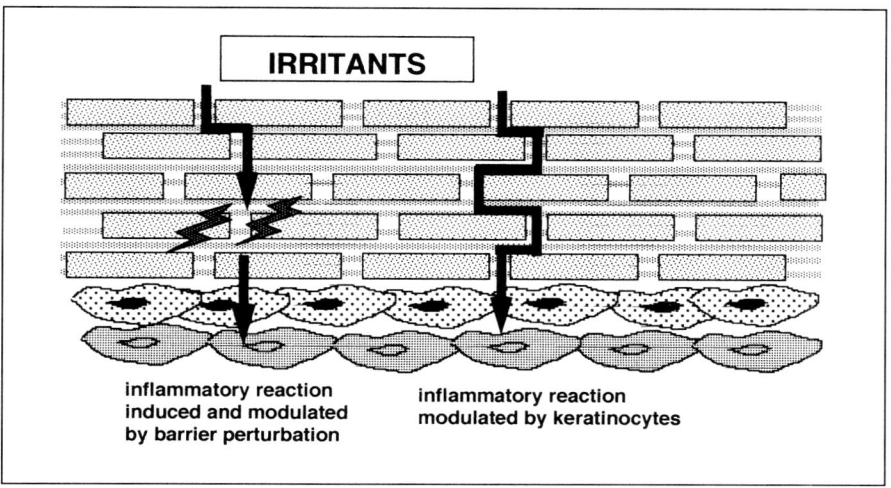

IRRITANTS

inflammatory reaction
induced and modulated
by barrier perturbation

inflammatory reaction
modulated by keratinocytes

*Fig. 1.* Mechanisms of skin irritation. The cascade of inflammatory events may be initiated via the stratum corneum or the keratinocytes.

increase DNA synthesis in epidermal basal cells. Occlusion with water-impermeable membranes inhibits the effect.

Stratum corneum has a role in the control of epidermal proliferation, but recent findings demonstrate that it may have an important role in the outbreak of skin inflammation, a factor that we have seen before it was thought to be mainly controlled by the irritant/keratinocyte interaction. Indeed, recently, Wood et al. [22] disrupted the stratum corneum barrier in several ways, using acetone, tape stripping or essential fatty acids in mice as a model of chronic damage and found a severalfold increase in TNF-$\alpha$, various ILs, and granulocyte-macrophage colony-stimulating factor. These results suggest that barrier disruption can stimulate cytokine production and promote the inflammatory skin reaction.

This reaction could be directed in two ways: the outbreak of inflammation, and the promotion of barrier recovery. TNF and ILs have been shown to increase lipogenesis in liver cells [23, 24]. In addition, TNF increases HMG-CoA reductase and thus increases cholesterol synthesis [23].

In conclusion, two distinct pathways may be correlated and might operate in connection to modulate irritant response: the direct effect of the irritants on keratinocytes on the one hand and the damage of the barrier on the other (fig. 1). Both of them may activate the biochemical signals which start the development of the irritant reaction as well as the repairing processes.

**Special Model – SLS Irritation**

In experimental dermatology there is a particular model of skin irritation that is generally used to study barrier function and irritant dermatitis: the surfactant-induced dermatitis model, i.e. the SLS-induced dermatitis model. SLS is widely used as a standard model to investigate skin irritation and we thought we knew almost everything about it. But some aspects of its mode of action need further investigation. For instance, the increase of TEWL after SLS has been always correlated to the delipidizing effect of surfactants on stratum corneum. But this is not totally correct, especially after short-term application. Recently, Leveque et al. [25] showed that the increase of TEWL after SLS is secondary to an increased s.c. hydration due to spongiosis; and, more interestingly, delipidization is minimal in the early phases of SLS-induced application. The study also showed that the organization of lipid bilayers rather than the extraction of lipids is responsible for the impairment of the barrier. The same findings were also reported by Wilhelm et al. [26]: they found an increase of both hydration and TEWL after short-term SLS. The effect is time- and concentration-dependent and proportional to carbon chain length and to the irritative potential of the solution. A possible explanation is that hydration is increased by disruption of the secondary and tertiary structures of keratin which exposes new water-binding sites [27].

These findings were confirmed in a study [28] showing that SLS leaves intact lipid bilayers in the upper s.c., but induces changes in viable epidermis with premature keratinization and damage of lipids of the lower stratum corneum: this means that the hypothetical model of SLS-induced irritation is mainly modulated by keratinocytes rather than by stratum corneum. Alterations of epidermal lipids and desquamation occur as a consequence of keratinocyte damage and impaired lipid synthesis.

Recent findings confirm this hypothesis and report normal lipid structure in the stratum corneum along with abnormal lamellar bodies secretion and cleft formation in the lower regions of the barrier [29]. Dry skin induced by SLS is therefore related to abnormalities of amino acids and proteins in corneocytes [30]. Another interesting hypothesis on the effects of surfactants on skin has recently been proposed [31]. In a three-dimensional reconstruction of the stratum corneum, the so-called domain mosaic model, stratum corneum lipids are not randomly distributed, but are organized in domains. Lipids with very long chain lengths are segregated in gel, a phase impermeable to water, separated by grain borders populated by lipids with relatively short chain lengths which are in a fluid phase, permeable to water. Surfactants, which usually have a chain length shorter than 18 carbon atoms, infiltrate primarily these fluid spaces increasing the width of the grain borders and therefore the TEWL.

In conclusion, it is clear that skin irritation has many faces and aspects. Further work is needed to investigate all these different mechanisms in order to better understand skin reactions and possible defense factors. Indeed, our ultimate goal is the development of efficient prevention and effective therapeutic regimens that could help occupational dermatologists and other colleagues in treating and preventing this common skin disorder.

## References

1 Lammintausta K, Maibach H: Contact dermatitis due to irritation; in Adams RM (ed): Occupational Skin Diseases. Philadelphia, Saunders, 1990, pp 1–15.
2 Lammintausta K, Maibach HI: Exogenous and endogenous factors in skin irritation. Int J Dermatol 1988;27:213–222.
3 Cua AB, Wilhelm KP, Maibach HI: Frictional properties of human skin: Relation to age, sex and anatomical region, stratum corneum hydration and transepidermal water loss. Br J Dermatol 1990; 123:473–479.
4 Berardesca E, Maibach HI: Racial differences in sodium lauryl sulphate induced cutaneous irritation: Black and white. Contact Derm 1988;18:65–70.
5 Agner T, Damm P, Skouby SO: Menstrual cycle and skin reactivity. J Am Acad Dermatol 1991;24: 566–570.
6 Cua AB, Wilhelm KP, Maibach HI: Cutaneous sodium lauryl sulphate irritation potential: Age and regional variability. Br J Dermatol 1990;123:607–613.
7 Lammintausta K, Maibach HI, Wilson D: Irritant reactivity in males and females. Contact Derm 1987;17:276–280.
8 Landman G, Farmer ER, Hood AF: The pathophysiology of irritant contact dermatitis; in Jackson EM, Goldner R (eds): Irritant Contact Dermatitis. New York, Marcel Dekker, 1990, pp 67–77.
9 Willis CM, Stephens CJM, Wilkinson JD: Epidermal damage induced by irritants in man: A light and electron microscopic study. J Invest Dermatol 1989;93:695–699.
10 Willis CM, Stephens CJM, Wilkinson JD: Selective expression of immune-associated surface antigen by keratinocytes in irritant contact dermatitis. J Invest Dermatol 1991;96:505–511.
11 Brasch J, Bugard J, Sterry W: Common pathogenetic pathways in allergic and irritant contact dermatitis. J Invest Dermatol 1992;98:166–170.
12 Larsen CG, Ternowitz T, Larsen FG, Zachariae, Thestrup-Pedersen K: ETAF/interleukin-1 and epidermal lymphocyte chemotactic factor in epidermis overlying an irritant patch test. Contact Derm 1989;20:335–340.
13 Hunziker T, Brand CU, Kapp A, Waelti ER, Braathen LR: Increased levels of inflammatory cytokines in human skin lymph derived from sodium lauryl sulphate-induced contact dermatitis. Br J Dermatol 1992;127:254–257.
14 Parish WE: Chemical irritation and predisposing environmental stress (cold wind and hard water); in Marks R, Plewig G (eds): The Environmental Threat to the Skin. London, Martin Dunitz, 1991, pp 185–193.
15 Grubauer G, Elias PM, Feingold KR: Transepidermal water loss: The signal for recovery of barrier structure and function. J Lipid Res 1989;30:323–333.
16 Grubauer G, Feingold KR, Elias PM: Relationship of epidermal lipogenesis to cutaneous barrier function. J Lipid Res 1987;28:746–752.
17 Fisher LB, Maibach HI: Effect of some irritants on human epidermal mitosis. Contact 1975;1: 273–276.
18 Wilhelm KP, Saunders JC, Maibach HI: Increased stratum corneum turnover induced by subclinical irritant dermatitis. Br J Dermatol 1990;122:793–798.

19  Willis CM, Stephens CJM, Wilkinson JD: Differential effect of structurally unrelated chemical irritants on the density of proliferating keratinocytes in 48 h patch test reactions. J Invest Dermatol 1992;99:449–453.

20  Marks F, Bertsch S, Fürstenburger G: Ornithine decarboxylase activity, cell proliferation, and tumour promotion in mouse epidermis in vivo. Cancer Res 1979;39:4183–4188.

21  Proksch E, Feingold KR, Mao-Quiang M, Elias PM: Barrier function regulates epidermal DNA synthesis. J Clin Invest 1991;87:1668–1673.

22  Wood LC, Jackson SM, Elias PM, Grunfeld C, Feingold KR: Cutaneous barrier perturbation stimulates cytokine production in the epidermis of mice. J Clin Invest 1992;90:482–487.

23  Feingold K, Grunfeld C: Tumor necrosis factor alpha stimulates hepatic lipogenesis in rat in vivo. J Clin Invest 1987;80:184–190.

24  Feingold KR, Soued RM, Serio MK, Moser AH, Dinarello CA, Grunfeld C: Multiple cytokines stimulate hepatic lipid synthesis in vivo. Endocrinology 1989;125:267–274.

25  Leveque JL, DeRigal J, Saint-Leger D, Billy D: How does sodium lauryl sulphate alter the skin barrier function in man? A multiparametric approach. Skin Pharmacol 1993;6:111–115.

26  Wilhelm KP, Cua A, Wolff H, Maibach HI: Surfactant-induced stratum corneum hydration in vivo: Prediction of the irritation potential of anionic surfactants. J Invest Dermatol 1993;101:310–315.

27  Rhein LD, Robbins CR, Ferne K, Cantore R: Surfactant structure effects on swelling of isolated human stratum corneum. J Soc Cosmet Chem 1986;37:125–139.

28  Fartasch M, Diepgen TL, Hornstein OP: Morphological changes of epidermal lipid layers of stratum corneum in SLS induced dry skin: A functional and ultrastructural study. J Invest Dermatol 1991;96:617.

29  Fartasch M, Bassukas ID, Diepgen T: Structural relationship between epidermal lipid lamellae, lamellar bodies and desmosomes in human epidermis: An ultrastructural study. Br J Dermatol 1993;128:1–9.

30  Denda M, Koyama J, Namba R, Horii I: Stratum corneum lipid morphology and transepidermal water loss in normal skin and surfactant-induced scaly skin. Arch Dermatol Res 1994;286:41–46.

31  Forslind B: A domain mosaic model of the skin barrier. Acta Derm Venereol 1994;74:1–6.

Enzo Berardesca, MD, Department of Dermatology, University of Pavia,
IRCCS Policlinico S. Matteo, I–27100 Pavia (Italy)

Elsner P, Maibach HI (eds): Irritant Dermatitis. New Clinical and Experimental Aspects.
Curr Probl Dermatol. Basel, Karger, 1995, vol 23, pp 9–17

..........................

# Papular and Follicular Contact Dermatitis: Irritation and/or Allergy?

*D. Perrenoud and the Swiss Contact Dermatitis Research Group*[1]

Departments of Dermatology, University Hospitals of Lausanne and Geneva,
Switzerland

The formulation of new topical products and the subsequent 'launching on the market' always includes a risk of unexpected cutaneous side effects in certain individuals. Meticulous premarketing studies of all the ingredients individually and the final formulations with in vitro, animal and human trials can markedly reduce the risk. However, when undesirable manifestations occur in relatively low quantities (under 1% of the users) and with atypical clinical features (with both irritant and allergic features), a rapid diagnosis constitutes a real challenge. The epidemic outbreak of papular and follicular contact dermatitis to a new line of cosmetics we have studied in the spring of 1992 in Switzerland illustrates this well and stresses the importance of a good postmarketing control for new topical formulations as well as for cosmetics and for pharmaceuticals. In this particular case, the dermatologists of the Swiss Contact Dermatitis Research Group worked as an efficient first-line network. The close collaboration of the manufacturer was particularly important in the identification of the offending agent.

In 1991, a Swiss manufacturer decided to change the base formulation of a cosmetic line. The ingredients were chosen for their safety after having been tested on animals or humans for potential irritant or allergenic properties. The final products passed through various human tests, such as open tests, use tests and Draize tests in the premarketing phase.

As the packaging claimed, the products offered good intrinsic qualities: (a) they were tested by dermatologists and presumed hypoallergenic; (b) were

---

[1] C. Hauser, P. Piletta, N. Hunziker, J.P. Grillet, Geneva; V. Emmenegger, Lausanne; M. Wyss, P. Elsner, Zürich; A. Bircher, Basel; B. Brand-Campbel, Th. Hunziker, Bern; F. Gillet, Bellinzona; G. Montaldi, S. Gilardi, Locarno; J. Roduner, Interlaken; H. Schibli, Fribourg; A. Suard, Monthey.

recommended for sensitive skin; (c) were only slightly perfumed; (d) contained no preservatives or dyes (the microbiological stability was guaranteed by a vacuum distribution system). In addition, full indication of the ingredients was given and the price was definitely competitive. In the 3 months following the placement of the products on the Swiss market, approximately 300,000 units were sold.

Quite unexpectedly, 4 months after the introduction of the products, a significant outbreak of dermatitis related to the use of the new line of cosmetics was observed in various regions of Switzerland. The severity and the frequent extent of the lesions seen in persons using these products permitted the manufacturer to rapidly withdraw all the products of the line. In order to isolate the origin of this dermatitis and to understand the physiopathological mechanisms of this unusual dermatitis, we conducted a collaborative study with the manufacturer, members of the Swiss Contact Dermatitis Research Group and other dermatologists in private practice [1].

## Materials and Methods

The clinical signs and epidemiology were studied on 263 observations reported by dermatologists. Additional cases (642) were directly reported by consumers to the manufacturer. The geographical localization and the independent batches of products involved were analyzed. Skin biopsies were performed in 12 cases for histological examination. Seventy-seven patients were submitted for a patch test with the body lotion and 26 patients with the individual ingredients (the latter were tested in 108 different combinations). Control patch tests were performed on 73 outpatients. The differences between patients and control subjects were analyzed statistically (Fisher's exact test). The relevance of the results was furthermore evaluated with a classical repetitive open application test on the outer-upper arm in 15 patients and on the anterior forearm and the upper back in 25 healthy volunteers. On these latter patients, the possible irritant effect of the base formulation of the cosmetics was evaluated in a repetitive twice-daily open test lasting 4 weeks. The final effect was assessed using bioengineering techniques including measurement of transepidermal water loss, skin erythema and capacitance.

*Fig. 1.* Knee of a 25-year-old woman 7 days after a single application of a body lotion containing vitamin E linoleate. The lesions appeared 2 days after the topical application of the cosmetic following a warm and humid day. Obvious sweating occurred in the patient but not UV exposure to the site.

*Fig. 2.* Abdomen of a 41-year-old woman who developed an urticarial papular and plaque eruption after a 6-day use of a vitamin E linoleate containing body lotion on the whole body with exception of the face. This photo represents 10 days following when the patient had stopped the applications and the associated pruritus and burning sensation was still intolerable.

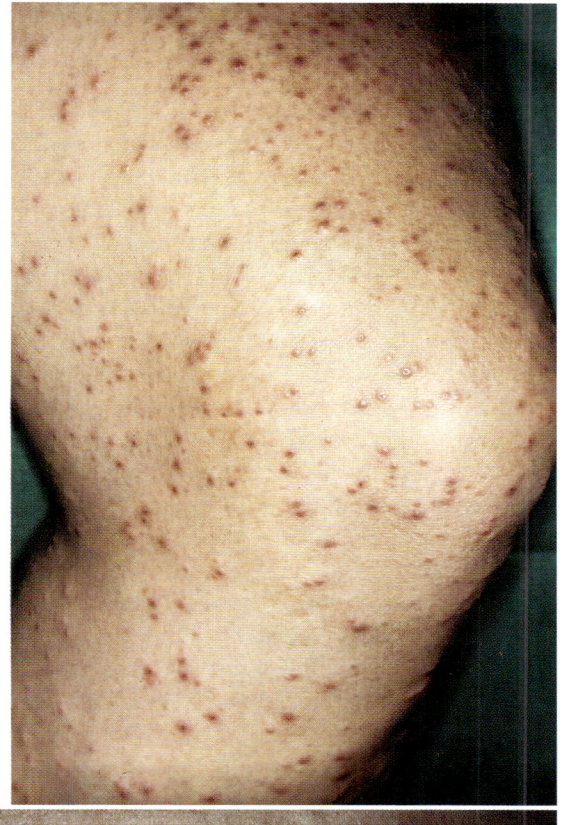

*1*

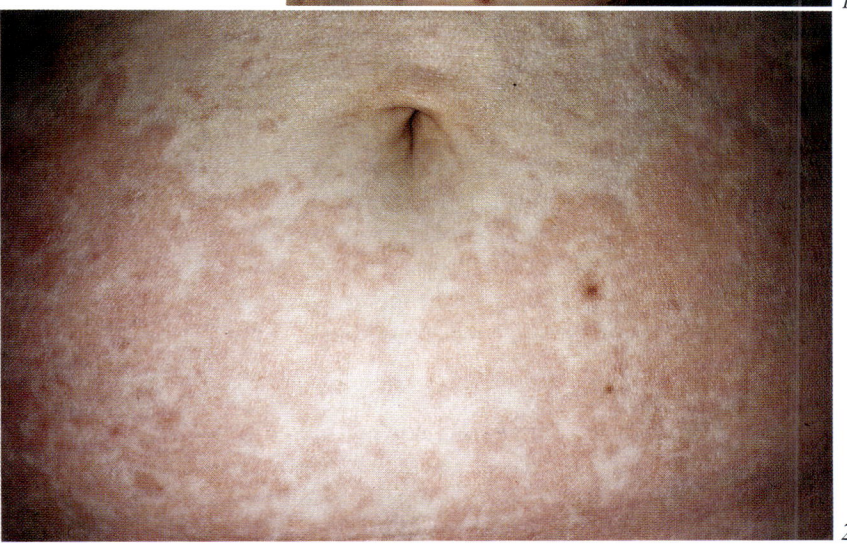

*2*

**Results**

*Clinical Features*

It was evident quite quickly that it was neither a geographical region nor a population nor a batch of the cosmetics which were particularly involved in the outbreak of dermatitis. Almost exclusively, women (96% out of the total patients; average age 36 years, range 9–70) were affected and the majority were previously free of skin diseases (less than 5% were atopic). The lesions presented as erythematous papules (75%), and/or as papules surrounded by a clear halo (38%), and/or as vesicles (21%) and were accompanied with facial erythema in 13% of the cases. Generally, the papules appeared after 2–3 weeks of use of the cosmetics, and were distributed mainly on the trunk and the extremities (fig. 1). The dermatitis was not only cosmetically unpleasant but was extremely uncomfortable with severe pruritus and in 25% of the cases a burning sensation. Most cases presented late onset of the lesions, as in classical cell-mediated contact dermatitis. In many cases, the rash spread and the pruritus increased several days after the end of the application of the cosmetic. In a few cases, a papular dermatitis beginning in the few hours following a single application of the cosmetic was reported. In 1/5 of the cases, a secondary extension to the face was seen where pronounced involvement of the ears, presenting as erythema with edema, was observed even though the cosmetics had not been applied to these sites. In a few cases, the papules were located on intensely erythematous, well-defined plaques, suggesting irritation rather than allergy (fig. 2). Heat, sweating and solar irradiation seemed to aggravate the lesions in many cases, while in other cases the lesions were located on body areas shielded from the sunlight. The eruption subsided with topical or oral corticosteroid treatment within 1–4 weeks. However, relapses were occasionally observed after sun exposure.

*Histology*

The histological pattern of the skin biopsies confirmed the preferential follicular involvement of the dermatitis. The follicular epidermis showed spongiosis with a perifollicular and perivascular dermal inflammatory infiltrate. The interfollicular epidermis was generally strikingly intact. In one severe case, the architecture of the follicule was destroyed by the inflammtory infiltrate, but again with very slight involvement of the overlying epidermis (fig. 3).

*Patch Testing*

Significant positive patch-test reactions are shown in table 1, which were mainly observed with the cosmetic products themselves and with a derivative of vitamin E, tocopheryl linoleate® (Roche, Basel, Switzerland), when applied undiluted (fig. 4). The number of positive reactions was increased by artificial aging of

*Table 1.* Significant positive patch test results of the individual ingredients of the cosmetic line

|  | Patients positive/tested | Controls positive/tested | Fisher's exact test, p |
| --- | --- | --- | --- |
| Body lotion (as is) | 5/26 | 0/63 | 0.002 |
| Night cream (as is) | 6/14 | 0/48 | <0.001 |
| Vitamin E linoleate (as is) | 4/11 | 0/49 | <0.001 |
| Vitamin E linoleate aged (as is) | 5/6 | NT | – |
| Isohexadecane (as is) | 2/26 | 11/55 | NS |
| Isohexadecane (10% pet.) | 0/19 | 0/56 | NS |
| Linoleic acid (as is) | 1/11 | 4/48 | NS |
| Linoleic acid (10% pet.) | 0/10 | 0/19 | NS |

$p > 0.05$ was considered not significant (NS). NT = Not tested.

vitamin E (by heating, or open-air mixing in thin layer, or UV irradiation). The differences of the results between patients and control subjects, analyzed with Fisher's exact test, clearly indicated that the positive reactions to the cosmetics and to vitamin E linoleate were highly specific, suggesting an allergic mechanism. The pattern of reactivity of isohexadecane (another ingredient of the cosmetics) in patients compared to control subjects suggested that it acts unspecifically as an irritant when undiluted. The same conclusions have been made with linoleic acid, a metabolite of tocopheryl linoleate (TL).

No patients reacted to various TL-free preparations of the cosmetics, confirming the responsibility of tocopheryl linoleate. Positive reactions with various derivatives of vitamin E other that linoleate (*D*-α-tocopherol, *DL*-α-tocopherol, *DL*-α-tocopheryl acetate, tocopheryl PEG-succinate 1000) suggested that cross-reactions were possible between vitamin E derivatives themselves. The patch tests performed with the remaining components of the cosmetics did not disclose any relevant results.

### Use Tests

Twelve patients out of 15 showed a positive repeated open application test to the body lotion. However, a TL-free formulation of the body lotion induced slight papular reactions in 4 patients, whereas this was not observed on 25 healthy volunteers. With the latter control group, open-controlled 4 weeks twice-daily application of TL-free formulations of the cosmetics did not reveal any irritant properties (stability of transepidermal water loss values). A favorable cosmetic effect could be assessed by measurement of increasing skin hydration and basal luminance.

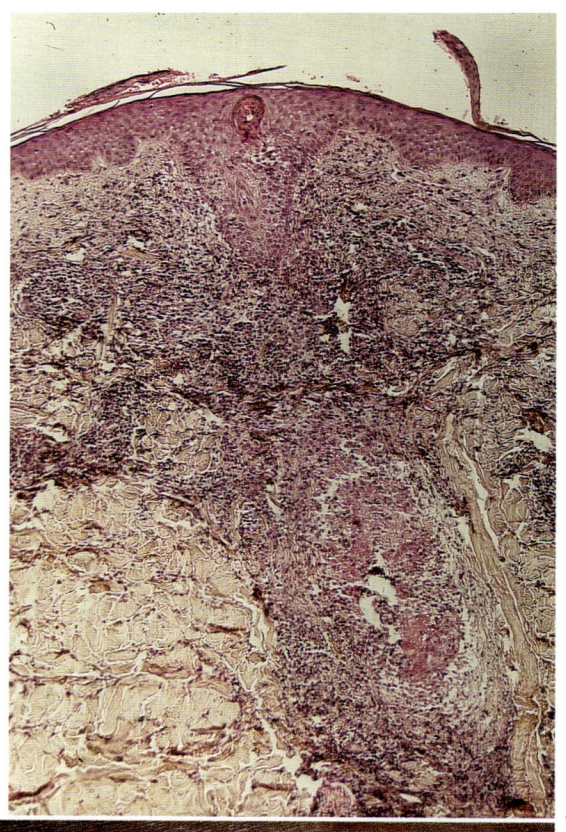

3

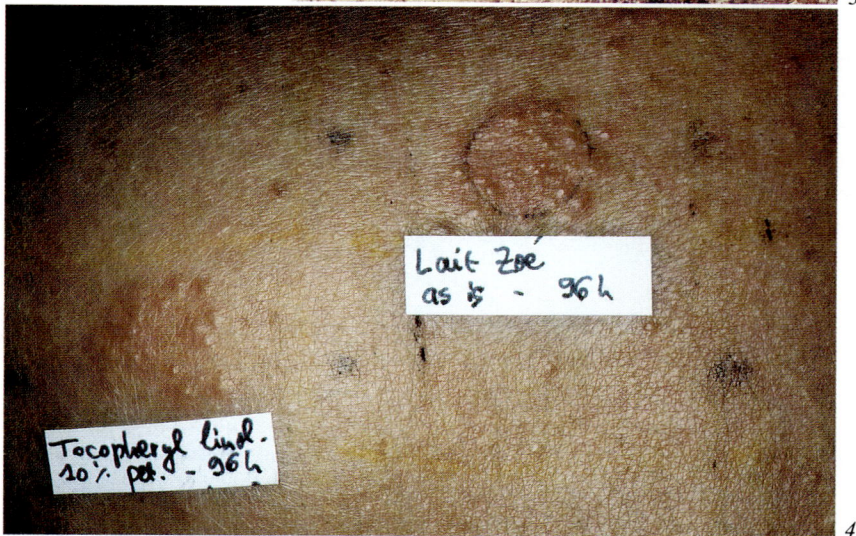

4

## Discussion

Contact dermatitis to vitamin E derivatives has only rarely been reported before our observations [1, 2]. However, the cases of vitamin E linoleate we have observed share features with the few reported cases of vitamin E contact dermatitis. In particular, an unusual papular morphology with long-lasting lesions and severe pruritus.

Although follicular patch-test reactions are usually considered a manifestation of irritancy [3], in our case, several observations suggest for a specific allergic mechanism in addition to a possible concomitant irritant mechanism.

The course of the dermatitis consisted of an asymptomatic initial period of 2–3 weeks (in the vast majority of the cases), followed by a sudden onset of a very pruritic papular rash. This two-phase evolution strongly suggested an allergic reaction, with its classical initial asymptomatic sensitization phase followed by the elicitation phase, rather than irritation. If there was the possibility of cumulative irritancy, a more progressive clinical course would have been expected. In a minority of cases, the eruption appeared after a few applications on discontinuous days or more rarely after a single application thus suggesting irritation rather than allergy. This coexistence of immediate-type reactions and of late reactions of the allergic type was formerly reported in vitamin E contact dermatitis by Aeling et al. [4] in 'Mennen E dermatitis' in 1973, whereas Fisher [5] interpreted them as contact urticaria.

Our investigations on healthy control subjects failed to detect any irritant effect of ingredients other than vitamin E-derivatives (at the concentrations used in the final products). Special vitamin E-free preparations prepared for use tests were similarly well tolerated which was in contrast with the original marketed products. Vitamin E linoleate positive patch test reactions were specifically observed in the patient population and not in the healthy control subjects, again allowing argument for a specific allergic mechanism. The observation that undiluted or only very slightly diluted vitamin E linoleate reproduced the lesions by

*Fig. 3.* Punch biopsy (HE) of a papular lesion on the abdomen of a 46-year-old women. The lesions appeared after 2 weeks of daily extensive application of a vitamin E linoleate containing body lotion. The biopsy was taken 10 days after the last application. The follicular structures are partially distroyed by an intense and localized inflammation whereas the overlying interfollicular epidermis is remarkably well preserved.

*Fig. 4.* Positive patch reactions (finished product and tocopheryl linoleate 10% in petrolatum) at day 4 on the upper back of a 41-year-old woman who was sensitized after only 15 applications of vitamin E linoleate containing cosmetics.

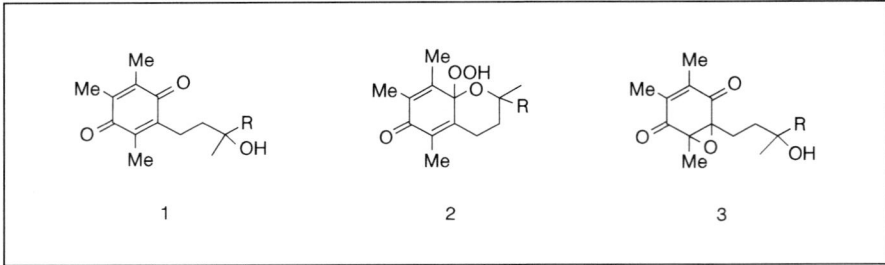

*Fig. 5.* Oxidation products of the tocol ring of tocopheryl linoleate that could act as haptens in vivo. Compound 1 is the normal product of oxidation of vitamin E; by action of oxygen, the hydroperoxide 2 can be formed, which itself rearranges to form the epoxide 3. The analogies of the clinical pictures with previously reported cases of contact dermatitis to other vitamin E derivatives favors this schematic. The linoleic acid lateral chain, which includes many unsaturated bonds that can form hydroperoxides, is probably less involved in an allergic mechanism while clearly having only irritant properties in patch testing.

patch testing suggested that a contaminant or a metabolite could be the offending agent (fig. 5).

Biopsies performed on lesional skin confirmed the unusual follicular local-ization of this contact dermatitis which showed very minor alterations of the interfollicular epidermis. The spongiotic aspect of the latter on skin biopsies per-formed on positive patch test reactions to vitamin E linoleate strongly suggest an allergic mechanism.

The long duration of the lesions, their relative resistance to topically applied potent steroids and their tendency to relapse following sunbathing weeks after the initial episode, suggest a persistence of vitamin E-related substances (metabolite or contaminant) in the skin.

To conclude, our observation of a large and totally unexpected outbreak of clinically atypical contact dermatitis emphasizes the importance of the postmar-keting control of every new topically applied pharmaceutical product, cosmetic or skin care formulation. In our case, the pre-existence of a network of dermatolog-ists specially trained in contact dermatitis and the efficient and immediate collab-oration of the manufacturer of the concerned products was of great help in promptly isolating the offending agent. This collaboration was of particular importance because the procedure described herein was laborious due to the pres-ence of clinical features of both the allergic and the irritant type simultaneously. The results of our compared investigations on patients and healthy control sub-jects support the hypothesis that a metabolite or a contaminant of vitamin E lino-leate could act as a hapten and/or as an irritant with possible synergistic effects.

## Acknowledgement

We thank H.P. Homberger, P.C. Auderset, R. Emmenegger (Mibelle AG Cosmetics, Buchs, Switzerland) for providing all the ingredients of their concerned products and giving us their active support. We also thank J.P. Lepoittevin (Laboratoire de Dermatochimie, Strasbourg, France) for helpful discussion on the possible pathogenic pathway of vitamin E derivatives in the skin.

## References

1   Perrenoud D, Homberger HP, Auderset PC, Emmenegger R, Frenk E, Saurat JH, Hauser C: An epidemic outbreak of papular and follicular contact dermatitis to tocopheryl linoleate in cosmetics. Dermatology 1994;189:225–233.
2   de Groot AC, Berretty PJ, van Ginkel CJ, den Hengst CW, van Ulsen J, Weyland JW: Allergic contact dermatitis from tocopheryl acetate in cosmetic creams. Contact Derm 1991;25:302–304.
3   Fischer T, Tystedt I: False-positive, follicular and irritant patch test reactions to metal salts. Contact Derm 1985;12:93–98.
4   Aeling JL, Panagotacos PJ, Andreozzi RJ: Allergic contact dermatitis to vitamin E aerosol deodorant. Arch Dermatol 1973;108:579–580.
5   Fisher AA: Three faces of vitamin E topical allergy. Cutis 1991;48:272–274.

For a complete reference list see Perrenoud et al. [1].

D. Perrenoud, MD, Department of Dermatology, CHUV, CH–1011 Lausanne (Switzerland)

Elsner P, Maibach HI (eds): Irritant Dermatitis. New Clinical and Experimental Aspects.
Curr Probl Dermatol. Basel, Karger, 1995, vol 23, pp 18–27

# What Can We Learn from Epidemiological Studies on Irritant Contact Dermatitis?

*Thomas L. Diepgen* [a], *Pieter-Jan Coenraads* [b]

[a] Department of Dermatology, Friedrich-Alexander University Erlangen-Nuremberg,
Germany;
[b] Occupational and Environmental Dermatology Unit, State University Groningen,
The Netherlands

Epidemiology of irritant contact dermatitis (ICD) is concerned with three closely interrelated components: distribution, determinants, and frequency of ICD in human populations. Investigating the patterns of ICD in groups of people we try to learn why certain individuals develop ICD and other people do not. Some individuals seem to be at 'high risk' by virtue of their personal characteristics, environment, and occupational activities. In order to prevent ICD in those subjects, we must first of all be able to identify endogenous and exogenous risk factors and then intervene to reduce that risk.

The epidemiological concept encompasses the following aspects: (1) disease surveillance; (2) searching for causes; (3) determining the natural history; (4) searching for prognostic factors; (5) testing strategies of prevention and new treatments.

Very few truly epidemiological studies have been published which were aimed at these aspects. Most of the studies are based on hospital patients and are not population based. Additionally, ICD is a multifactorial disease and it is necessary to assess the validity of any observed statistical association by excluding possible alternative explanations, such as chance, systematic error in collecting and interpreting data (bias), as well as the effects of additional variables that might be responsible for the observed association (confounding). In the context of methodological aspects, epidemiological studies on ICD will be discussed.

*Table 1.* Prevalence of hand eczema in population-based studies (selection)

| Author | Method of case ascertainment | Measures of prevalence years | Prevalence of hand eczema, % | | |
|--------|------------------------------|------------------------------|-------|---------|-------|
| | | | males | females | total |
| Coenraads et al. [3] | E | 3 | 4.6 | 8.0 | 6.2 |
| Kavli and Forde [4] | Q | 1 | 4.9 | 13.2 | 8.9 |
| Meding and Swanbeck [6] | Q, E | 1 | 8.8 | 14.6 | 10.6 |
| Smit et al. [7] | Q | 1 | 5.2 | 10.6 | 8.2 |
| Funke et al. [8] | E | 1 | 5.6 | 10.5 | 6.7* |

E = Dermatological examination; Q = questionnaire.
* 940 males and 256 females.

## Incidence and Prevalence

One of the most basic questions in occupational dermatology is the frequency with which contact dermatitis and sensitizations against various allergens occur. The availability of such data is a prerequisite for any systematic investigation of patterns of disease occurrence in human populations and to find strategies to prevent irritant and allergic contact dermatitis. Measures of disease frequencies are 'prevalence', which is the amount of disease that is already present in a population, 'incidence', which refers to the number of new cases of contact dermatitis during a defined period in a specified population, and 'incidence rate', which is the number of nondiseased persons who become diseased within a certain period of time, divided by the number of person-years in the population. All measures of disease frequency consist of the number of cases in the numerator, and the size of the population under study in the denominator. But, very often the sample of the numerator is biased and the size of the denominator unknown. These terms are often misused, too [1].

Studies among patient populations from dermatology clinics are not adequate for estimating prevalence or incidence rates. Estimates of the prevalence of hand eczema (HE) in the general population vary between 2 and 11% [2–8]. In table 1 the results of some population-based studies are given. But, the way in which persons with HE were identified differs, the definition of the disease is not the same, and different prevalence rates are reported. The study of Meding and Swanbeck [6] demonstrates the difference between point and period prevalence: in the same population the point prevalence of hand eczema was 5.4 and the 1-year period prevalence 10.6. In all the studies listed in table 1, the prevalence among women was higher than among men. Irritant HE is more often than allergic HE and HE occurred twice as often among females as among males. The highest prevalence of ICD was found in females aged between 20 and 29 years and in males 30–39 years of age. For females the prevalence of ICD is decreasing but for males it increases between the age of 20 and 39 [6]. The influence of a control group on the prevalence ratio was demonstrated by Smit et al. [7]. The prevalence of hand eczema in female and male nurses was quite similar between 29 and 32%. But because of the fact that the prevalence of HE in the general population was twice as high among women as among men, in male nurses the prevalence ratio was two times as high as in female nurses (PR of 5.7 and 3.0, respectively), and the age-adjusted prevalence ratio was approximately 4 times as high (PR of 9.3 and 2.3, respectively). Age-adjusted prevalence ratios were calculated using the pooled Mantel Haenszel estimator for relative risks, which pools stratum-specific estimates of the prevalence ratio under the null hypothesis (PR=1). ICD is often a chronically relapsing disease. The point prevalence includes only subjects with actual ICD and is therefore less informative than the period prevalence. On the other hand, the accuracy of recall will decrease with time, because persons who did not have complaints recently will more often forget to report their earlier ICD.

Prevalence is in fact a measure of the product of incidence and duration of disease. In examining the changes in the biology of a disease, incidence data are preferable. The period prevalence includes subjects with long-lasting ICD as well as relatively recent cases and thus possesses all interpretional difficulties that are inherent to a period prevalence [9]. No inference can be made between exposure and ICD because the exposure may have changed over time, past exposure may be over- or underestimated, and preventive measures may have been taken after complaints occurred. Given these considerations, incidence figures are preferred for analyzing risk factors for ICD.

In Germany, occupational skin diseases (OSD) constitute 34% of all registered occupational diseases. A population-based prospective study was performed to investigate all closed cases of OSD in North Bavaria [10]. Between 3/1990 and 3/1993 in 2,567 cases a work-related skin disease was stated. In cooperation with the German State Institute of Labour and Occupation the number of all employees of this referral area were evaluated for the same time period. Because all cases of OSD have been recorded and the number of employees during the same time period in the same referral area is known, incidence rates (IR) could be calculated (for a 3-year period) as follows:

$$\text{Incidence rate} = \frac{\text{number of closed cases with OSD}}{\text{number of all employees}} \, 10{,}000.$$

*Table 2.* Incidence rates (in 3 years per 10,000 employees) of ICD and ACD in different occupational groups according to a population-based study in North Bavaria [Diepgen, in prep.]

| Occupational group | Employees | Incidences | |
|---|---|---|---|
| | | ICD | ACD |
| Hairdressers | 11,560 | 270 | 401 |
| Bakers | 5,611 | 135 | 70 |
| Electroplaters | 1,773 | 17 | 102 |
| Metalworkers | 89,674 | 21 | 13 |
| Health services | 93,853 | 19 | 17 |
| Bricklayers | 48,287 | 8 | 15 |

In table 2, the incidence rates of selected occupations according to the diagnoses of ICD and allergic contact dermatitis (ACD) are given. The incidence rates (IR) were sex and age related, too. For example, in the food industry [11] bakers had a higher risk of OSD with an IR of 191 (95% CI: 156–226) compared with confectioners (IR = 84; 95% CI: 55–113) and cooks (IR = 34; 95% CI: 28–40). Females had a considerably higher risk to develop OSD than men. The incidences of OSD were highest between the ages of 15 and 24 years. ICD was the main reason of OSD in bakers, confectioners and cooks.

**Errors and Case Ascertainment**

Appreciation of the issue related to sources of errors and difficulties in case ascertainment in epidemiological studies on ICD is important for interpretation and appropriate application of research findings in the clinical setting. The main errors that interfere with research inferences are random error and systematic error (table 3). Components of systematic error are bias and confounding, which must always be considered in assessing the presence of a valid statistical association in an epidemiological study. There are two types of bias: information and selection bias.

In epidemiological studies on ICD, the ascertainment of cases of hand eczema varied from intensive efforts by a medical examination of the complete study population [3, 5, 8] to the relatively easy-to-apply method of self-administered questionnaires [4, 6]. The size of the differences in prevalence estimates that may

*Table 3.* The five explanations when an association between a history of metal sensitivity (HMS) and ICD is observed

| Explanation | Type of association | Basis for association | Causal model |
|---|---|---|---|
| Chance | spurious | random error | HMS and ICD not related |
| Bias | spurious | systematic error | HMS and ICD not related |
| Effect-cause | real | | ICD is a cause of HMS<br>HMS → ICD |
| Effect-effect | real | confounding | HMS and ICD are both caused by a 3rd factor<br>factor X<br>↓   ↓<br>HMS  ICD |
| Cause-effect | real | | HMS is cause of ICD<br>HMS → ICD |

arise as a result of differences in the definition and method of diagnosing hand eczema (information bias) has been investigated by Smit [12]. Two types of questionnaire diagnoses, a 'symptom-based' and a 'self-reported' diagnosis were compared with the medical diagnosis of hand eczema. The prevalence of HE according to the medical diagnosis was 18%, but according to the symptom-based diagnosis 48%, and to the self-reported diagnosis 17%. The sensitivity and specificity of the symptom-based diagnosis were 100 and 64%, and of the self-reported 65 and 93%. That means that with the self-reported questionnaire only 65% of the hand eczema patients could be detected, and contrary with the symptom-based questionnaire only 64% of subjects without the disease are thought to be healthy. The symptom-based diagnosis of HE overestimated, the self-reported diagnosis underestimated the prevalence of HE according to the medical diagnosis.

The distinction between ACD and ICD can be difficult, because in many instances, the simultaneous exposure to irritant and sensitizing agents plays an essential part in the development of contact dermatitis. Additionally, ICD can be considered as a subcategory of eczema. Beside ACD other categories of eczema are, for example, atopic, nummular, dyshidrotic, and tylotic eczema [13]. The classifications are based upon a combination of morphological, etiological, constitutional and other factors, and lead to overlapping categories [1].

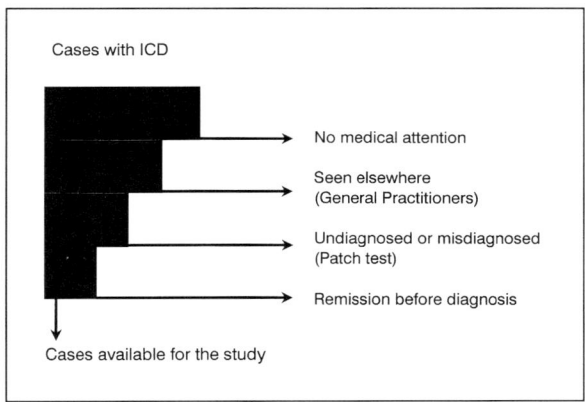

Cases with ICD

No medical attention

Seen elsewhere
(General Practitioners)

Undiagnosed or misdiagnosed
(Patch test)

Remission before diagnosis

Cases available for the study

*Fig. 1.* Selection bias in epidemiological studies on ICD.

The important question 'how do we know who has the disease and who does not' can cause selection bias (fig. 1). Ideally, the sample of cases would be a random sample of everyone who had the disease. Especially in case-control studies the cases were sampled from patients in whom contact dermatitis had already been diagnosed, and who were available for study. This sample is not representative of all patients with the disease because those who are undiagnosed, misdiagnosed, or lost to follow-up are less likely to be included.

In an epidemiological study on the prevalence of HE in different occupational groups [7], it was noted that in population-based studies the symptoms of ICD are relatively mild in the majority of cases, judged from the proportion of cases that resulted in sick leave or medical attention. Out of all persons with hand eczema in different occupational groups, the majority reported improvement of their symptoms after weekends or holidays. Only between 15 and 36% of the cases had consulted a physician for treatment of their symptoms, and sick leave due to the symptoms results in only 4–9%.

## Risk Factors

When searching for causes of irritant contact dermatitis, the association between personal and environmental characteristics and the occurrence of ICD can occur by coincidence, by noncausal linkage to other features, or by cause-and-effect relationships (table 3). Of course we are primarily interested in determinants of disease development, also known as risk factors. Identification of risk factors can result in a better understanding of the pathways leading to disease acquisition and thereby suggest preventive strategies.

Apart from exposure to irritating agents there are many factors that may influence the development of ICD, such as atopic constitution, psychological factors, age, gender, humidity and weather conditions, etc. If these factors are not properly controlled for either in the design or in the analysis, they may act as confounders in the study [1].

Because we lack a test to determine whether an irritant is relevant to a subject's contact dermatitis, it remains a clinical decision to state the etiological role of irritants in ICD. Well-known irritants are water and wet work, detergents and cleansing agents, hand cleaners, unspecific chemicals, cutting fluids, and abrasives. In a study on HE [13] at least one of those irritants were always involved in irritant HE but also in 60% in atopic HE, in 84% in allergic HE and in 63% in other types of HE. A complex interplay of exogenous risk factors, such as exposure to irritants, and the endogenous disposition (atopy) is believed to be responsible for the occurrence and the course of ICD in humans.

Atopy and especially atopic eczema (AE) are well-known factors influencing ICD, but the role of atopic features in the developing of the disease is still unclear. In comparing the figures quoted in the literature one is faced with the same difficulties of selection and interpretation that have been mentioned before [14]. Additionally, the definition of atopy itself differs considerably. Some authors include a family history as well as a personal history of atopy, others divide their subjects in those with atopic eczema and those with respiratory allergy, some would only accept positive prick tests as evidence for the atopic diathesis. Even if in epidemiological terms atopic eczema is an effect modifier, it can be argued that the observed ICD associated with atopy is in fact an exacerbation of AE, rather than ICD. Since atopic dermatitis is associated with respiratory atopy, one would expect that respiratory atopy is a surrogate risk factor as well.

Especially AE in childhood are risk factors for hand eczema in adults [15, 16]. However, these studies also found that a considerable number of subjects with a personal history of AE managed to work in risk occupations without developing HE. Therefore, a reduced resistance to irritants does not occur in all subjects with AE and may occur in subjects with respiratory atopy and in non-atopics.

Lammintausta and Kalimo [15] introduced the term atopic skin diathesis (ASD) as a prognostically useful definition of the skin condition which might be involved in the development of HE. In order to establish a diagnostic score of ASD we evaluated basic and minor features of AE systematically in established cases of AE and in subjects randomly collected from the Caucasian normal population of young adults (NP) in a prospective computerized study [17, 18].

*Table 4.* Score of atopic skin diathesis (ASD score) based on $\chi^2$ values without laboratory investigations (according to Diepgen and co-workers [17, 18])

| Atopic feature | Points | $\chi^2$ | OR | 95% CI of OR |
|---|---|---|---|---|
| Xerosis | 3 | 429 | 27.9 | 23.2–33.8 |
| Itch when sweating | 3 | 410 | 25.4 | 21.1–30.1 |
| White dermographism | 3 | 357 | 19.3 | 16.2–23.2 |
| Wool intolerance | 3 | 355 | 15.8 | 13.4–18.5 |
| Pityriasis alba | 2 | 304 | 60.1 | 41.6–87.0 |
| Infraorbital fold | 2 | 292 | 11.0 | 9.4–12.7 |
| Hertoghe sign | 2 | 282 | 44.8 | 32.1–62.6 |
| Palmar hyperlinearity | 2 | 242 | 11.7 | 9.8–13.9 |
| Ear rhaghade | 2 | 236 | 19.2 | 15.2–24.4 |
| Perlèche | 1 | 201 | 7.0 | 6.1–8.2 |
| Cradle cap | 1 | 184 | 10.6 | 8.7–12.9 |
| Family history of atopy | 1 | 69 | 2.9 | 2.6–3.3 |
| Facial pallor/erythema | 1 | 117 | 5.3 | 4.5–6.3 |
| Keratosis pilaris | 1 | 103 | 4.9 | 4.2–5.8 |
| Food intolerance | 1 | 85 | 4.7 | 4.0–5.7 |
| Allergic rhinitis | 1 | 65 | 3.1 | 2.7–3.6 |
| Allergic asthma | 1 | 55 | 4.8 | 3.4–6.0 |
| Metal sensitivity | 1 | 55 | 2.7 | 2.4–3.1 |
| Photophobia | 1 | 41 | 2.6 | 2.3–3.1 |

Atopy score: $\chi^2 > 350$: 3 points; $350 > \chi^2 > 220$: 2 points; $\chi^2 < 220$: 1 point.
OR = Odds ratio; 95% CI = 95% confidence interval of OR.
The statistical analysis is based on 428 AE patients and 628 non-eczematous controls.

Anamnestic and clinical atopic basic and minor features were investigated in all test subjects by two investigators to obtain a good interobserver agreement. On the base of statistical modelling a diagnostic score system was constructed, which should be based on anamnestic and clinical features without laboratory investigations. The presence of an itching flexural dermatitis was not included since this was the selection base. For practical use every atopic feature obtained a value between 1 and 3 points according to its statistical significance (table 4). Based on this score system patients with more than 10 points should be considered to have an ASD, patients with more than 6 points are suspicious of ASD.

Other possible risk factors are age, gender, race, contact sensitization, condition of the skin. There is therefore no hard epidemiological evidence on the relevance of these factors. Some of these may be just proxy variables: age, for example, is within the time window spanning the employment years not a major risk factor in itself. Age-related exposure characteristics are for the most part responsible for the alleged effect of age. The same may be true for gender, which is probably a proxy variable for exposure to irritants in the household environment. The term condition of the skin, which may include dry skin, is rather ambiguous and therefore subject to random error and bias.

From this very brief discussion on risk factors for ICD, it may be concluded that the concept of a 'sensitive skin', whether in subjective or in objective terms, still eludes a meaningful definition, although considerable progress has been made in outlining atopy as a major component of this concept.

Key points:

(1) Any study that includes cases from clinical material or registries suffers from selection bias, which should be accounted for in the analysis.

(2) Prospective, incidence type studies are scarce, but should have priority.

(3) Skin atopy is a major risk factor, or atopic dermatitis activated by exposure.

(4) Within the time span of employment years, age is not a risk factor.

(5) Gender is not a risk factor, but proxy variable for exposure.

(6) Exposure is the most important determinant of risk, but exposure quantification techniques (for example, job-exposure matrices) are underdeveloped.

## References

1   Coenraads PJ, Smit J: Epidemiology; in Rycroft RJG, Menné T, Frosch PJ, Benezra C (eds): Textbook of Contact Dermatitis. Berlin, Springer, 1992, pp 133–149.

2   Agrup G: Hand eczema and other dermatoses in South Sweden. Acta Derm Venereol (Stockh) 1969;61:1–91.

3   Coenraads PJ, Nater JP, Lende van der R: Prevalence of eczema and other dermatoses of the hands and arms in the Netherlands. Association with age and occupation. Clin Exp Dermatol 1983;8: 495–503.

4   Kavli G, Forde OH: Hand dermatoses in Tromso. Contact Derm 1984;10:174–177.

5   Lantinga H, Nater JP, Coenraads PJ: Prevalence, incidence and course of eczema on the hands and forearms in a sample of the general population. Contact Derm 1984;10:135–139.

6   Meding B, Swanbeck G: Prevalence of hand eczema in an industrial city. Br J Dermatol 1987;116: 627–634.

7   Smit HA, Burdorf A, Coenraads PJ: The prevalence of hand dermatitis in different occupations. Int J Epidemiol 1993;22:288–293.

8   Funke U, Diepgen TL, Fartasch M: Identification of high-risk groups for irritant contact dermatitis by occupational physicians; in Elsner P, Maibach HI (eds): Irritant Dermatitis: New Clinical and Experimental Aspects. Curr Probl Dermatol. Basel, Karger, 1995, vol 23, pp 64–72.

9   MacMahon B, Pugh TF: Epidemiology: Principles and Methods. Boston, Little, Brown, 1970.

10  Diepgen TL, Schmidt A, Schmidt M, Fartasch M: Demographic and legal characteristics of occupational skin diseases. Allergologie 1994;17:84–89.

11  Tacke J, Schmidt A, Fartasch M, Diepgen TL: Occupational contact dermatitis in bakers, confectioners and cooks. A population-based study. Contact Derm, in press.

12  Smit HA, Coenraads PJ, Lavrijsen APM, Nater JP: Evaluation of a self-administered questionnaire on hand dermatitis. Contact Derm 1992;26:11–16.
13  Diepgen TL, Fartasch M: General aspects of risk factors in hand eczema; in Menné T, Maibach HI (eds): Hand Eczema. Boca Raton, CRC Press, 1993, pp 141–156.
14  Diepgen TL, Tepe A, Pilz B, Schmidt A, Hüner A, Huber A, Hornstein OP, Frosch PJ, Fartasch M: Occupational skin diseases in hairdressers and nurses during apprenticeship: Design of a prospective epidemiological study. Allergologie 1993;10:396–403.
15  Lammintausta K, Kalimo K: Atopy and hand dermatitis in hospital wet work. Contact Derm 1981; 7:301–308.
16  Rystedt I: Hand eczema and long-term prognosis in atopic dermatitis; thesis. Acta Derm Venereol (Stockh) 1985;117:1–59.
17  Diepgen TL, Fartasch M, Hornstein OP: Criteria of atopic skin diathesis. Dermatosen 1991;39: 79–83.
18  Diepgen TL, Fartasch M: Recent epidemiological and genetic studies in atopic dermatitis. Acta Derm Venereol (Stockh) 1992(suppl 176):13–18.

Priv.-Doz. Dr. Thomas L. Diepgen, Department of Dermatology, Dermato-Epidemiology Unit, Hartmannstrasse 14, D–91052 Erlangen (Germany)

Elsner P, Maibach HI (eds): Irritant Dermatitis. New Clinical and Experimental Aspects.
Curr Probl Dermatol. Basel, Karger, 1995, vol 23, pp 28–40

..........................

# Statistics on Occupational Dermatoses in Finland[1]

*Lasse Kanerva, Riitta Jolanki, Jouni Toikkanen, Kyllikki Tarvainen, Tuula Estlander*

Finnish Institute of Occupational Health, Helsinki, Finland

The Department of Social Research of the Ministry of Social Affairs and Health started compiling statistics on occupational diseases in Finland in 1926. Since 1974, the Act on the Supervision of Labour Protection has obligated doctors to report every case of occupational disease. In 1975, the Finnish Institute of Occupational Health assumed the responsibility for compiling these statistics. The collection of data is illustrated in figure 1. All the cases of occupational disease have been diagnosed by a physician. The insurance companies provide the Register with data on every case reported to them, irrespective of the final decision with regard to compensation. Statistics on new cases of occupational diseases are published annually in Finnish [1, 2]. In this study we have collected data from the Register to get more information about occupational dermatoses in Finland. We have earlier used the Register to collect data on occupational allergic rhinitis [3], occupational allergic diseases [2], occupational dermatoses during 1974–1983 [4] and occupational dermatoses caused by man-made mineral fibers [5].

## Material and Methods

Only the main cause was considered as the source of an occupational dermatosis. If the disease is diagnosed as an occupational allergic disease, it is recorded as an occupational allergic disease despite the fact that probably also irritant factors have been involved. A patients' occupational skin disease is registered only once.

[1] This research is part of the Allergy and Work programme of the Finnish Institute of Occupational Health. The Allergy and Work programme is headed by one of the authors (L.K.).

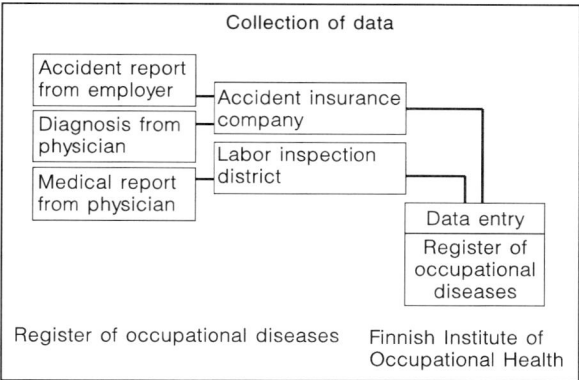

*Fig. 1.* Collection of data to the Register of Occupational Diseases in Finland.

Data from the Register for the years (1989) 1990–1993 were collected as follows:

(i) The total number of cases with occupational dermatoses, namely (a) irritant contact dermatoses; (b) allergic contact dermatoses; (c) protein contact dermatitis (PCD)/contact urticaria (CU), were collected from the years 1989–1993. In the following, CU and the related PCD are dealt with together under the heading of contact urticaria.

(ii) The number of individual causes of occupational dermatoses were compiled for the years 1990–1993.

(iii) The number of occupational irritant and allergic dermatoses by industry per 100,000 employed workers were compiled from the latest statistics, i.e. 1993 [1].

**Results and Discussion**

*Occupational Diseases in Finland in 1993*

A total of 7,036 cases of occupational diseases were registered in all of Finland with a population of about 5 million and a total work force of 2,041,000 (men 1,048,000/women 993,000). Of these 16% were occupational dermatoses [1]. Occupational dermatoses were the fourth most common cause of occupational diseases, the three most common being musculoskeletal diseases, hearing loss and diseases caused by asbestos.

*Frequency of Occupational Dermatoses*

During 1990–1993, 5,177 occupational skin diseases were reported to the Register. A total of 4,368 cases were classified as irritant contact dermatitis, allergic contact dermatitis or contact urticaria. Of these, 4,368 cases, irritant dermatoses comprised 2,181 cases (49.9%), allergic contact dermatitis 1,567 cases

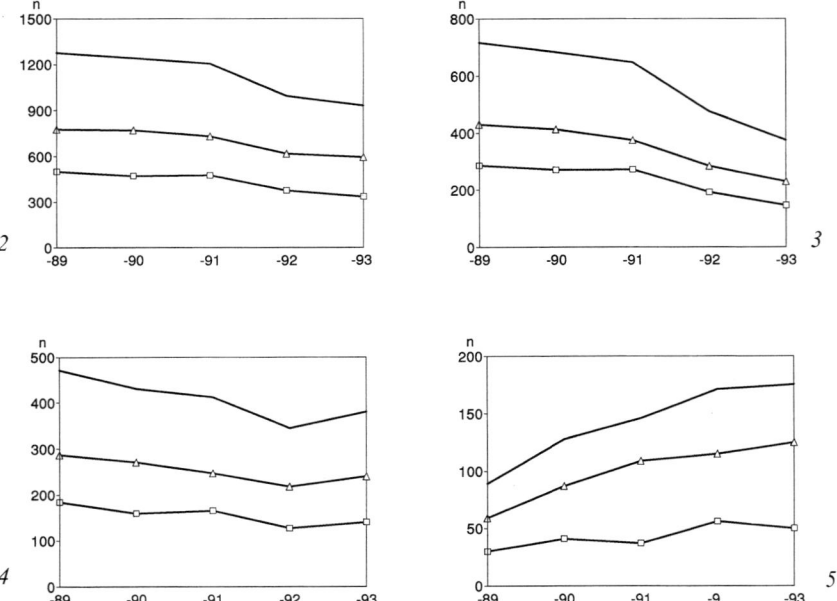

*Fig. 2.* Number of occupational dermatoses by gender during 1989–1993. ——— = All; —△— = females; —□— = males.

*Fig. 3.* Number of occupational irritant contact dermatoses by gender during 1989–1993. For symbols see figure 2.

*Fig. 4.* Number of occupational allergic contact dermatoses by gender during 1989–1993. For symbols see figure 2.

*Fig. 5.* Number of occupational cases of protein contact dermatitis and contact urticaria by gender during 1989–1993. For legend see figure 2.

(35.9%) and contact urticaria 620 cases (14.2%). Accordingly, the number of allergic cases was about the same (50.1%) as the number of irritant cases (49.9%). This differs from most other reports in which 60–75% of the occupational dermatoses are irritant [6]. Interestingly, Diepgen and Fartasch [7] also encountered about equal numbers of occupational irritant and allergic dermatoses in Germany.

### Total Number of Occupational Dermatoses during 1989–1993

The frequency of occupational dermatoses during 1989–1993 is given in figures 2–5. The greatest decrease has occurred in irritant dermatoses. The reason is probably the following: Firstly, the work force has significantly decreased in Finland during 1991–1993 because of an economic recession resulting in unemployment. Secondly, it seems feasible to assume that because of the high unemployment rate, workers do not seek medical advice in cases of less serious disorders, e.g. irritant contact dermatitis. The reason for this is that if work force in a work place has to be diminished, an occupational disease may be one criterion for selecting workers who will be laid off.

Occupational contact urticaria has been compiled as a separate entity in Finland since 1989; earlier all allergic contact dermatoses, i.e. allergic contact dermatitis and contact urticaria, were grouped together. The number of cases with occupational contact urticaria has nearly doubled during the observation period of 1989 to 1993 (fig. 5).

*Occupational Irritant Contact Dermatoses during 1990–1993*

The cause of an irritant dermatitis is usually multifactorial. As there is no test to verify an irritant dermatitis, the statistics reflect what the physicians, based on their knowledge of exposure, have considered to be the main cause. Occupational irritant dermatoses are usually not caused by pure chemicals but my mixes, the detergents being the most common in our statistics (table 1; fig. 6). The causative agents are in accordance with those that have been reported elsewhere, including textbooks [6, 8–11]. The following brief review gives the most common causes of occupational irritant contact dermatoses in the current statistics:

*Detergents (33%).* The most irritant detergents are those containing solvents, alkalis or acids, irritating surfactants or disinfectants. Added abbrasives, such as sand or silica, may cause further mechanical damage [8].

*Wet Work (10%).* A wide range of occupations include wet work. Prolonged contact with water causes the skin to dry. This is facilitated by prior damage to the lipids of the horny layer [6, 8].

*Oils, Greases, Cutting Fluids, Organic Solvents and Dirty Work (18% when compiled).* Mostly synthetic coolants are used today. They may contain mineral or vegetable oil, and in addition emulsifiers, antioxidants, anticorrosive agents, antimicrobials, dyes, perfumes and water [8]. Organic solvents, in particular aromatic solvents, are irritants. Organic solvents may also cause diseases of the nervous system and have partly been replaced by water-soluble chemicals, which on the other hand may be more allergenic [12].

*Handling of Foods, Organic and Materials (8%).* Wet work connected with the handling of raw meat, viscera, fish, shellfish, raw vegetables, fresh fruits and berries is irritant. Kitchen maids, cooks, cold buffet managers, processed food factory workers, chefs and other workers who process and prepare food are the occupations mostly involved. Recently, it has become evident that dermatitis in this group is often accompanied by immediate allergy [2, 13].

Table 1. The causes of occupational irritant contact dermatoses during 1990–1993 (2,181 cases) according to the Finnish Register of Occupational Diseases

| Cause | Cases | Total | % |
|---|---|---|---|
| *Individual chemicals* | | 64 | 3 |
| Inorganic acids and alkalis | | 12 | 1 |
| Metals | | 9 | <1 |
| Organic acids | | 7 | <1 |
| Acrylate compounds | | 5 | <1 |
| Formaldehyde | | 5 | <1 |
| Other chemicals | | 26 | 1 |
| *Mixes* | | 1,859 | 85 |
| Detergents | | 729 | 33 |
| Wet work | | 211 | 10 |
| Oils, greases, cutting fluids | | 133 | 6 |
| Dirty work | | 120 | 6 |
| Organic solvents | | 120 | 6 |
| Handling of foods | | 103 | 5 |
| Organic dusts and materials | | 68 | 3 |
|     Animal-derived products | 17 | | |
|     Vegetables | 11 | | |
|     Flours, grains and feed | 8 | | |
|     Wood | 8 | | |
|     Ornamental plants | 6 | | |
| Cement | | 59 | 3 |
| Glues | | 37 | 2 |
|     Plywood glues | 7 | | |
|     Phenol formaldehyde glues | 6 | | |
| Synthetic mineral fibres | | 29 | 1 |
|     Glass fibre | 17 | | |
|     Rockwool fibre | 9 | | |
| Dusts from plywood, plastics, etc. | | 29 | 1 |
| Cosmetics | | 23 | 1 |
|     Hairdressers' chemicals | 20 | | |
| Synthetic resins and plastics | | 17 | 1 |
| Medicaments | | 16 | 1 |
| Paints and lacquers | | 13 | 1 |
| Disinfectants | | 10 | <1 |
| Dyes | | 8 | <1 |
| Rubbers and elastomers | | 7 | <1 |
| Barrier creams, udder salves, other creams | | | |
|     (not medicaments) | 6 | <1 | |
| Fuels | | 7 | <1 |
| Pesticides | | 5 | <1 |
| Photographic chemicals | | 5 | <1 |
| Other chemicals | | 104 | 5 |
| *Physical causes* | | 30 | 1 |
| Humidity | | 13 | 1 |
| Maceration | | 10 | <1 |
| *Biological causes, physical and* | | | |
|    *psychological causes* | | 54 | 2 |
| Abnormal friction of the skin | | 47 | 2 |
| *Cause not given* | | 174 | 8 |

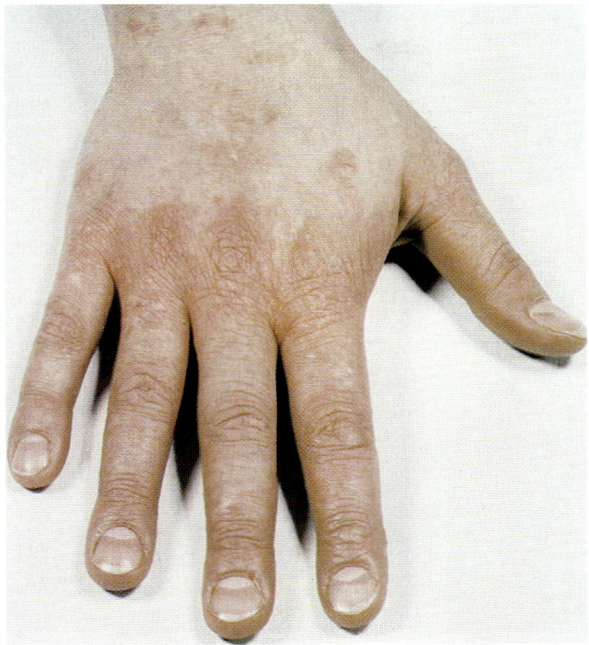

*Fig. 6.* Typical irritant dermatitis on the backs of the hands caused by detergents.

*Cement (2%).* Cement is one of the alkalis that induces irritant dermatitis. Its role is diminishing because the workers have been taught to use protective equipment, e.g. protective gloves [14].

*Synthetic Mineral Fibres (1%).* This well-known cause of occupational skin disease [5] (fig. 7) was less common in the present statistics than would have been expected. Probably the recession in the building industry in Finland has lowered the frequency, and underreporting may partly be another cause for the low numbers.

*Physical Causes (below 2%).* These are rather rarely reported possibly because of underreporting [15, 16].

*Occupational Allergic Skin Diseases*

The most common causes of ACD are given in table 2. The most common causes were rubber chemicals (26%; i.e. the rubber chemical group and the individual rubber chemicals in table 2), synthetic resins, plastics, glues and paints

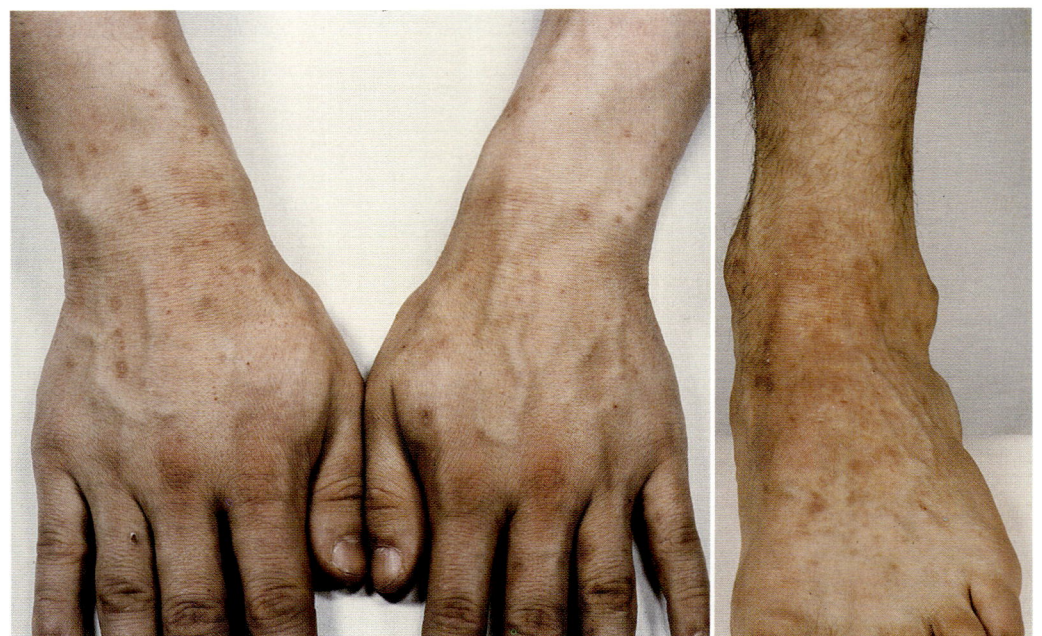

*a*

*b*

*Fig. 7. a* Typical irritant dermatitis on the wrists caused by mineral wool dust. *b* The same patient also had dermatitis on the ankles, caused by 'airbone' mineral wool dust.

*Table 2.* The causes of occupational allergic contact dermatoses (excluding contact urticaria) during 1990–1993 (1,567 cases) according to the Finnish Register of Occupational Diseases

| Cause | Cases | Total | % |
|---|---|---|---|
| *Individual chemicals* | | 612 | 39 |
| Metals | | 247 | 16 |
|   Nickel | 132 | | |
|   Chromium | 84 | | |
|   Cobalt | 19 | | |
|   Mercury | 5 | | |
| Aldehydes | | 86 | 5 |
|   Formaldehyde | 75 | | |
|   Glutaraldehyde | 11 | | |
| Thiuram sulfides | | 73 | 5 |
| Acrylate compounds | | 67 | 4 |
|   Aliphatic acrylates | 33 | | |
|   Aliphatic methacrylates | 31 | | |
| Amines | | 52 | 3 |
|   Phenylene diamines | 17 | | |
|   4-Isopropylaminodiphenyl (IPPD) | 7 | | |

*Table 2* (continued)

| Cause | Cases | Total | % |
|---|---|---|---|
| Glycidylethers | | 22 | 1 |
|     Diglycidylether of bisphenol A | 18 | | |
| Glycerylmonothioglycolate | | 10 | 1 |
| Mercaptobenzothiazole (MBT) | | 6 | <1 |
| Thioureas | | 5 | <1 |
| Ammonium thioglycolate | | 5 | <1 |
| Other individual chemicals | | 39 | 2 |
| *Mixes and chemicals classified according* | | | |
|    *to their usage* | | 895 | 57 |
| Rubber chemicals | | 328 | 21 |
| Synthetic resins and plastics | | 147 | 9 |
|     Epoxy resins and plastics | 82 | | |
|     Phenolformaldehyde resins and | | | |
|       plastics | 27 | | |
|     *p*-Tertiary butylphenolformal- | | | |
|       dehyde resins and plastics | 12 | | |
| Organic dusts and materials | | 71 | 5 |
|     Ornamental plants | 22 | | |
|     Wood | 13 | | |
|     Spices | 10 | | |
|     Flour, grains and feed | 5 | | |
| Preservatives and antimicrobials | | 60 | 4 |
|     Isothiazolinones | | | |
|       (e.g. Kathon CG) | 36 | | |
|     Benzalkonium chloride, | | | |
|       benzethonium chloride | 6 | | |
| Natural resins | | 49 | 3 |
| Glues | | 42 | 3 |
|     Epoxy glues | 16 | | |
|     Phenolformaldehyde glues | 15 | | |
| Dyes | | 26 | 2 |
|     Textile dyes | 12 | | |
|     Hair dyes | 7 | | |
| Medicaments | | 26 | 2 |
|     Antimicrobials | 6 | | |
| Paints and lacquers | | 17 | 1 |
|     Epoxy, paints | 9 | | |
| Barrier creams, udder salves, other creams | | | |
|     (not medicaments) | | 16 | 1 |
| Cosmetics | | 14 | 1 |
|     Hairdressers' chemicals | 10 | | |
| Fragrances | | 13 | 1 |
| Detergents | | 13 | 1 |
| Oils, lubricants and cutting fluids | | 12 | 1 |
| Other chemicals | | 62 | 4 |
| *Physical and biological causes* | | 2 | <1 |
| *Cause not given* | | 59 | 4 |

(19%), and metals (16%). Occupational rubber chemical allergy is most commonly caused by protective gloves [14]. Thioureas are an unusual cause of rubber chemical allergy; most cases in Finland have been diagnosed at our Institute [17]. Interestingly, no cases of occupational allergic dermatoses were detected in rubber industry during 1993 (fig. 9). Chromate used to be the most common metal causing occupational skin disease [4, 18], but nickel has now surpassed it. The relatively high number of nickel allergy cases reflects the change in the Finnish law: since 1988 also work-related aggravation of a pre-existing disease can be considered as an occupational disease. Accordingly, it is probable that a part of the nickel allergy cases represent pre-existing allergy from, e.g., jewellery, aggravated by the work of, e.g., hairdressers and cashiers. Synthetic resins and plastics have remained a common cause of occupational dermatoses in Finland since the 1960s and 1970s [4, 18]. The most common cause has been the epoxy resins [19] but the share of acrylate compounds [20] is increasing. Among the antimicrobials, formaldehyde is still a common occupational allergen although isothiazolinones (e.g. Kathon CG) need to be remembered as occupational allergens. The number of allergy cases due to thioglycolates is small compared to some other countries [21].

*Contact Urticaria.* The main cause of CU was cow epithelium (70%; table 3) in dairy farming. Natural rubber latex was the second most common cause (n = 139; 22%; table 3). Natural rubber latex is an important cause of occupational diseases, especially in hospital and health care workers [22] and laboratory work in general, but it was also the most common cause of occupational allergic skin disease in, e.g., boat builders [23]. Flours, grains and feeds are the third most common group.

*Occupational Dermatoses by Industry per 100,000 Employed Persons*
Leather and leather goods manufacture, followed by agriculture were the most common industries in which occupational irritant dermatoses occurred during 1993 when calculated per 100,000 workers (fig. 8). Similarly, the manufacture of refined petroleum products, followed by leather and leather goods manufacture, and beverage manufacture were the most common industries in which allergic skin diseases occurred during 1993 when calculated per 100,000 workers (fig. 6). Interestingly, in several industries no cases of either allergic or irritant dermatoses were reported during 1993 (fig. 8, 9), including, e.g. forestry and logging. There has probably been underreporting of occupational dermatoses. It should also be emphasized that the number of workers in one field may be small, and accordingly the figures may be misleading. Our numbers per 100,000 workers are much lower than those reported by Diepgen and Fartasch [7]; they reported, e.g., as many as 2,440 occupational skin diseases per 100,000 employed hairdressers.

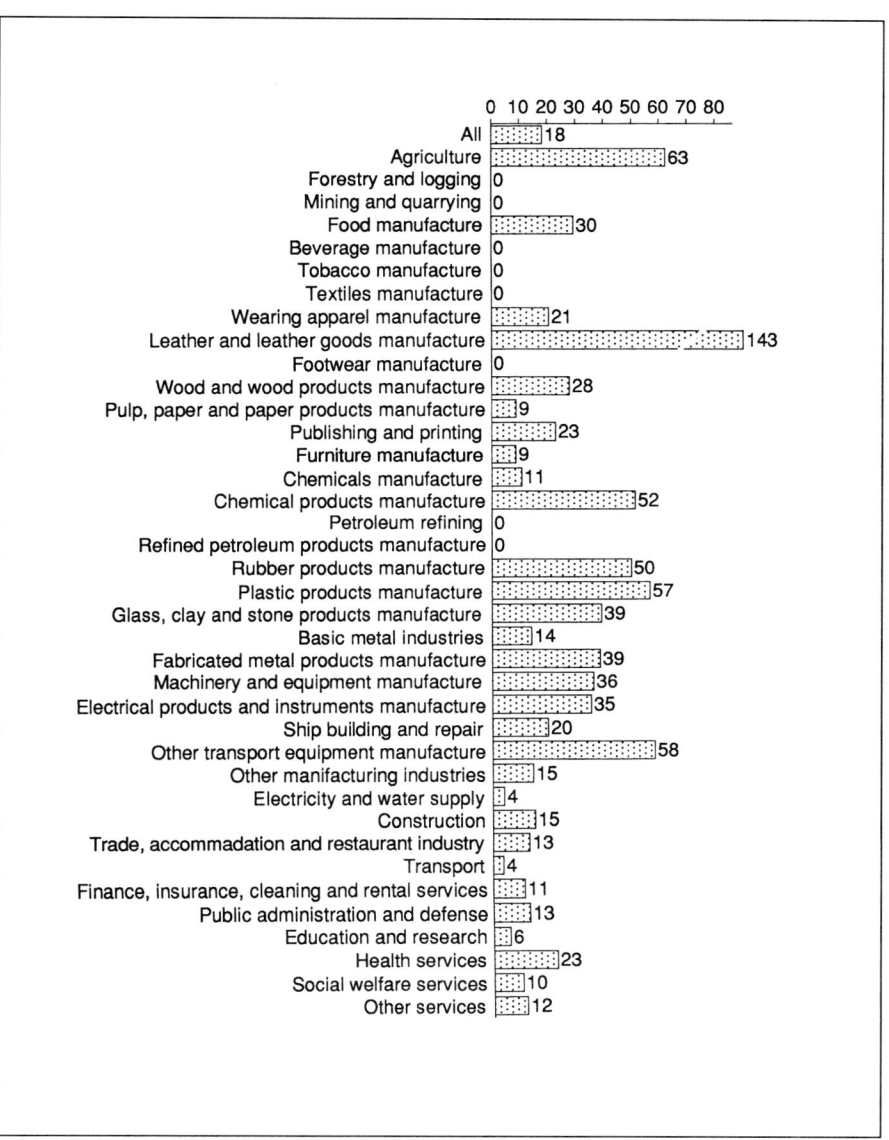

*Fig. 8.* Occupational irritant contact dermatoses by industry per 100,000 employed persons in 1993.

*Table 3.* The causes of occupational protein contact dermatitis and contact urticaria during 1990–1993 (620 cases) according to the Finnish Register of Occupational Diseases

| Cause | Cases | Total | % |
|---|---|---|---|
| *Individual chemicals* | | 11 | 2 |
| 2-Ethylhexylacrylate | | 5 | 1 |
| Other chemicals | | 6 | 1 |
| *Mixes and chemicals classified according* | | | |
| *to their usage* | | 598 | 96 |
| Organic dusts and materials | | 433 | 70 |
|    Cow epithelium | 276 | | |
|    Flour, grains and feed | 71 | | |
|    Enzymes | 11 | | |
|    Decorative plants | 10 | | |
|    Spices | 10 | | |
|    Roots | 9 | | |
|    Vegetables | 6 | | |
|    Pork | 5 | | |
|    Egg | 4 | | |
|    Onion | 4 | | |
|    Fish, fish meal | 3 | | |
|    Wood | 3 | | |
| Natural rubber, latex | | 139 | 22 |
| Handling of foodstuffs | | 19 | 3 |
| Other chemicals | | 7 | 1 |
| *Biological causes* | | 6 | 1 |
| Storage mites | | 4 | 1 |
| *Cause not given* | | 5 | 1 |

**Concluding Remarks**

The significance of irritant factors should not be underestimated when drawing conclusions from the present results. It is evident that irritant factors are involved in most of the allergic dermatoses, but they were not registered in the present statistics if a relevant occupational allergen was revealed.

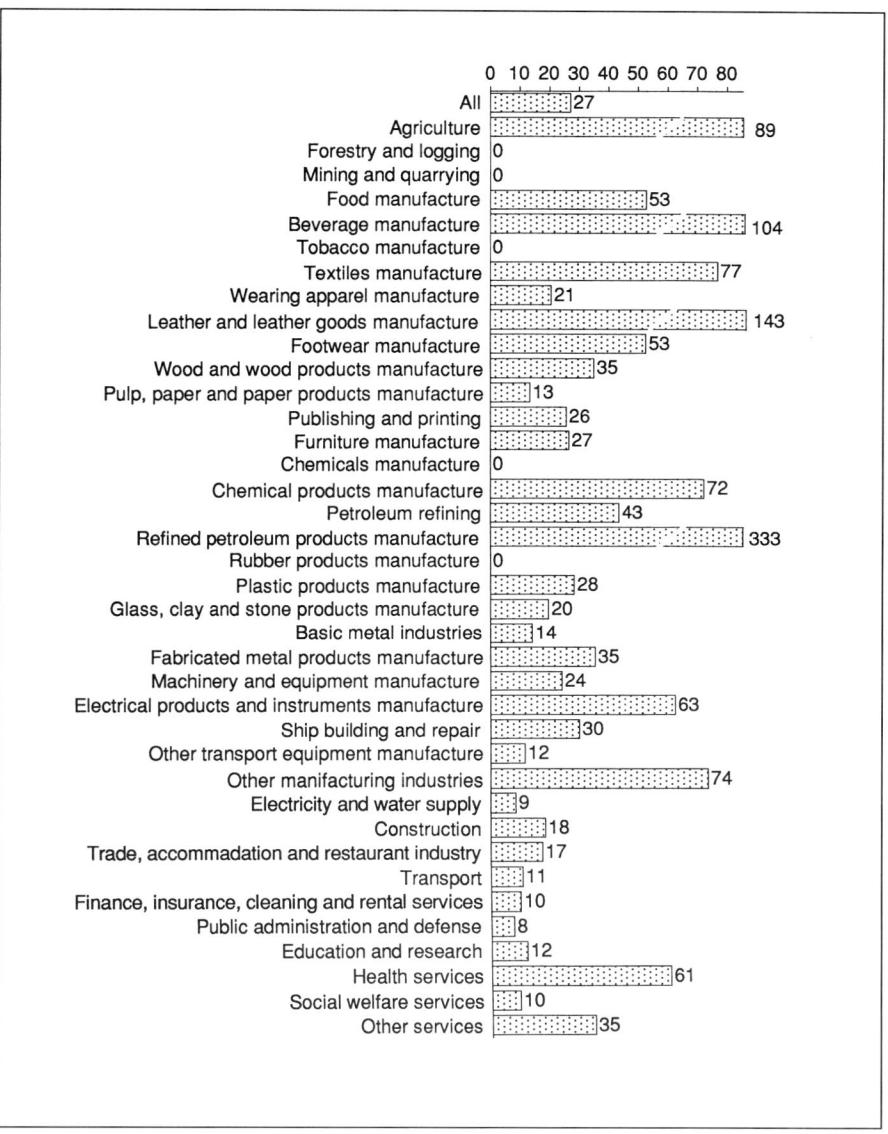

*Fig. 9.* Occupational allergic skin diseases (allergic contact dermatitis and contact urticaria) by industry per 100,000 employed persons in 1993.

# References

1 Kauppinen T, Vaaranen V, Vasama M, Toikkanen J, Jolanki R: Ammattitaudit 1993. Katsauksia 130 (in Finnish). Työterveyslaitos, Helsinki, 1994.

2 Kanerva L, Jolanki R, Toikkanen J, Keskinen H: Allergiset ammattitaudit v. 1992; in Kanerva L, Jolanki R, Toikkanen J (eds): Työterveyslaitos, Allergia ja työ-ohjelma (in Finnish), Helsinki, 1993.

3 Kanerva L, Vaheri E: Occupational allergic rhinitis in Finland. Int Arch Occup Environ Health 1993;64:565-568.

4 Kanerva L, Estlander T, Jolanki R: Occupational skin disease in Finland. An analysis of 10 years of statistics from an occupational dermatology clinic. Int Arch Occup Environ Health 1988;60:89-94.

5 Tarvainen K, Estlander T, Jolanki R, Kanerva L: Occupational dermatoses caused by man-made mineral fibers. Am J Contact Derm 1994;5:22-29.

6 Goldner R: Work-related irritant contact dermatitis; in Nethercott JR (ed): Occupational Medicine: State of the Art Reviews. 1994, Philadelphia, Hanley & Belfus, 1994, vol 9, pp 37-44.

7 Diepgen TL, Fartasch M: General aspects of risk factors in hand eczema; in Menné T, Maibach HI (eds): Hand Eczema, Boca Raton, CRC Press, 1994, pp 141-156.

8 Bruze M, Emmett EA: Occupational exposures to irritants; in Jackson EM, Goldner R (eds): Irritant Contact Dermatitis. New York, Marcel Dekker, 1990, pp 81-106.

9 Bason M, Lammintausta K, Maibach HI: Irritant dermatitis; in Marzulli FN, Maibach HI (eds): Dermatotoxicology, ed 4. New York, Hemisphere, 1991, pp 223-253.

10 Rycroft RJG, Wilkinson JD: Irritants and sensitizers; in Champion RH, Burton JL, Ebling FJG (eds): Rook-Wilkinson-Ebling, Textbook of Dermatology, ed 5. Oxford, Blackwell, 1992, pp 717-754.

11 Frosch PJ: Cutaneous irritation; in Rycroft RJG, Menné T, Frosch PJ, Benezra C (eds): Textbook of Contact Dermatitis. Berlin, Springer, 1992, pp 28-61.

12 Kanerva L, Estlander T, Jolanki R, Tarvainen K: Occupational allergic contact dermatitis and contact urticaria caused by polyfunctional aziridine hardener. Contact Derm, in press.

13 Hjorth N, Roed-Petersen J: Occupational protein contact dermatitis in food handlers. Contact Derm 1976;2:28-42.

14 Estlander T, Jolanki R, Kanerva L: Dermatitis and urticaria from rubber and plastic gloves. Contact Derm 1986;10:201-205.

15 Susten AS: The chronic effects of mechanical trauma to the skin: A review of the literature. Am J Ind Med 1985;8:281-288.

16 Kanerva L: Physical causes of occupational skin disease; in Adams RM (ed): Occupational Skin Disease, ed 2, Philadelphia, Saunders, 1990, pp 41-65.

17 Kanerva L, Estlander T, Jolanki R: Occupational allergic contact dermatitis caused by thiourea compounds. Contact Derm 1994;31:242-248.

18 Förström L, Pirilä V: 27 years of occupational dermatology in Finland. Derm Beruf Umwelt 1975; 23:207-213.

19 Jolanki R, Kanerva L, Estlander T, Tarvainen K, Henriks-Eckerman M-L: Occupational dermatoses from epoxy resin compounds. Contact Derm 1990;23:172-183.

20 Kanerva L, Estlander T, Jolanki R, Tarvainen K: Dermatitis from acrylates in dental personnel; in Menné T, Maibach HI (eds): Hand Eczema. Boca Raton, CRC Press, 1994, pp 231-254.

21 Frosch PJ, Burrows D, Camarasa JG, et al: Allergic reactions to a hairdressers' series: Results from 9 European countries. Contact Derm 1993;28:180-183.

22 Turjanmaa K: Latex glove urticaria; thesis, University of Tampere, 1988.

23 Tarvainen K, Jolanki R, Forsman-Grönholm L, Estlander T, Pfäffli P, Juntunen J, Kanerva L: Exposure, skin protection and occupational skin diseases in the fiberglass reinforced plastics industry. Contact Derm 1993;29:119-127.

Prof. Lasse Kanerva, MD, PhD, Section of Dermatology,
Finnish Institute of Occupational Health, Topeliuksenkatu 41 aA, FIN-00250 Helsinki (Finland)

Elsner P, Maibach HI (eds): Irritant Dermatitis. New Clinical and Experimental Aspects.
Curr Probl Dermatol. Basel, Karger, 1995, vol 23, pp 41–48

..........................

# Irritant Contact Dermatitis of the Hands in Housewives

*C.L. Meneghini*[a], *A. Sertoli*[b], *C. Nava*[c], *G. Angelini*[a], *S. Francalanci*[b],
*C. Foti*[a], *P. Moroni*[c, 1]

[a] Department of Dermatology, University of Bari;
[b] Department of Dermatology, University of Florence;
[c] Institute of Occupational Medicine, Milan;
  with the co-operation of Assocasa, Federchimica, Milan, Italy

Contact dermatitis (CD) caused by irritants and/or sensitizing agents in housewives (HW) and cleaners (Cl) is still a current occupational and social problem, especially related to the introduction and use of new detergents and hygiene products in the domestic and extradomestic environment. The purpose of this study, started in 1988, was to gather information furthering the knowledge of CD in HW and Cl.

The aspects considered were: (a) the type and composition of common household cleaning and hygiene products; (b) their irritant or sensitizing incidence in CD; (c) predisposing factors, particularly atopy, and (d) clinical patterns and other features.

## Methods and Materials

Three University Centers participated in this study, with the cooperation of 'Assocasa' (member of the Italian Federchimica), an association of manufacturers of household cleaning and hygiene products, which represents about 100 Italian and multinational companies operating in Italy. The first step was to collect generic data on the use of the products and possible complaints by means of a questionnaire published in a well-known weekly magazine addressed to HW and Cl. Nearly 2,000 women answered, indicating 824 different products as alledged noxious agents for the skin.

[1] We thank the Executive Board and the manufacturers of Assocasa (Italy) for providing information and household products, and for their kind support.

The second part of the investigation was carried out at the dermatology centers involved in the study according to a standard design, as follows:
(1) Obtaining informed consent.
(2) Eliciting the history of occupational and nonoccupational exposures.
(3) Medical and dermatological history.
(4) Clinical examination with special regard to the patterns of hand eczema and other cutaneous features.
(5) Diagnostic patch tests, combined with the above examination, to differentiate irritant (I) from allergic (A) CD.

*Table 1.* Substances contained in household cleaning and hygiene products and concentrations for patch testing

| | |
|---|---|
| Na alkyl-benzene sulfonate | 0.1 % aq. |
| Na toluene sulfonate | 0.5 % aq. |
| Alkylamine oxide | 1 % aq. |
| Diethanolamide (coccoil) | 1 % pet. |
| Tetra-acetylethylenediamine | 2 % pet. |
| Na perborate | 0.5 % aq. |
| Na tripolyphosphate | 0.5 % aq. |
| Ethoxylated fatty alcohols | 0.1 % aq. |
| Ethoxysulfate fatty alcohols | 0.1 % aq. |
| Distearyl-dimethyl ammonium chloride | 0.1 % aq. |
| Na dichloroisocyanurate | 2 % aq. |
| Na ethylenediamine tetraacetate | 1 % aq. |
| Na xylene sulfonate | 1 % aq. |
| Polyacrylate | 1 % aq. |
| DDT | 1 % pet. |
| Fluorescent brightener 119 | 2 % pet. |
| 4,4-bis-(2-Benzoazolyl) stilbene | 2 % pet. |
| 3-Phenyl-7-naphthotriazol-2-cumarine | 2 % pet. |
| Waxes mix | 30 % pet. |
| Triethanolamine sulfate | 1 % pet. |
| Color Index 42735 | 1 % pet. |
| Color Index 45170 | 1 % pet. |
| Color Index 61585 | 1 % pet. |
| Color Index 59040 | 1 % pet. |
| Color Index 74160 | 1 % pet. |
| 2-Bromo-2-nitropropane-1-3-diol | 0.2 % |
| Fatty alcohol phosphonate | 10 % aq. |
| Tetrametryne | 2 % pet. |
| Bioalletryne | 2 % pet. |
| Piperonilbutoxide | 3 % pet. |
| Propoxur | 1 % pet. |

Data on atopy, cutaneous or mucosal, were obtained according to the criteria of Hanifin and Rajka [1].

The majority of subjects were patch tested with: (a) a standard series of allergens used by GIRDCA (Italian Contact and Environmental Dermatitis Research Group), with some additional allergens such as fragrances mix, phenol formaldehydic resin, garlic, primin, Vioform and (b) another series of substances, prepared with components of household cleaning and hygiene products suggested and provided by the experts of manufacturers coworking in the investigation (table 1). In some cases 'open' and 'usage' tests were made.

The method of recording patch test reactions was as recommended by the International Contact Dermatitis Research Group [2].

The study included 1,719 female subjects selected on the basis of their working activity as HW (full time or part time) or professional Cl, exposed to contact with household and hygiene products: 1,100 of them were patients affected with hand contact dermatitis referred by their general practitioners or dermatologists for allergological investigation, or female workers with suspected hand dermatitis from hospitals, schools, cleaning firms and other kinds of employment.

Another group of 619 women, HW or not, with other mild diseases, cutaneous (noneczematous type) or mucosal, were considered as controls. Not all the 1,719 subjects who agreed to the study completed it (loss due to not returning for follow-up, lack of cooperation, or other reasons).

### Results

Table 2 summarizes the history of mucosal or cutaneous atopy in our patients with CD (hand eczema): atopy is slightly prevalent in the ICD group. Table 3 shows that more than 60% of patients preferred protective gloves to barrier creams: in comparison, only 26.6% of controls used gloves and 22.2% creams. Table 4 underlines the high frequency of subjective symptoms (itching, burning, smarting) and the chronic patterns of dry and scaling lesions. Clinical and allergological findings are summarized in tables 5–9. Patch test results are negative in 55.6% and positive to 1 or more allergens in 44.4% of the 1,100 patients. The main incidence of ICD is in subjects aged between 16 and 45 years, with a percentage of about 80%; approximately the same incidence is found for ACD in this age group. The primary site of CD of the hands is obviously local in the majority of cases but onset can be also observed in some patients (9%) on the forearms, face or neck.

As regards the patch test results with the Italian standard series of allergens (table 8), the most frequent positive reactions are caused by nickel, in 28.7% of the patients and in 10.1% of the controls. Lower percentages are attributable to chromium and cobalt and, in decreasing order, to fragrances, thiurams, Kathon CG, balsam of Peru, other rubber components, p-phenylenediamine, garlic, and other allergens.

*Table 2.* Positive history of mucosal atopy in 1,100 patients with CD (hand eczema)

|  | ICD, % | ACD, % |
|---|---|---|
| In childhood | 6.2 | 4.5 |
| In adult age | 12.2 | 10.5 |
| Positive history of 'dermatitis' in childhood | | |
|    (in most cases atopic dermatitis) | 10.4 | 8.5 |

*Table 3.* Protective clothing or creams adopted by patients with hand eczema and controls in cleaning work

|  | Gloves % | Barrier creams % |
|---|---|---|
| Patients (n = 1,100) | 60.6 | 7.6 |
| Control (n = 619) | 26.6 | 22.2 |

*Table 4.* Symptoms and lesions in 1,100 patients with CD (hand eczema)

|  | n | % |
|---|---|---|
| Itching, stinging, smarting | 857 | 77.9 |
| Erythema | 841 | 76.4 |
| Scaling | 757 | 68.8 |
| Fissures | 403 | 36.6 |
| Vesicles | 357 | 32.4 |
| Edema | 132 | 12 |

*Table 5.* Positive patch tests (PT) to the standard series allergens of GIRDCA (Italian Research Group of Contact and Environmental Dermatitis) and to the additional series with household cleaning, hygiene and domestic environmental substances

|  | PT positive | | PT negative | |
|---|---|---|---|---|
|  | n | % | n | % |
| Patients (n = 1,100) | 488 | 44.4 | 612 | 55.6 |
| Controls (n = 619) | 73 | 11.8 | 546 | 88.2 |

Table 6. Incidence of ICD of the hands with respect to age (n = 556)

| Age, years | % |
|---|---|
| >16 | 0.7 |
| 16–30 | 47.8 |
| 31–45 | 32.1 |
| 46–60 | 14.8 |
| >60 | 4.3 |

Table 7. Location at onset of CD of the hands in 1,100 patients

| | ICD, % | ACD, % |
|---|---|---|
| Hands (fingers, palms, dorsa, nails) | 91.7 | 85.3 |
| Face | 4.1 | 6.5 |
| Forearm | 3.4 | 5.5 |
| Neck | 2.0 | 1.5 |
| Other sites | 4.6 | 5.9 |

Table 8. Positive patch tests (PT) in patients with CD (hand eczema) and controls with allergens of the standard series used by GIRDCA and other additional allergens

| Allergens | Patients | | Controls | |
|---|---|---|---|---|
| | PT positive | % | PT positive | % |
| Ni sulfate | 316/1,100 | 28.7 | 66/619 | 10.1 |
| Co chloride | 71/1,100 | 6.4 | 12/619 | 1.9 |
| K bichromate | 53/1,100 | 4.8 | 4/619 | 0.6 |
| Fragrance mix | 34/1,100 | 3.1 | 8/619 | 1.3 |
| Thiuram mix | 31/1,100 | 2.8 | 1/619 | 0.2 |
| Kathon CG | 24/925 | 2.6 | 2/459 | 0.4 |
| Balsam of Peru | 28/1,100 | 2.5 | 6/619 | 1.0 |
| DDM | 24/1,100 | 2.2 | 7/619 | 1.1 |
| Carba mix | 15/980 | 1.5 | 0/619 | – |
| PPD | 17/1,100 | 1.4 | 5/619 | 0.8 |
| Ethylenediamine | 16/1,100 | 1.4 | 2/619 | 0.3 |
| Benzocaine | 13/1,100 | 1.2 | 2/619 | 0.3 |
| PPD mix | 12/1,100 | 1.1 | 2/619 | 0.3 |
| Mercaptobenzotiazole | 12/1,100 | 1.1 | 1/619 | 0.2 |
| MBT mix | 10/1,100 | 0.9 | 1/619 | 0.2 |
| Primin | 3/804 | | 1/459 | |
| Garlic | 11/788 | | 2/451 | |
| Phenolform.resin | 6/798 | | 3/455 | |
| Vioform | 2/798 | | 0/456 | |

*Table 9.* Positive patch tests (PT) to the additional series of household substances tested in groups of consenting patients and controls

| Substances | PT positive patients | PT positive controls |
|---|---|---|
| DDT, 1% | 2/980 | 0/619 |
| Na tripolyphosphate, 0.5% aq. | 2/980 | 0/619 |
| Alkylamine oxide, 1% aq. | 1/980 | 0/619 |
| Triethanolamine sulfate, 1% | 2/802 | 0/454 |
| Diethanolamide coccoil, 1% | 2/980 | 0/619 |
| Color Index 45170, 1% | 1/475 | 0/316 |
| Fluorescent brightener 119, 2% | 1/980 | 0/619 |
| Tetrametryne, 2% | 2/980 | 0/619 |

Table 9 summarizes the positive patch test reactions to components of household cleaning and hygiene products, which were very few.

## Discussion

A history of atopy, either cutaneous or mucosal, is present in slightly more patients with ICD of the hands than in those with ACD or the general population.

Patients with CD of the hands preferred gloves to barrier creams in a higher percentage than controls: this is presumably related to protective measures adopted by patients only after the onset of skin manifestations. Chronic patterns of dry and scaling lesions are more common in the ICD than in the ACD group, the latter presenting erythematous and vesicular features more frequently.

As regards primary locations of CD in HW and Cl on the face or neck, these could be considered as airborne dermatitis.

The high incidence of sensitivity to nickel in women, even without clinical manifestations, has already been pointed out [3, 4]. The relatively high number of positive reactions to nickel in the control group may be explained by women having forgotten/not noticed a mild past dermatitis. The metals and its salts are ubiquitous allergens, while the sources of contact and potential sensitization are usually but not always recognizable. With regard to household detergents and hygiene products, recent analytical data on the content of nickel, chromium or cobalt in some household products do not seem to support an important sensitizing role of these metals contained in trace levels as impurities in some detergents [5]. However, we cannot completely exclude that the potential sensitizing capacity of nickel, chromium and cobalt may increase proportionally with the occasions of contact with minimal amounts of allergen, coming from many sources and acting on an altered skin for a long time.

With reference to nickel sensitivity without apparent cutaneous manifestations, an interesting recent study [6] on the effect of repeated open nickel application (up to 1 ppm) combined with repeated treatment with anionic surfactant sodium dodecyl sulfate on the hands of four nickel-allergic subjects, elicited negative results.

The question of how many cases of sensitivity to metals are really related to household materials and environment is open to doubt. The same consideration may be made for fragrances, Kathon CG and balsam of Peru, contained in household hygiene products or cosmetics.

The relatively high frequency of positive reactions to rubber components can be imputed to the use of protective gloves, as is well known. Sweating and macerating are important contributing factors in irritation and contact allergy. Gloves of different material are therefore recommended.

Among the positive results to various allergens, those to garlic are not negligible. The handling of these vegetable can cause either irritant or, to a lesser degree, allergic dermatitis of the finger tips. Other vegetables, such as celery, onion, carrot, and lettuce can also induce the same reactions.

Sensitivity to components of household cleaning and hygiene products is rare and scarcely significant. The action of some of them is presumably more irritant than sensitizing.

Thus, HW and Cl are most likely to be affected by CD during household wet activities, especially when using detergents without adopting correct preventive measures.


### Conclusion

The study of a large population of housewives and cleaners with CD of the hand shows: a slight prevalence of positive history of atopy in the ICD in comparison with the ACD group; itching and stinging is the major complaint in patients; patterns of chronic, dry, scaling and fissured eczema are prevalent especially in the ICD group; the main incidence of ICD is in subjects between 16 and 45 years of age, approximately the same as for ACD; patch test results are negative in about 55% of patients; a high incidence of sensitivity to nickel, about 29% in the patient group, and 10% in controls; lower positive reactions are observed to chromium and cobalt; high frequency of positive results to rubber components, presumably caused by gloves, and to fragrances; significant reactions to garlic, found to be not only an irritant but also a sensitizing agent; very rare sensitivity to the main compounds contained in household and hygiene products.

The major risk for housewives and cleaners of being affected with CD remains the high frequency of household wet activities, especially using detergents, and the lack of correct preventive measures.

### References

1   Hanifin JM, Raika G: Diagnostic features of atopic dermatitis. Acta Derm Venereol 1980; 92(suppl):44–47.
2   Fregert S, Hjorth N, Magnusson B, et al: Epidermiology of contact dermatitis. Trans St John's Hosp Derm Soc 1969;55:17–34.
3   Nava C, Meneghini CL, Sertoli A, et al: Indagine sulla dermatite da contatto delle mani delle casalinghe. Dati preliminari di uno studio multicentrico. G Ital Med Lavoro 1989;11:109–112.
4   Peltonen L: Nickel sensitivity in the general population. Contact Derm 1979;5:27–32.
5   Fedler R, Stromer K: Nickel sensitivity in atopics, psoriatics and healthy subjects. Contact Derm 1993;29:65–69.
6   Basketter DA, Briatico-Vangosa G, Keastner W, Lally C, Bontinec WJ: Nickel, cobalt and chromium in consumer products: A role in allergic contact dermatitis? Contact Derm 1993;28:15–25.
7   Allenby CF, Basketter DA: The effect of repeated open exposure to low levels of nickel on compromised hand skin of nickel-allergic subjects. Contact Derm 1994;30:135–138.

Prof. C.L. Meneghini, Clinica Dermatologica Università Bari,
Via A. Gimma, 99, I–70122 Bari (Italy)

Elsner P, Maibach HI (eds): Irritant Dermatitis. New Clinical and Experimental Aspects.
Curr Probl Dermatol. Basel, Karger, 1995, vol 23, pp 49–55

..........................

# Occupational Dermatitis in Hairdressing Apprentices

## Early-Onset Irritant Skin Damage[1]

*Wolfgang Uter*[a], *Olaf Gefeller*[b], *Hans Joachim Schwanitz*[a]

[a] University of Osnabrück/AGW, Osnabrück, and
[b] University of Göttingen, Department of Medical Statistics, Göttingen, Germany

Wet work has been identified as a major cause of irritant hand eczema in several occupations, including the hairdressing trade. Hand eczema can result in significant morbidity or necessitate a change of job. Because of the great individual significance and the socioeconomic impact primary prevention has to be improved. This may be achieved by both: (1) adequate preplacement dermatological examination and advice, and (2) the conception and realization of skin-safe working conditions [1, 2].

Prospective studies [3, 4] directed towards evaluating constitutional and occupational risk factors for the development of hand eczema in hairdressing apprentices are currently under way. First results of the study centered in Osnabrück will be presented and discussed.

## Methods

The study presented involves vocational training schools in 14 cities of different size in Northern Germany, where hairdressing apprentices are examined soon after the start of their training (a *pre*-employment examination is rarely possible for logistic reasons). Follow-up examinations are scheduled at the end of the first year and after 3 years (at the end of the training). Two waves have been started in summer 1992 and 1993, and a third and last wave commenced in summer 1994.

Prior to the first examination, a standardized lecture on structure, function and protection of the skin is delivered and discussed. The compact intervention is supplemented by

---

[1] Without the help of the observers J. Raguz, I. Jánossy, K. Keller, I. Kirbach and S.M. John, this extensive study would not have been possible.

*Table 1.* Professional tasks for hairdressing novices

| | Average time min | Done by % | Times per day | |
|---|---|---|---|---|
| | | | mean | median |
| Shampooing | 9 | 96 | 12.5 | 10 |
| Fixation | 2 | 90 | 5.0 | 4 |
| Acid perm | 20 | 51/80[1] | 2.8 | 2 |
| Alcaline perm | 20 | 57/74 | 2.7 | 2 |
| Coloring | 20 | 58/85 | 2.5 | 2 |
| Bleaching | 5 | 34 | 0.8 | 0 |

[1] Percentages at right are for rinsing off.

educational units developed by educationalists of the Osnabrück team available for routine use in the schools [5]. A short self-administered questionnaire relating to family and personal history (atopy, intolerances), leisure activities and specific tasks and skin protection at work is checked by the observer, who additionally records physical findings on a standardized form.

Data analysis was performed using the SAS™ (version 6.08) software. Estimated average times per occupational task (table 1) were used to approximatively calculate the individual time of unprotected wet work per day by multiplication with the number of times the respective task was performed per day *without* gloves (and similarly *with* protective gloves). Logistic regression analysis was employed to analyze the relationship between skin damage and a variety of potential risk factors. Adjusted odds ratios and their accompanying 95% confidence intervals were derived from the primary logistic regression model [6].

## Results

In 1993, 859 participants were examined at the start of their training. Many apprentices showed slight skin changes (23.5%) or frank eczema (14.7%), mostly localized interdigitally (66.5%), at the back of the hand (13.0%) or both sites (14.2%), sometimes at other sites. Morphologically, scaling (76.7%) and erythema (77.3%) prevailed, while infiltration (10.2%), vesicles (8.3%), erosions (3.8%), papules (2.7%) or fissuring (1.5%) were observed only occasionally. From the clinical picture and test results (if available), *allergic* contact dermatitis was considered only as an exception, i.e. in 5 of 328 (1.5%), compared to (slight) *irritant* contact dermatitis.

At that initial stage of their training, hairdressing apprentices' tasks (table 1) mainly include frequent hairwashing, fixation and rinsing off of colorings or per-

*Table 2.* Use of protective gloves (% values)

|  | Regularly | Sometimes | Never |
|---|---|---|---|
| Shampooing | 18.5 | 21.1 | 60.4 |
| Fixation | 39.0 | 11.5 | 49.5 |
| Acid perm | 39.5 | 11.6 | 48.9 |
| Alcaline perm | 33.7 | 13.1 | 53.2 |
| Coloring | 87.1 | 5.7 | 7.3 |
| Bleaching | 69.8 | 7.3 | 22.9 |

No gloves at all: 17.1%; no active work yet: 3.5%.

*Table 3.* Material of protective gloves

|  | Alone | With other materials | Total |
|---|---|---|---|
| Latex | 7.2 | 11.8 | 19.0 |
| Vinyl | 8.0 | 11.4 | 19.4 |
| PE | 23.9 | 22.1 | 46.0 |
| 'Thick' rubber[1] | 10.0 | 21.3 | 31.3 |

[1] Multiple-use, long sleeve, 0.3-mm-thick gloves (natural latex) with a thin lining of cotton fleece.

manent waving solutions, often supplemented by cleaning jobs both dry and wet (45.4% up to 1 h, 43.3% 1–4 h/day).

Protective gloves are used by nearly 80% of apprentices (table 2), but those who perform the respective work mostly do not use them regularly, except for the application of coloring solutions. Glove materials used are shown in table 3. Moisturizing creams are recommended to counteract the washout of epidermal lipids which is still to be expected to some extent even if protective gloves are worn (e.g. by handwashing with water). While most apprentices (82.2%) stated to use creams at work, relatively few applied them often, e.g. 13.2% more than 10 times per day. At home, 83.5% used creams (mean 3.1 times/day).

Results of a multifactorial analysis relating the outcome ('mild-to-severe skin changes' with a frequency of 38.2%) or 'hand eczema' (i.e. more pronounced findings, with a frequency of 14.7%), respectively, to (risk) factors included in the model are presented in table 4.

*Table 4.* Primary logistic regression analysis of potential risk factors

| | % of all | 'Skin changes' | | 'Eczema' | |
|---|---|---|---|---|---|
| | | OR | 95% CI | OR | 95% CI |
| Sex (M) | 7.3 | 1.38 | 0.79–2.44 | 1.58 | 0.76–3.27 |
| Score | | | | | |
| 3.5–5.0 | 20.4 | 0.98 | 0.61–1.60 | 0.71 | 0.34–1.44 |
| 5.5–7.0 | 19.0 | 1.22 | 0.74–2.00 | 0.87 | 0.43–1.76 |
| 7.5–9.5 | 19.6 | 1.72* | 1.06–2.79 | 1.61 | 0.84–3.06 |
| >9.5 | 22.0 | 1.11 | 0.67–1.81 | 0.99 | 0.51–1.95 |
| Past hand eczema | | | | | |
| (pompholyx or eczema) | 8.5 | 2.51* | 1.47–4.26 | 1.85 | 0.98–3.49 |
| Flexural eczema | | | | | |
| (past or present) | 6.5 | 1.71 | 0.95–3.08 | 1.25 | 0.58–2.69 |
| Duration of training | | | | | |
| >3–6 weeks | 35.0 | 0.66 | 0.39–1.11 | 2.25 | 0.95–5.36 |
| >6–9 weeks | 22.0 | 1.15 | 0.66–2.01 | 2.34 | 0.94–5.82 |
| >9–12 weeks | 5.2 | 2.60* | 1.17–5.77 | 5.72* | 1.98–16.54 |
| >12 weeks | 23.7 | 0.93 | 0.53–1.63 | 1.59 | 0.63–3.97 |
| Observer[1] | | | | | |
| 1 | 39.1 | 1.35 | 0.57–3.17 | 0.38 | 0.12–1.23 |
| 2 | 10.1 | 1.94 | 0.74–5.05 | 0.28 | 0.07–1.01 |
| 3 | 12.6 | 0.54 | 0.21–1.41 | 0.29 | 0.08–1.09 |
| 4 | 29.8 | 1.02 | 0.43–2.43 | 0.68 | 0.21–2.20 |
| Wet work | | | | | |
| >1–2 h | 16.8 | 1.10 | 0.68–1.77 | 1.36 | 0.67–2.72 |
| >2–4 h | 31.5 | 2.28* | 1.54–3.39 | 2.07* | 1.17–3.66 |
| >4 h | 22.2 | 2.51* | 1.63–3.88 | 2.49* | 1.36–4.55 |
| Smoking (>5/day) | 47.8 | 1.36* | 1.01–1.83 | 1.92* | 1.27–2.91 |
| *Alternatives* | | | | | |
| For wet work (more than median) | | | | | |
| Shampooing | 29.7 | 1.50* | 1.07–2.11 | 1.10 | 0.70–1.73 |
| Acid perm | 36.2 | 1.61* | 1.12–2.31 | 1.56 | 0.98–2.51 |
| Alkaline perm | 36.9 | 1.11 | 0.77–1.59 | 1.46 | 0.91–2.35 |
| Fixation | 23.4 | 0.82 | 0.56–1.22 | 0.82 | 0.49–1.38 |

OR = Odds ratio; CI = confidence interval.
* Significant (p < 0.05).
[1] Only those included who participated in more than one school (91.6%).

## Discussion

The frequency of skin damage is high in German hairdressing apprentices after a relatively short period of heavy professional exposure (e.g. 51% within 3 months [4]), and an association to the duration of training can be seen at this stage (table 4). (The apparent decline in risk seen for a duration of more than 12 weeks is mostly a geographical or organizational 'artifact': 75% of this group have been seen in one particular school where trainees had not been working for 2–4 weeks prior to examination, thus allowing for resolution of possible previous skin damage.)

A retrospective analysis appears adequate to evaluate risk factors of this type of 'early onset irritant skin damage' mostly presenting as interdigital web space eczema. This very typical irritant eczema should not be neglected, as it may be a 'soft spot' with regard to the development of contact allergy in a second phase.

### Constitutional Risk Factors

A history of 'childhood eczema' [7], and hand eczema in particular [8], or pompholyx ('dyshidrosis') [9], has been identified as a risk factor for the development of hand eczema, particularly in wet work occupations. The OR of 1.85 found in our study is in agreement with these previous investigations, but may nevertheless underestimate the true deleterious effect of this risk factor, as persons with such a history who developed hand eczema very early in their training may have dropped out before they could be included in the study population. For a relatively unequivocal criterion of atopic dermatitis such as (past or present) pruritic flexural eczema (encountered here in 6.5%, about as often as in a comparable study: 6% [4]) presently an OR of 1.71 has been found for the global risk of 'irritant skin damage' and of 1.25 for hand eczema, although this was not significant (p > 0.05). Clinical and anamnestic atopic parameters have been operationalized with a score system according to Diepgen et al. [10]. This score only had a significant effect for a subgroup of persons with a value between 7.5 and 9.5 with regard to 'irritant damage' (but not those with a *higher* score). This 'decline in risk' for the highest quintile of the score is hard to explain and probably due to a yet unidentified confounding factor which has not been included in the model, or an effect modifier (i.e. an interaction between parameters included).

Smoking (more than 5 cigarettes per day, which is the median in the whole group) could be identified as a marginally significant risk factor for 'irritant damage', with increasing importance for 'hand eczema' (table 4). It may be argued, however, that this association does not reflect a true 'pharmacological', but rather a confounding effect: smoking could be an indicator for a careless attitude towards health.

*Exposure and Protection*

Prolonged, repeated contact with water, in particular when combined with detergents or other irritants, is a well-known cause of cumulative irritant skin damage [11]. This has been confirmed by the analysis presented, where even some dose-response gradient could be observed (table 4). It is noteworthy that 34.4% 'normal' persons (i.e. arbitrarily, those with an atopy score of 7 or less) had irritated skin, two thirds due to wet work of more than 2 h. It seems unlikely that the significant association between wet work and skin damage is due to differential misclassification (i.e. those with damaged skin 'explaining' this fact by exaggerating their self-reported workload), because in the majority of cases (minor) skin damage had either not been noticed or was considered 'normal' (and temporary) by the novices.

The analysis of single, contributing wet work tasks is hampered by imprecisions introduced by general or individual over- or underestimation of times needed per task or the number of times a task is performed per day, respectively. Thus, only marginally significant ORs have been found for shampooing more than ten times per day or acid permanent waving more than two times per day for global irritant damage (table 4).

*Role of the Observer*

Considerable effort had been devoted to the training of the observers (dermatologists or residents in dermatology) and the conception and application of operational definitions to achieve standardization for findings relating to atopic signs and skin damage of the hands. Despite this, considerable variability (nonsignificant in primary analysis) was partly found. Although we expect optimum training to further reduce variability, it will remain inevitable to a certain degree for the respective parameters as long as visual evaluation is concerned which cannot be objectively quantified or even qualified. The aspect of interobserver variability should therefore be considered in *any* multiobserver assessment.

**Outlook**

Further follow-up of the study cohort will allow for the evaluation of the long-term protective effect of glove wearing according to current legal guidelines [2]. The *need* for such preventive measures is already emphasized by the cross sectional results presented. A subgroup of persons with (atopic) risk factors who protect their skin adequately will be of particular interest to either reject or confirm the notion that these persons may do as well as others, which respective consequences for pre-employment advice.

# References

1   Schwanitz HJ: Prävention chronischer Friseurekzeme. Allergologie 1993;16:408–412.
2   TRGS 530: Technische Regeln für Gefahrstoffe 530. Bundesarbeitsblatt 1992;9:41-45.
3   Keller K, Köhn R, John SM, Wulfhorst B, Budde-Wamhoff U, Schöbel K, Steinebrunner B, Schwanitz HJ: Handekzeme bei Friseurauszubildenden. Erste Ergebnisse einer prospektiven ausbildungsbegleitenden Kohortenstudie. Allergologie 1993;16:404–407.
4   Diepgen TL, Tepe A, Pilz B, Schmidt A, Hüner A, Huber A, Hornstein OP, Frosch PJ, Fartasch M: Berufsbedingte Hauterkrankungen bei Auszubildenden im Friseur- und Krankenpflegeberuf. Konzept einer prospektiven Längsschnittstudie. Allergologie 1993;16:369–403.
5   Budde U: Unterrichtseinheit über Hautschäden im Friseurhandwerk. Osnabrück, Universität Osnabrück, 1992.
6   Hosmer DW Jr, Lemeshow S: Applied Logistic Regression. New York, Wiley, 1989.
7   Meding B, Swanbeck G: Predictive factors of having hand eczema. Contact Derm 1990;23:154–161.
8   Rystedt I: Factors influencing the occurrence of hand eczema in adults with a history of atopic dermatitis in childhood. Contact Derm 1985;12:185–191.
9   Hornstein OP, Bäurle G, Kienlein Kletschka B: Prospektivstudie zur Bedeutung konstitutioneller Parameter für die Ekzemgenese im Friseur- und Baugewerbe. Derm Beruf Umwelt 1985;33:43–49.
10  Diepgen TL, Fartasch M, Reindl H, März A, Hornstein OP: Die Bedeutung atopischer Major- und Minorkriterien für die ärztliche Atopie-Diagnostik. Akt Beitr Umwelt Berufskrankh Haut 1991;3:77–101.
11  Kienlein Kletschka B: Feuchtarbeit als konditionierender Faktor bei der Genese berufsbedingter Dermatosen. Derm Beruf Umwelt 1984;32:14–16.

Dr. med. W. Uter, University of Osnabrück/AGW, Albrechtstrasse 28,
D–49069 Osnabrück (Germany)

Elsner P, Maibach HI (eds): Irritant Dermatitis. New Clinical and Experimental Aspects.
Curr Probl Dermatol. Basel, Karger, 1995, vol 23, pp 56–63

..........................

# Expression of Atopic Criteria in a Population of Medical Nurses and Hairdressers at the Beginning of Vocational Training[1]

*M. Gebhardt*[a], *A. Seidel*[b], *R. Bartsch*[b]

Departments of
[a] Dermatology, and
[b] Occupational and Social Medicine, University of Jena, Germany

Within the last decades, the number of notifications of occupational dermatoses in Germany is constantly increasing. Therefore, prophylactic procedures in or even before vocational training are of substantial interest. Whereas predictive patch testing, e.g. testing before the first contact to the substance, was used to prognosticate the development of occupational dermatoses in the past, we know about the absurdity of such measures in the prediction of diseases which appear during occupation today. Nevertheless, prediction of occupational tolerance still remains a problem and reliable instruments are required in this field.

Atopic diathesis was proved to be a serious risk of development of irritant or allergic skin diseases due to occupational influences [1, 2]. Therefore, recording atopics at the beginning of job training and follow-up of their skin condition during the training might lead to the diagnosis of occupational skin disorders at an earlier stage. So prophylactic or therapeutic procedures directed against the disorder could act more effectively and the trainees are able to stay in their job.

Jobs most probably associated with hazards for development of occupational dermatoses are health care workers in general, hairdressers, bricklayers, metal workers, food handlers, painters, gardeners and cleaning staff [3–6]. There were 6,170 (32.9%) hairdressers and health care workers among the nearly 21,000

[1] The study was supported by the 'Hauptverband der Gewerblichen Berufsgenossenschaften', St. Augustin, Germany.

notifications of occupational dermatoses in Germany in 1990 [7]. The aim of our study was to assess a cohort of 187 hairdressers and 266 health care workers (nurses, doctor's receptionists, dentist's assistants, midwives) concerning the skin state using a combination of questionnaire and medical examination supervised by a dermatologist. In the present paper, we present the data of all persons included in the study at the beginning of vocational training. We focus on atopic criteria of the population as a basic value for follow-up studies on the same population which we are about to perform at the moment.

## Methods

Examinations were performed in 2 vocational schools (Jena, Gera) in the east of the Thuringian region (Germany). At a first visit, pupils were informed about the aim of the study. Moreover, after they had given their informed consent, they were given a questionnaire, established by Diepgen's group, concerning the history of atopic diseases, skin diseases and the presence of atopic criteria.

In a second step, pupils consulted the dermatologist in the school for case history and examination of the skin, especially regarding minimal changes or manifestations of hand eczema. The examination schedule included information on previous jobs, history of occupational skin problems, atopic criteria, family case history and a documentation of present skin disturbances.

The aim of the study was to screen employees for an atopic disposition and to follow up their skin condition during the period of job training by means of repeated examinations.

All the examinations were performed during the first weeks of job training. (Before beginning vocational training, the prospective trainees were not available, although that time point would be better for a first examination.) This period is characterized by a fortnightly rotation between theoretical education and training on the job. We have seen the employees in the theoretical education phase only. Evaluating the data of skin problems in training combined with the atopic case history, we had to unite both groups in order to reach statistically assessable numbers.

## Results

### Evaluation of Case Histories

454 trainees, 267 health care workers (266 were evaluated) and 187 hairdressers were included in the study until now. The mean age of hairdressers was 18.2 years and that of health care workers 19.9. 32.0% of health care trainees had had a former job compared with 8.0% of the hairdressers.

19.3% (= 36) of the hairdressers and 14.3% (= 38) of the health care workers complained of skin problems during the first weeks of their training. Figure 1

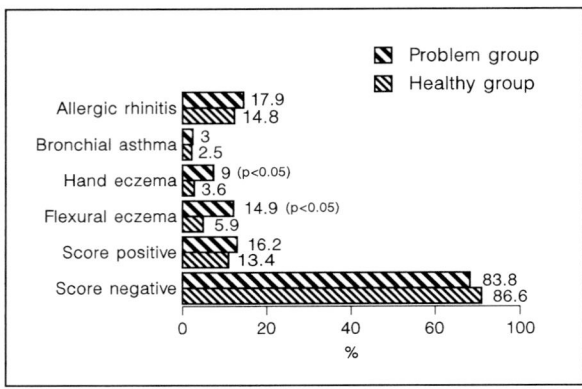

*Fig. 1.* Frequency of atopic diseases. Comparison between the group with skin problems and the healthy group.

shows a comparison of a group with skin problems in training and the group without problems concerning the incidence of atopic disease in their case histories.

32.1%/32.8% of trainees in nursing/hairdressing complained of dry skin. 86.8% of the health care trainees reported the use of skin protection ointments whereas 92.5% of the hairdressers used such products. The role of 'milky crusts' in the prediction of atopic diseases is controversial. 10.2% of medical trainees and 7.5% of hairdressers had a positive history. An intolerance to textiles, mostly wool, was confirmed by 32.7%/30.5% of interviewees in nursing/hairdressing. 39.1% of health care workers but only 29.9% of hairdressers complained of metal intolerance, appearing as jewellery eczema. 90.6% of health care trainees had had their ears pierced compared to 95.7% of hairdressers. 31.7%/41.5% of them had acquired an intolerance to fashion jewellery by the age of 12 years and 61.5%/71.7% by the age of 15. There was not a single case who had the complaint without previous ear piercing.

Food intolerances were reported in 9.4%/9.1% of nursing/hairdressing trainees. Some cases were pollen-associated food allergies and some citrus fruit irritations of the perioral region.

Allergies which led to medical consultation (to drugs, food, pollens, animals, house dust, etc.) were reported in 20.7% of the health care workers and in 11.2% of the hairdressers.

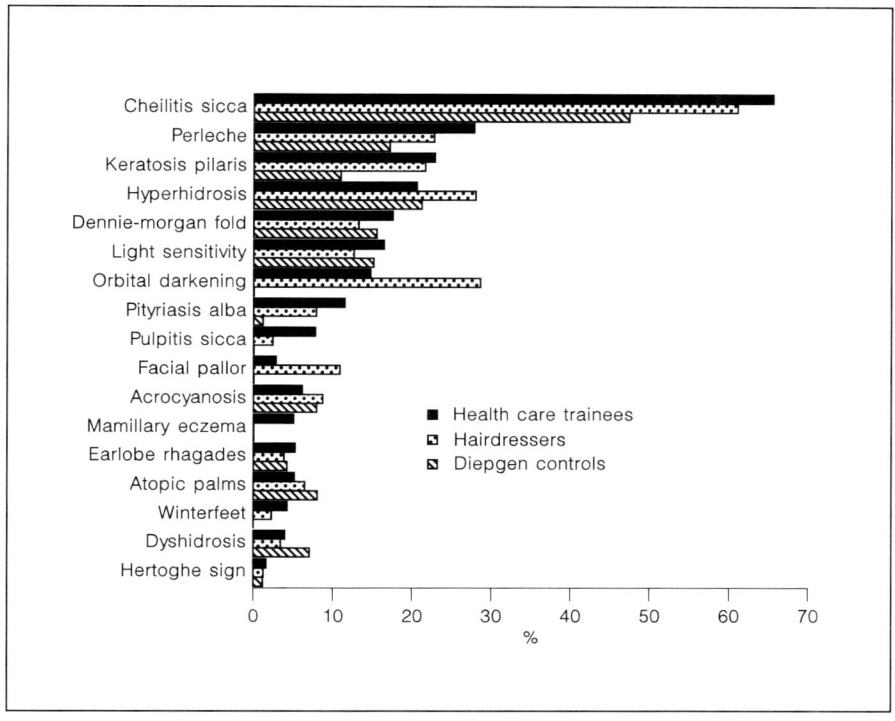

*Fig. 2.* Criteria of atopy. Comparison between the data obtained in hairdressers, health care workers and the data of noneczematous population of Diepgen [10].

*Evaluation of Examination Findings*

Every trainee was examined concerning the existence of atopic criteria by a dermatologist. In figure 2, we present positive data without regarding questionable data. Most of the criteria are showing a good correlation between both groups except for pulpitis sicca, mammillary eczema, facial pallor and orbital darkening. Some facts occurred rather frequently, e.g. cheilitis sicca, hyperhidrosis, perleche, keratosis pilaris and orbital darkening, which resulted in a frequency of more than 20%.

*Examination of Questionnaires in Comparison to Examination Data*

Not all the applicants responded to the questionnaire. Therefore, we have to accept a number of 128 missing cases from a total of 454. In figure 3, we present the data concerning atopic case histories.

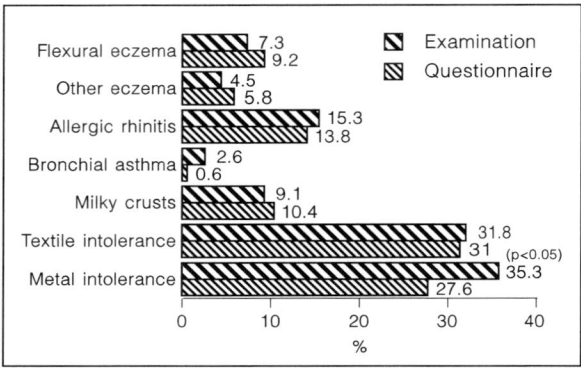

*Fig. 3.* Comparison between examination and questionnaire data. Other eczema = 'hand eczema' in examination and 'nonflexural eczema' elsewhere in questionnaire.

## Discussion

The assessment of skin conditions, especially the decision whether the applicant in any profession is atopic or not is one of the most subjective and therefore critical points in occupational dermatology. The definition of major and minor criteria for atopic dermatitis by Hanifin and Rajka [8] was the first effort to reach an objective diagnosis of atopic dermatitis. However, the Hanifin criteria can only be applied for manifest dermatosis. The diagnostic criteria recommended by Sampson [9] are another possibility to classify an eczema as atopic.

That is why the examination and statistical evaluation of more than 1,000 people concerning the presence of atopic criteria by Diepgen's group [10, 11] was a breakthrough in defining the term *'atopic skin diathesis'*. Since then, atopic diathesis was no longer established by IgE detection alone. The advantage of the Diepgen criteria was that the term diathesis describes the genetic aspect, e.g. atopics without manifestation, too. We have shown a rather good correlation between the Diepgen criteria for normals and both our groups of trainees concerning the anamnestic or clinical criteria of atopy (fig. 3). Przybilla et al. [12] found a much higher incidence of some features in their normal population, probably resulting from a different definition of the criteria.

The more frequent manifestations of atopic diseases in younger members of the family is a typical feature of the increasing frequency of atopic disorders in general in the last decades. Some extensive studies have shown this trend for any atopic disease by comparing previous data in a selected group with examinations in comparable groups today [13, 14].

We cannot explain the difference in frequency of metal intolerances, but we suppose, that the nursing trainees are more likely to be allergic because of their higher age in some cases and therefore their higher probability to acquire an allergy during their previous occupations.

It is of great importance that nearly 1/5 of all hairdressers and 1/6 of all health care applicants complained of skin problems during the first weeks (fig 1). Frequent shampooing is the most important irritant factor during the early years of young hairdressers (trainees) as noted by Maibach and Engarser [15]. Of course, we cannot conclude the predictive value of atopic diseases for the development of occupational skin disorders [16] regarding only the first time of practical work. But using a small number of participants, we found a higher incidence of history of all atopic diseases in the group of trainees who complained of skin problems during that period. The data are statistically significant in the case of atopic skin diseases (hand eczema, flexural eczema) and probably accidental in the case of atopic airways disease. However, regarding the Diepgen score (cut-off point = 10 points), we did not find a significant difference between atopics and nonatopics in the problem/nonproblem group during the first weeks. But even due to that period of training our data confirm the statement of Rystedt [17] that atopic airways disease is not a risk for the development of occupational skin disorders. On the contrary, a history of atopic skin disease is likely to provoke an occupational sentinel health event under workplace exposure to irritants and allergens [6, 18, 19]. Therefore, 'skin atopics' should not be placed in a job with skin hazards [20]. If they want to work in such a job they should be monitored carefully and taught to recognize early skin disturbances as well as develop an attitude to industrial safety, protective measures and preventive practices [21].

As we expected, the comparison of examination- and questionnaire-based results in our study shows a difference of up to 8% (27 vs. 35% in the case of metal intolerance) between the information given by the trainees (fig. 3). The reliability of the data is actually better in examination-based information since the questionees are able to explain their answers and so the examiner is able to reduce misunderstandings. Moreover, the motivation of the questionees is higher when there is direct contact with the interviewer. It is important that Bakke et al. [22] reported a higher agreement between self-reported diagnoses and the physician's diagnoses in hay fever patients than in eczema patients. However, the great interest of social insurances, trade unions, health care organizers and occupational physicians is to find an easy and less expensive way to obtain the data. Based on our study, we conclude that questionnaires could be a useful instrument in screening populations for atopic diathesis. Some better explanations in the Diepgen questionnaire might be necessary to improve the quality of the answers.

Although questionnaire- or clinical-examination-based examinations are an effective and inexpensive way to screen a lot of employees, applicants or patients, modern bioengeneering methods could be more comparable, evaluable and therefore reliable [23]. The increased irritant reactivity of atopic skin is proved to be a risk factor for development of occupational dermatoses. Methods to assess the irritancy of skin have been shown to be a sensitive and specific predictive marker of irritant susceptibility [24]. So we do not expect a decrease in use of such techniques following an increased use of questionnaire examinations. A combination of epidemiological screening examinations in large groups of employees with irritancy studies in some characteristic individuals should lead to a better understanding of irritant occupational skin diseases.

## Conclusion

Questionnaire-based evaluations alone cannot be recommended in studying mild skin problems in a population [25]; minor criteria of skin atopy are not estimated as harmful by prospective employees. Moreover, real existing skin problems could be negated by the trainees who are afraid to lose their positions. Therefore, predictive preoccupational examinations in skin-risk professions should be performed under the responsibility of dermatologists. The examiner should concentrate his efforts on registration and documentation of atopic criteria which are proved to be markers for development of skin disorders during vocational training.

The aim of predictive examinations in this field is the prevention of occupational dermatoses. Applicants showing skin risk factors (atopic diathesis and/or dry skin) should be warned against all risk occupations and should not start a vocational training in a high risk occupation [17]. Follow-up examinations are necessary to monitor the individual tolerance to occupational hazards.

Moreover, repeated contact between the dermatologist and the employee as well as employer might promote the awareness of potential health hazards and should lead to increased motivation for the use of protective measures.

## References

1  Lammintausta K, Maibach HI: Irritant dermatitis syndrome. Immunol Allergy Clin North Am 1989;9:435–446.
2  Menghini CL, Angelini G: Atopic dermatitis and occupational and contact allergy. Immunol Allergy Clin North Am 1989;9:523–534.
3  Cronin E, Kullavanijaya P: Hand dermatitis in hairdressers. Acta Derm Venerol 1979;59(suppl 85): 47–50.
4  Kavli G, Angell E, Moseng D: Hospital employees and skin problems. Contact Derm 1987;17: 156–158.

5    Rajka G: Occupational choice for persons with common chronic dermatoses and pathological skin function. Acta Derm Venerol 1967;47:15–19.

6    Rystedt I: The role of atopy in occupational skin disease; in Adams RA (ed): Occupational Skin Disease, ed 2. Philadelphia, Saunders, 1990, pp 215–222.

7    BK-DOK '90: Dokumentation des Berufskrankheitengeschehens in der Bundesrepublik Deutschland. St. Augustin, Schriftenreihe des Hauptverbandes der gewerblichen Berufsgenossenschaften, 1992, pp 1–86.

8    Hanifin JM, Rajka G: Diagnostic features of atopic dermatitis. Acta Derm Venerol 1980;92(suppl): 44–47.

9    Sampson HA: Pathogenesis of eczema. Clin Exp Allergy 1990;20:459–467.

10   Diepgen TL: Die atopische Hautdiathese. Gentner 1991, pp 1–128.

11   Diepgen TL, Fartasch M: Statistische Evaluierung klinisch-diagnostischer Kriterien beim atopischen Ekzem. Allergologie 1991;14:301–306.

12   Przybilla B, Ring J, Enders F, Winkelmann H: Stigmata of atopic constitution in patients with atopic eczema or atopic respiratory disease. Acta Derm Venerol 1991;71:407–410.

13   Kunz B, Ring J, Ueberla K: Do allergies increase? Allergologie 1990;13:238.

14   Wüthrich B: In Switzerland pollinosis has really increased in the last decade. ACI News 1991;3: 41–44.

15   Maibach HI, Engasser PG: Dermatitis due to cosmetics; in Fisher AA (ed): Contact Dermatitis. Philadelphia, Lea & Febiger, 1986, pp 368–393.

16   Rystedt I: Hand eczema and long-term prognosis in atopic dermatitis. Acta Derm Venerol 1985; 117(suppl):1–59.

17   Rystedt I: Work-related hand eczema in atopics; in Maibach HI (ed): Occupational and Industrial Dermatology, ed 2. Chicago, Year Book, 1987, pp 158–167.

18   Lammintausta K, Kalimo K: Atopy and hand dermatitis in hospital wet work. Contact Derm 1981; 7:301–308.

19   Svensson A: Hand eczema: An evaluation of the frequency of atopic background and the difference in clinical pattern between patients with and without atopic dermatitis. Acta Derm Venerol 1988; 68:509–513.

20   Wilkinson DS: Careers advice to youths with atopic dermatitis. Contact Derm 1975;1:11.

21   Mathias CGT: Prevention of occupational contact dermatitis. J Am Acad Dermatol 1990;23:742–748.

22   Bakke P, Gulsvik A, Eide GE: Hay fever, eczema and urticaria in southwest Norway. Allergy 1990; 45:515–522.

23   Pinagoda J, Tupker RA, Agner T, Serup J: Guidelines for transepidermal water loss (TEWL) measurement. Contact Derm 1990;22:164–178.

24   Taylor JS: Chair's summary: Occupational dermatoses; in Burgdorf WHC, Katz SI (eds): Dermatology – Progress and Perspectives. New York, Parthenon, 1993, pp 1121–1125.

25   Berg M: Evaluation of a questionnaire used in dermatological epidemiology. Discrepancy between self-reported symptoms and objective signs. Acta Derm Venereol 1991;156(suppl):13–17.

Dr. M. Gebhardt, Department of Dermatology, University of Jena, D–07740 Jena (Germany)

Elsner P, Maibach HI (eds): Irritant Dermatitis. New Clinical and Experimental Aspects.
Curr Probl Dermatol. Basel, Karger, 1995, vol 23, pp 64–72

# Identification of High-Risk Groups for Irritant Contact Dermatitis by Occupational Physicians[1]

*Ulrich Funke*[a], *Thomas L. Diepgen*[b], *Manigé Fartasch*[b]

[a] Gesundheitswesen Audi AG, Ingolstadt, and
[b] Department of Dermatology, University of Erlangen-Nürnberg, Germany

The most work-related skin disease is contact dermatitis of the hands due to irritants. Certain factors related to the workplace conditions and certain factors intrinsic to the employees have to be present that, when combined, result in irritant contact dermatitis (ICD). On the one hand, a major task of the occupational physician is the prevention of ICD by identification of job-related factors like the frequent and/or intense exposure to irritants. He or she can then try to minimize that risk for example by suggesting ways to the management how to reduce risks under the condition of a specific workplace [1]. On the other hand, knowledge of host-related predisposing factors and of 'risk groups' is an important tool of prevention [2]. Identification of an employee at high risk for work-related ICD enables the physician to give accompanying guidance already at the job applicant stage either by educational campaign of the employee and/or preventive measures like preemployment hand care advice (e.g. use of barrier creams, provision and use of mild cleansing agents) and consideration of the appropriateness and choice of protective gloves, which might have an important impact on the prevention of ICD. In the last years, research in the field of occupational dermatology has given a lot of hints to identify risk groups, especially when epidemiological methods were used [3, 4]. But most studies were either based on clinical populations (in- and out-patients) or questionnaire studies on population surveys. In the hospital-based studies, research settings and samples are biased and differ from the general population. The degree and severity of already existing skin diseases are different

[1] This study is financially supported by grants from the German research minister (BMFT, Förderkennzeichen 01HK919/5 und 01HK899/3).

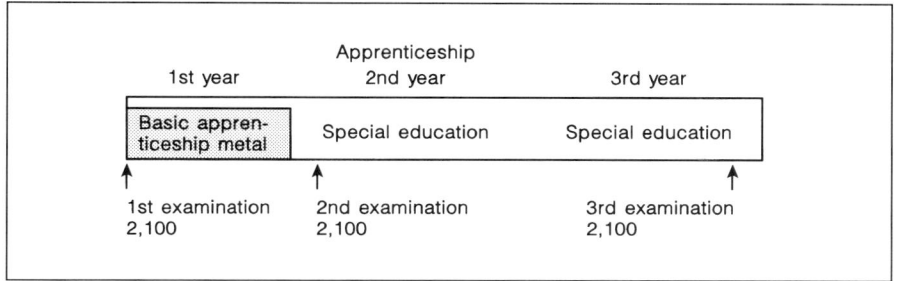

*Fig. 1.* Study design of the prospective cohort study of apprentices in the automobile industry.

and criteria of high-risk groups, won by special occupational dermatosis clinics may lead to false especially 'overprotective' decisions in preemployment screenings. In questionnaire-based population surveys, it is impossible to find exact diagnoses of the skin diseases and slight variations of skin conditions or minor dermatosis that still might predispose to work-related skin diseases cannot be detected.

To date, epidemiological studies of work-related skin disease [2–6] have shown that a history of atopic dermatitis is an important factor in the development of hand eczema (HE) in general and work-related skin disease in particular. The atopic non-eczematous skin seems to be more vulnerable because of its diminished threshold for irritation and the skin disease may flare up under certain workplace conditions.

The aims of our prospective cohort study were to identify intrinsic (especially by anamnestic and existing cutaneous atopic symptoms and signs [5, 7]) and extrinsic risk factors which might lead to ICD in apprenticeships in the automobile industry with special regard to the situation of the occupational physician in preemployment screening and periodical medical examinations.

## Materials and Methods

*Study Design*

In close long-term cooperation with the Department of Dermatology of the University of Erlangen-Nürnberg, a prospective cohort study of all apprentices of the AUDI AG in Ingolstadt and Neckarsulm was started in 1990 (fig. 1). Before the start, after the first year, and at the end of the apprenticeship, anamnestic and cutaneous atopic symptoms and signs [5, 7], domestic risk factors (questions about irritant exposure at home), use of barrier creams, history of HE, and other parameters were evaluated. Irritant exposure at the work-

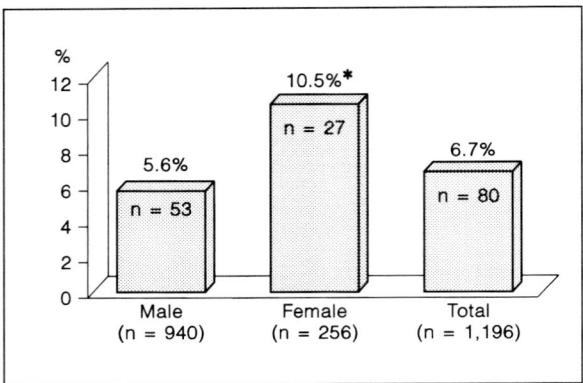

*Fig. 2.* The 1-year incidence of HE in all apprentices (n = 1,196). * = RR female vs. male 95% CI 1.2–2.9.

place was recorded according to already existing standardized descriptions of jobs and working processes during the apprenticeship.

Currently, half of the study has been performed with 3,543 examinations in total (1st examinations: 2,054; 2nd examinations: 1,196; 3rd examinations: 293).

The first examination (at the job application stage) was part of the regular preemployment examination of the prospective apprentices and was performed by specially trained occupational physicians. At this time, the subjects of our study were not aware of a study situation and the results involved all possible problems concerning knowledge of former health problems and reporting them in the preemployment situation. There was no employment or job selection in respect of skin susceptibility by occupational physicians.

The second and third examinations were performed by dermatologists, and participants were then aware of the study. No examination was refused by the apprentices.

In the first year during the basic apprenticeship as metalworker, nearly two third of all apprentices have the same exposures and perform the same work: e.g. metal filing, grinding, drilling, milling and turning.

### Studied Cohort

At the second examination, the study cohort consisted of 1,196 subjects (79% males, 21% females, median age 15.7 and 16.0 years, respectively, at the time of the 1st examination). The drop out rate was 5.3% (67 of 1,263 subjects). To our knowledge, no one quit their job due to skin problems. Actually, there were only 2 drop outs without information about their skin condition at the time of ceasing apprenticeships.

### Parameters, Outcome Variables and Statistics

The reported parameters are the results of the first examination, which was performed by occupational physicians. The outcome variables were the diagnoses of HE and ICD performed by dermatologists at the time of the second examination. The conditions for the outcome variable ICD were:

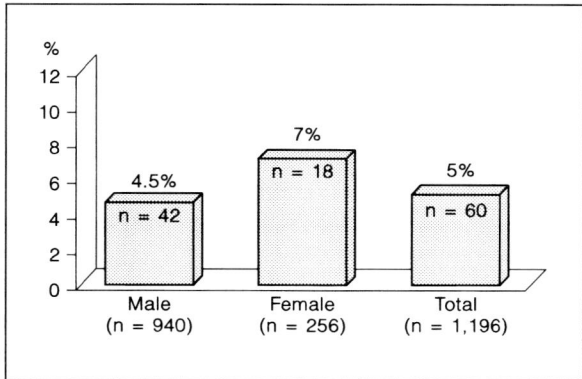

*Fig. 3.* The 1-year incidence of ICD in all apprentices (n = 1,196).

(1) History, e.g. skin problems during the first year of apprenticeship on the hands and/or forearms. This was corroborated by the information available from the 2nd medical examination (including standardized questionnaire and documentation of findings) and files of the occupational physicians at the time HE was acute. In certain cases, the fulfillment of the outcome variable was discussed between observers. Only minimal skin symptoms like redness and scaling did not fulfill the ICD criteria.
(2) The requirement of job-related worsening during exposure to irritants at the workplace (common irritants were cutting and machine fluids, metal dust/chips and wet work).
(3) Exclusion of mycosis, other not-job-related eczema, or allergic contact dermatitis. Relative risks (RR) and 95% confidence intervals (95% CI) were calculated by SPSS.

## Results

### Incidence of HE and ICD

The 1-year incidences of HE are shown in figure 2. The relative risk (RR) was 1.9 (95% CI 1.2–2.9) for women compared to men. The overall 1-year incidence of ICD was 5%. This represents 75% of all HE in this sample (fig. 3). In the basic apprenticeship as metalworker, the incidences of ICD (male: 4.8%; female: 7.6%; total: 5.2%) were only slightly increased compared to the overall apprenticeship.

### RR of Parameters Associated with ICD

For preventive risk group identification, RR and predictive values associated with different parameters are needed (fig. 4, 5). The highest RR of 6.7 (95% CI 2.1–21.8) was calculated for a history of HE plus a history of flexural eczema

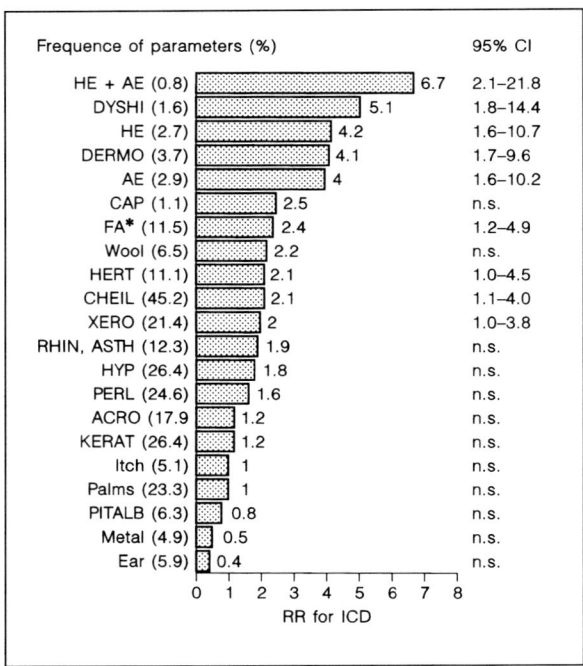

Fig. 4. Frequencies (%) of atopic symptoms and signs in apprentices (basic apprentice-ship as metalworkers, n = 736) according to the first examination by occupational physicians. The RRs and 95% CIs for ICD during the first year of the apprenticeship are given. * = Respiratory atopy only.

(AE). Other risk factors were dyshidrosis [5.1], history of HE (4.2), white dermographism (4.1), and history of AE (4.0). For cradle cap a RR of 2.5 was calculated which, however, was not statistically significantly increased. The RR of a family history of respiratory atopy was 2.4 (95% CI 1.2–4.0), whereas the RR of a personal history of respiratory atopy was only 1.9 (not statistically significantly increased). Hertoghe sign, cheilitis and xerosis had significant RRs between 2.1 and 2.0. Other parameters were not statistically significantly increased in our sample. It has to be pointed out that a history of metal sensitivity was not associated with an elevated risk for ICD and HE during the apprenticeship as metalworker.

*Predictive Values of Parameters Associated with ICD*

Predictive values (for 1-year incidence of ICD) were as follows (fig. 5): history of HE plus history of AE 33.3%, dyshidrosis 25.0%, history of HE 20.0%, history of AE 19.0%, white dermographism 18.5%. But the frequencies of these

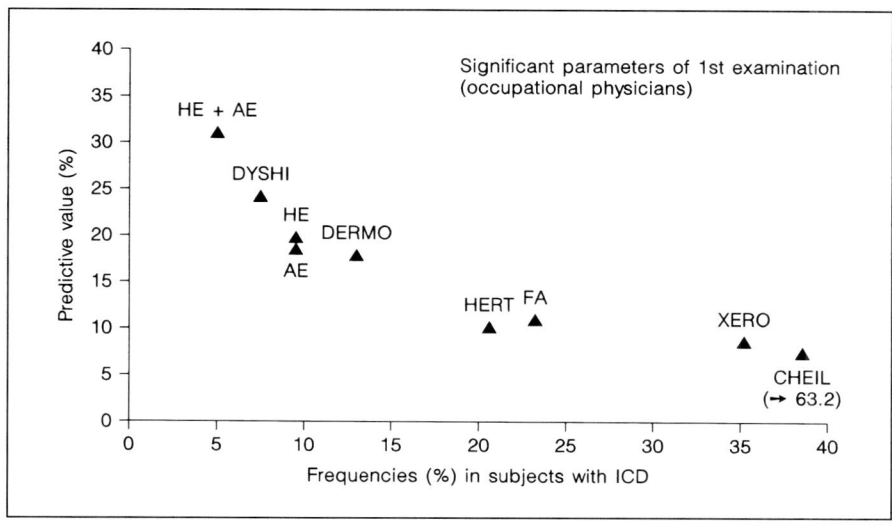

*Fig. 5.* Predictive values of atopic features associated with ICD.

parameters in subjects who developed ICD were low (between 5.3 and 13.5%). Other criteria like Herthoge sign (21.1%) and family history of respiratory atopy (23.7%) appeared to be more frequent. Although xerosis (35.1%) and cheilitis (63.2%) were often diagnosed in subjects with ICD, the predictive values of these both parameters were low, and therefore they seem to be less helpful for the identification of high-risk groups.

*Use of Barrier Creams*

A relevant part of the prevention of ICD is the proper use of barrier creams. But only 45.5% of apprentices with HE (in the last year) and only 22.5% without HE have used barrier creams effectively (fig. 6).

## Discussion

To date, there is a lack of population-based studies dealing with occupational skin diseases (OSD). Metal workers are believed to be at high risk for developing of ICD. In a population-based study of OSD in North Bavaria it could be demonstrated that 41% of male subjects with OSD were employed in the metal industry [8].

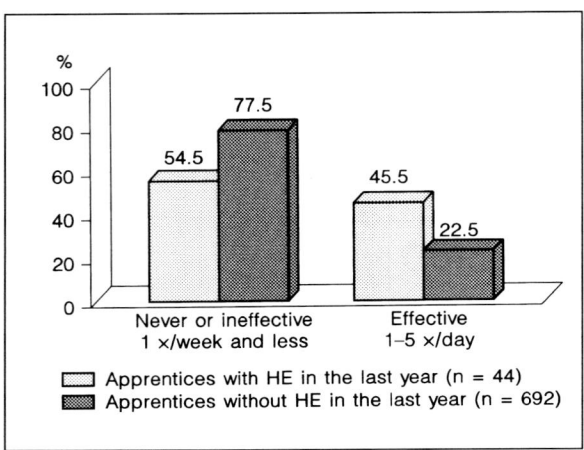

*Fig. 6.* The use of barrier creams (%) in apprentices with and without hand eczema during the last year of the apprenticeship.

In our prospective cohort study the role of endogenous and exogenous risk factors are investigated in the automobile industry. The quality of this epidemiological study profits from the intensive cooperation between dermatologists and occupational physicians. Another advantage is the structure of a large automobile company with an integrated department of occupational medicine. From the epidemiological point of view further advantages of our study are: (i) the results were not affected by refusing of examinations or drop outs; (ii) nearly identical exposures to irritants at the workplace within 1 year (basic apprenticeship metal), and (iii) in general no former exposures to irritants at the workplace are given.

These circumstances allow a valid calculation of characteristics for the identification of high-risk groups for ICD. But the study results are representative only for moderate exposure to irritants at the workplace (metalwork apprenticeship).

Previous studies have shown that HE is more common in women than in men [9, 10]. The possibility and extent of more extensive and frequent exposure to irritants in day-to-day activities, wet work, the distribution of atopic dermatitis, and female sex characteristics are still being discussed. In the observed population of female apprentices without children and only moderate housekeeping activities, it is not probable that higher domestic risk factors are responsible for the main difference. In the subsample of women itself, there was no correlation between washing dishes (indicating domestic wet work) and the occurrence of HE. This finding is corroborated by the study of Kristensen [11], who also investigated new employees in an automobile manufacturing industry and found a significantly higher prevalence of HE among female employees. But the degree of wet work performed at home had not been investigated in this study.

As three quarters of all HE are associated with exposure to irritants at the workplace, ICD represents the most important occupational dermatological risk in the reported agegroup. The reason for only moderately raised incidence of ICD in the basic apprenticeship as metalworker may be that in the other studied groups the exposure to irritants at the workplace may be heterogeneous and can be more (e.g. cooks) or less intensive (e.g. office employees) than in metalworker occupations. Self-selection of atopics who choose jobs without exposure to irritants must also be mentioned.

Even though the study design involved the difficulties of reporting parameters in a preemployment examination and investigators were trained but not dermatologists, our results correspond with other studies [5, 6, 9, 11]. Previous HE combined with AE was the predominant predictor for the 1-year incidence of ICD (33.3%) followed by dyshidrosis (25.0%), history of HE (20.0%) and AE (19.0%), white dermographism (18.5%), family history of respiratory atopy (10.7%) and Hertoghe sign (9.9%). However, it must be mentioned that there was some correlation between the different parameters, and therefore the use of multivariate statistical analysis is planned. The use of multivariate statistical techniques (logistic regression) will be performed for the evaluation of the parameters when the follow-up of all subjects has been completed. A further increase of the proportion of identified subjects which develop ICD, up to 50% with sufficient predictive values for ICD by just combining history and clinical criteria, seems to be possible.

### Conclusions

A valid risk group identification for ICD by (well-trained) occupational physicians is possible, necessary and worthwhile. A history of HE, atopic skin disease and associated parameters has the highest predictive value for ICD during an apprenticeship as metalworker, whereas respiratory atopy is only moderately associated with ICD. Skin protection even in risk groups for ICD is not performed sufficiently, nevertheless the majority of high-risk groups are able to work without HE with irritants at the workplace. Especially considering the possibilities of good occupational medical care and accompanying guidance *one must be careful not to be overprotective in respect of job selection or job restriction even in high-risk groups.*

Taking into account the performance of special periodical medical examinations for high-risk groups including their guidance to minimize skin contact with irritants and enforce use of protection equipment and barrier creams, job selection in respect of ICD remains necessary only for persons with extreme susceptibility of the skin demanding jobs with long-term unavoidable exposure to irritants at the workplace.

## References

1    Shmunes E: Predisposing factors in occupational skin diseases. Dermatol Clin 1988;6:7–13.
2    Lammintausta K, Maibach HI: Dermatologic considerations in worker fitness evaluation. Occup Med State Art Rev 1988;3:341–350.
3    Rystedt I: Atopy, hand eczema, and contact dermatitis: Summary of recent large scale studies. Semin Dermatol 1986;5:290–300.
4    Diepgen TL, Tepe A, Pilz B, Schmidt A, Hüner A, Huber A, Hornstein OP, Frosch PJ, Fartasch M: Occupational skin diseases in hairdressers and nurses during apprenticeship-design of a prospective epidemiological study. Allergologie 1993;10:396–403.
5    Diepgen TL, Fartasch M: Recent epidemiological and genetic studies in atopic dermatitis. Acta Derm Venerol (Stockh) 1992;(suppl 176):13–18.
6    Nilsson E, Mikaelsson B, Andersson S: Atopy, occupation and domestic work as risk factors for hand eczema in hospital workers. Contact Derm 1985;13:216–223.
7    Hanifin JM, Rajka G: Diagnostic features of atopic dermatitis. Acta Derm Venerol (Stockh) 1980; (suppl 92):44–47.
8    Diepgen TL, Schmidt A, Schmidt M, Fartasch M: Demographic and legal characteristics of occupational skin diseases. Allergologie 1994;17:84–89.
9    Diepgen TL, Fartasch M: General aspects of risk factors in hand eczema; in Menné T, Maibach HI (eds): Hand Eczema. Boca Raton, CRC Press, 1994, pp 141–156.
10   Patil S, Maibach HI: Effect of age and sex on the elicitation of irritant contact dermatitis. Contact Derm 1994;30:257–264.
11   Kristensen O: A prospective study of the development of hand eczema in an automobile manufacturing industry. Contact Derm 1992;26:341–345.

Dr. med. U. Funke, Gesundheitswesen Audi AG, Postfach 10 02 20,
D–85045 Ingolstadt (Germany)

Elsner P, Maibach HI (eds): Irritant Dermatitis. New Clinical and Experimental Aspects.
Curr Probl Dermatol. Basel, Karger, 1995, vol 23, pp 73–76

..............................

# The Long-Term Prognosis in Irritant Contact Hand Dermatitis

*David A. Fitzgerald, John S.C. English*

Department of Dermatology, North Staffordshire Hospital, Stoke-on-Trent, UK

Hand eczema is a major clinical problem and a significant cause of social and occupational disability. A Dutch study found a population prevalence of hand eczema of 3.4% [1], while 22% of Danish women reported current or previous hand eczema [2]. Many cases of hand eczema are multifactorial in origin, which gives rise to difficulties in both diagnosis and management. The morphology of the eruption is not always a reliable pointer to its cause. Exogenous factors include irritants and allergens, but both may coexist. The major endogenous predisposing factor is atopic dermatitis, but this frequently interacts with allergic and irritant influences.

In routine clinical practice irritant contact dermatitis is often diagnosed in a patient with a chronic superficial dermatitis affecting particularly the dorsa of the hands and the finger webs and a history of repeated exposure to mild or moderate irritants, without recourse to patch testing. Ideally, however, the diagnosis should be made only after patch testing has failed to reveal a relevant allergen [3], although this may in fact only indicate that the responsible allergen was not present in the test battery. Allergic contact dermatitis and atopic dermatitis may also be present and it may be impossible to determine which is the 'primary' cause.

Several large studies have investigated the natural history of irritant contact dermatitis in different populations with varying results [1, 4, 5]. We wished to ascertain the prognosis for irritant contact dermatitis of the hands in our own local population.

## Patients and Methods

We identified from our records 53 patients who had attended our department in 1988 and been diagnosed as having irritant contact dermatitis of the hands after patch testing. As mentioned above, logistic considerations preclude the patch testing of all patients with clinically typical irritant dermatitis and our sample therefore represents only a subset of our patients with this diagnosis. All had been tested to the European Standard Series as well as to any additional potential allergens to which individual patients had been exposed. Application and interpretation of tests was performed in accordance with the guidelines of the International Contact Dermatitis Research Group [6]. In all cases the primary diagnosis was felt to be irritant contact dermatitis, although some patients also had atopic dermatitis or allergic contact dermatitis.

A postal questionnaire was sent to these patients in 1993. This enquired about the current severity of their dermatitis with the possible responses being 'cleared', 'improved', 'same' or 'worse', and also about any change in employment.

## Results

Replies were received from 36 patients, giving a response rate of 68%. Twenty-four (66%) patients were female and the ages ranged from 22 to 73 years (mean 40 years) (table 1).

## Discussion

We acknowledge that our sample size is suboptimal; our primary intention was to establish whether the natural history of irritant contact dermatitis in our own local population was comparable with that reported in larger studies which have mostly been carried out in other countries. Our response rate of 68% is relatively high, which may in part result from the very stable population in the Stoke-on-Trent area. Postal surveys are subject to various sources of bias and it is likely that the response rate is higher in patients with persistent symptoms, so the true outcome may in fact be more favourable than suggested by studies of this type. The fact that 66% of our respondents were female reflects the greater incidence of irritant contact dermatitis in women as a result of different exposure patterns [1].

For the reasons mentioned above, irritant contact dermatitis of the hands can be a difficult condition to diagnose precisely and follow-up studies are correspondingly difficult to perform and interpret. Numerous studies have been reported; some consider all patients with irritant contact dermatitis, some patients with hand eczema and other patients with 'pure' irritant contact dermatitis of the hands only [4, 7, 8].

*Table 1.* Current severity of dermatitis

| | Cleared | | Improved | | Same | | Worse | | Total |
|---|---|---|---|---|---|---|---|---|---|
| | n | % | n | % | n | % | n | % | |
| All ICD | 11 | 30.6 | 19 | 52.8 | 5 | 13.8 | 1 | 2.8 | 36 |
| ICD alone | 8 | 36.4 | 10 | 45.4 | 4 | 18.2 | 0 | | 22 |
| ICD/secondary ACD | 3 | 33.3 | 4 | 44.4 | 1 | 11.1 | 1 | 11.1 | 9 |
| ICD/atopic dermatitis | 0 | | 5 | | 0 | | 0 | | 5 |
| Occupational history | | | | | | | | | |
|   Changed job | 4 | 21.1 | 13 | 68.4 | 2 | 10.5 | 0 | | 19 |
|   Unchanged job | 7 | 41.2 | 6 | 35.3 | 3 | 17.6 | 1 | 5.9 | 17 |

ICD = Irritant contact dermatitis; ACD = allergic contact dermatitis.

Most investigators report complete healing in between 25 and 42% of patients over follow-up periods ranging from 6 months to 16 years [4, 7, 8]. A recent large Australian study of occupational contact dermatitis considered rates of improvement as well as of full healing and found an overall improvement rate of 67% in irritant contact dermatitis, although only 30.3% of patients were completely clear of eczema at follow-up [5]. The results of our small study are similar; while only 30.6% of patients had completely recovered, 82.4% reported at least some improvement in their dermatitis.

Our sample revealed no significant difference in prognosis between those patients with irritant contact dermatitis alone and those who also had allergic contact dermatitis or atopic dermatitis. This is likely to result from the limited sample size as previous larger studies have revealed a significantly worse prognosis in the latter groups of patients [1, 5].

For the same reason we observed no significant difference in outcome between patients who changed their occupation and those who did not. Rosen and Freeman [5] reported an improvement in occupational contact dermatitis in patients who were able to change their working pattern and an even greater improvement in those who left the original industry altogether. Previous series had not detected such a difference [4, 8].

In conclusion, it is important to recognise, as emphasised by Rosen, that although the prospects for complete healing of irritant contact dermatitis are relatively poor, the majority of patients will experience at least some reduction in the severity of their condition with time, and it may be valuable for patients to be made aware of this fact.

## References

1  Lantinga H, Nater JP, Coenraads PJ: Prevalence, incidence and course of eczema on the hands and forearms in a sample of the general population. Contact Derm 1984;10:135–139.
2  Kaaber K, Veien NK: The significance of chromate ingestion in patients allergic to chromate. Acta Derm Venereol 1977;57:321–323.
3  Malten KE: Thoughts on irritant contact dermatitis. Contact Derm 1981;7:238–247.
4  Keczkes K, Bhate SM, Wyatt EH: The outcome of primary irritant hand dermatitis. Br J Dermatol 1983;109:665–668.
5  Rosen RH, Freeman S: Prognosis of occupational contact dermatitis in New South Wales, Australia. Contact Derm 1993;29:88–93.
6  Wilkinson DS, Fregert S, Magnusson B, et al: Terminology of contact dermatitis. Acta Derm Venereol 1970;50:287–292.
7  Agrup G: Hand eczema with other dermatoses in South Sweden. Acta Derm Venereol 1969; 49(suppl 61):61.
8  Fregert S: Occupational dermatitis in a 10-year material. Contact Derm 1975;1:96–107.

Dr. David A. Fitzgerald, Department of Dermatology, Royal Liverpool Hospital,
Prescot Street, Liverpool L8 7XP (UK)

Elsner P, Maibach HI (eds): Irritant Dermatitis. New Clinical and Experimental Aspects.
Curr Probl Dermatol. Basel, Karger, 1995, vol 23, pp 77–86

..........................

# Metalworking Fluid Dermatitis: A Comparative Follow-Up Study in Patients with Irritant and Non-Irritant Hand Dermatitis[1]

*Peter Elsner*[a, b]*, Friederike Baxmann*[b]*, Hans-Martin Liehr*[b]

Departments of Dermatology,
[a] University of Zurich, Switzerland, and
[b] University of Würzburg, Germany

Hand dermatitis in metalworkers is one of the most frequent occupational dermatoses. Metalworkers in the tool machine industry are the group of workers with the third-highest incidence of occupational dermatoses in Switzerland. Their annual incidence rate reaches 600 cases per 100,000 person years (fig. 1) [2], with irritant reactions being the predominant diagnosis. In a population-based study in Northern Bavaria, incidence rates (in 3 years per 10,000 employees) in metalworkers were 21 for irritant contact dermatitis and 13 for allergic contact dermatitis [3]. In an unselected population of 286 metalworkers examined at the workplace, the vast majority of hand dermatitis cases (>97%) were of the irritant type [4].

The risk for hand dermatitis in metalworkers is closely linked to metalworking fluid (MWF) exposure. In situations of high MWF exposure, the prevalence of hand dermatitis may reach 30% in the workforce [5]. Two main types of MWF (synthetic cooling fluids, cutting fluids) should be distinguished: water-based MWF with a higher irritating and sensitizing potential and oil-based MWF (synthetic neat oils) with a significantly lower risk of irritation and sensitization. In addition to MWF, microlesions induced by metal dust, friction, pressure, environmental temperature and humidity play an etiological role in the development of hand dermatitis in metalworkers.

---

[1] This paper is partially based on data of the dissertation of Baxmann [1].

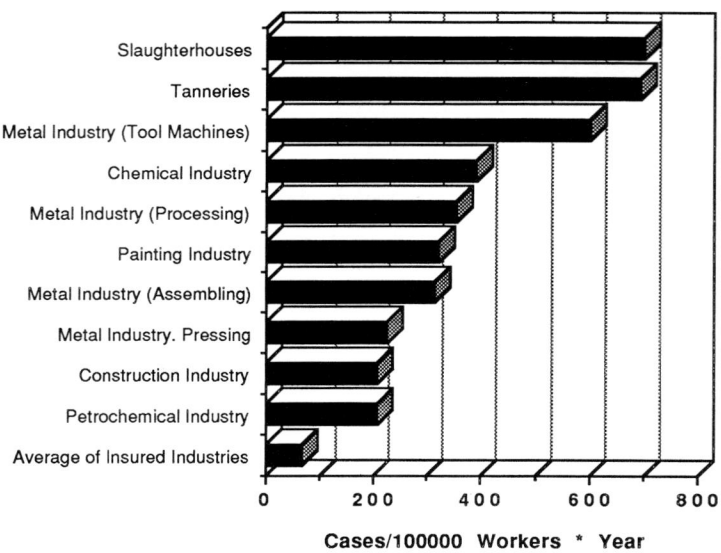

*Fig. 1.* Incidence of occupational dermatoses in Switzerland [2].

Since the potential of water-based MWF to induce acute irritation is rather low [6], the majority of MWF-associated cases of irritant contact dermatitis (ICD) are of the cumulative or chronic type. Once cumulative ICD is established, healing may be very slow even if contact to irritants is avoided, and recurrences are frequent. Consequently, the reason behind the recommendation of a job change in workers with chronic hand dermatitis has been questioned [7, 8]. It is therefore important to study the factors involved in clearing or persistence of hand dermatitis in metalworkers in order to optimize strategies for treatment and consultation of these patients.

In order to learn more about the history of hand dermatitis in metalworkers, we performed a questionnaire-based follow-up study in metalworkers patch tested in the Department of Dermatology, University of Würzburg, Germany.

## Patients and Methods

*Patient Population*
Between January 1, 1980, and December 31, 1989, 138 metalworkers with hand dermatitis and a history of exposure to cutting fluids were referred to the Allergology Unit of the Department of Dermatology, Julius Maximilian University of Würzburg for patch testing.

*Table 1.* Demographic and clinical data of 138 patients with MWF dermatitis

| | | |
|---|---|---|
| Sex | 134 (97.1%) male | 4 (2.9%) female |
| Age | mean 35.8 years | median 34.0 years |
| Time in job | mean 10.5 years | median 7.0 years |
| Time since onset of skin disease | mean 3.0 years | median 1.0 years |
| Diagnosis of hand dermatitis based on patch testing | 96 (69.6%) allergic and possibly combined allergic/irritant | 42 (30.4%) only irritant |

Referrals were mainly from dermatologists or from insurers for expert opinion. Demographic data and history of the disease are shown in table 1.

*Patch Testing*

The patients were patch tested with the standard series, a MWF series, a metal series, a sulfur compound series, a polyamine and an ethanolamine series, and with their own used and fresh metal working fluids. Patch testing and evaluation of patch test reactions were performed according to the recommendations of the ICDRG.

*Case Definitions*

The dermatitis of any patient with one or more positive allergic patch test reactions to any of the above allergens was categorized as allergic or allergic plus possibly irritant (ACD), whereas the dermatitis of patients with irritant or no reactions to the allergens was considered as solely irritant (ICD). This categorization probably overestimates the proportion of patients with allergic and possibly irritant contact dermatitis (ACD) and underestimates the patient group with irritant dermatitis (ICD), but it minimizes the number of false-positives in the irritant dermatitis group.

*Follow-Up Questionnaire Study*

In 1990, a six-item questionnaire was sent to 127 of the 138 patients of whom mailing addresses were available. The items included: history of the hand dermatitis; duration of the dermatitis until clearing; compensation by the insurers; change of job and/or profession; continued MWF exposure, and handling of MWF if exposure was continued.

Four weeks after the first mailing, a reminder with a second questionnaire was sent to those individuals who had not returned the first questionnaire. Furthermore, we tried to contact them by phone.

*Data Acquisition and Analysis*

The data of all returned questionnaires were entered into a spreadsheet on an Apple Macintosh personal computer and analyzed statistically using the SPSS® software. Differences in frequencies between groups were tested for significance using the $\chi^2$ test. Differences between means were tested for significance using the t test.

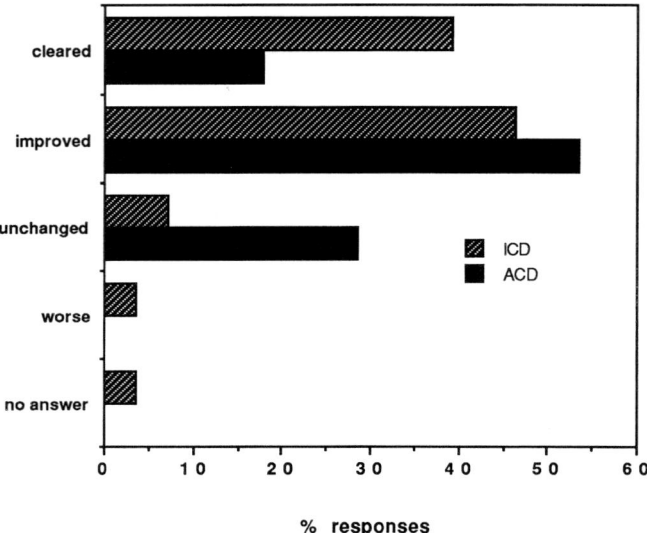

*Fig. 2.* Hand dermatitis at follow-up (n = 56, 28 in each group, matched for time of follow-up). ICD = Metalworkers with MWF-associated irritant contact dermatitis only; ACD = metalworkers with MWF-associated allergic or combined allergic/irritant contact dermatitis.

## Results

One hundred and one of 127 patients (79.5%) returned the questionnaire. The mean follow-up time since the presentation for patch testing was $5.7 \pm 0.4$ years (mean ± SEM), median 5.0 years. Seventy-three (72.3%) of the responders had been grouped as cases of ACD, whereas 28 (27.7%) had been grouped as ICD patients. Since the two groups differed regarding age and follow-up period with the ACD group being significantly older ($43.6 \pm 1.5$ vs. $37.0 \pm 2.4$ years, $p \leq 0.05$) and the follow-up being significantly longer ($6.3 \pm 0.5$ vs. $4.1 \pm 0.6$ years, $p \leq 0.01$), a one-to-one matching of ACD and ICD cases was performed adjusting for follow-up. If more than one case met the criteria for matching, selection was by chance. This resulted in a dataset of 56 cases, 28 ACD and 28 ICD, with an identical distribution of follow-up times ($4.1 \pm 0.6$ years). Results calculated from these cases will be reported in the following.

At follow-up, the dermatitis had cleared or was improved in 86% of the ICD and in 71% of the ACD patients (fig. 2). While this seems to indicate a more favorable prognosis in ICD patients, the difference was not significant in the $\chi^2$

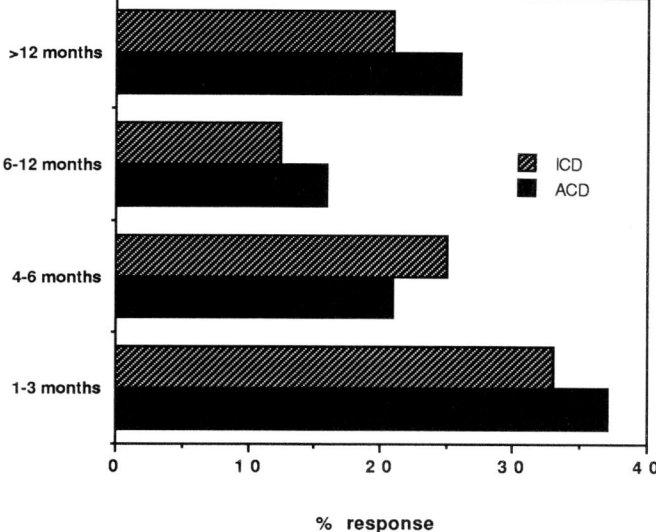

*Fig. 3.* Time to clearing or improvement of hand dermatitis.

test. Clearing or improvement did not seem to be correlated with the duration of follow-up: there was a clearing/improvement rate of 81% in patients with a follow-up of up to 2 years and 77% in patients with a longer follow-up period.

The majority of patients both in the ACD and the ICD group who cleared or improved did so within the first few months (fig. 3). In only 26% of the patients with ACD and 21% of those with ICD clearance or improvement occurred after more than 12 months. Healing or improvement was slightly more frequent in workers below the age of 30 (82%) compared to those above that age (76%).

Changes of occupation in the study groups are shown in figure 4. 46% of the patients in both groups continued the same work they had done before their skin disease. The most frequent alteration was to continue to work in the metal industry, but to change the job, followed by a new occupation in another field.

Clearing or improvement of hand dermatitis was not influenced by a job change in ICD patients. In ACD cases, the paradoxical finding was encountered of slightly more cases healing or improving if the patients stayed in the job compared to those who had left it without reaching significance (fig. 5).

Looking more specifically at MWF exposure, more patients healed or improved when avoiding MWF in the ACD group (80 vs 64%), whereas slightly more patients who continued to work with MWF got better in the ICD group (fig. 6).

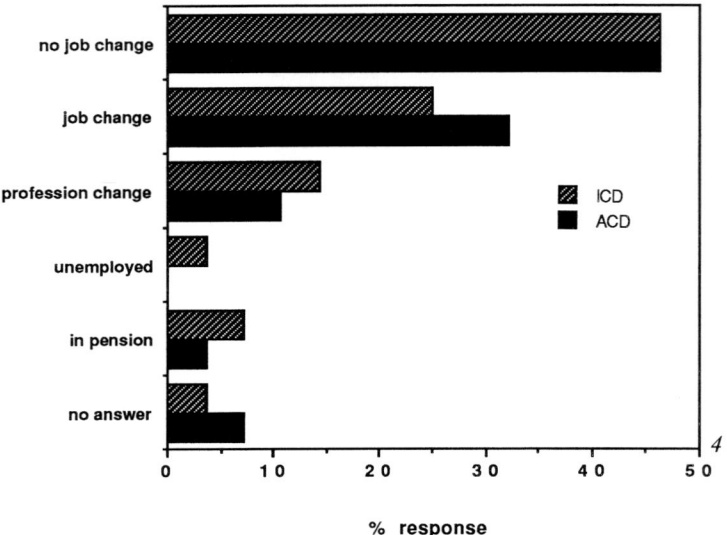

% response

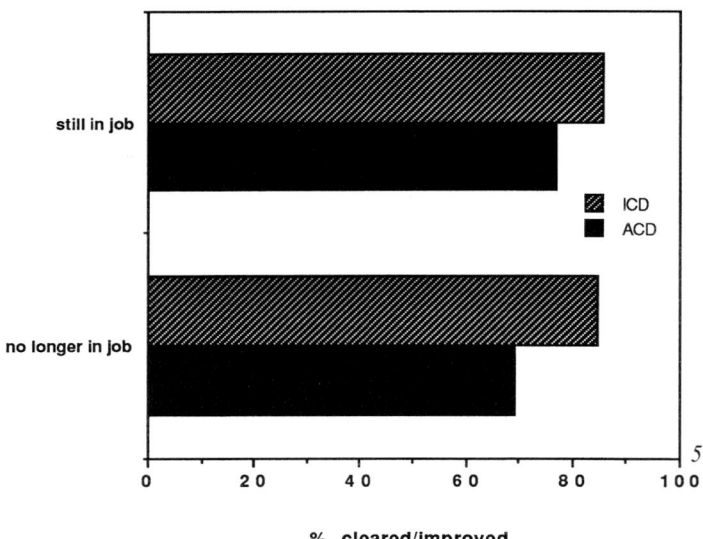

% cleared/improved

*Fig. 4.* Changes in occupation in patients with MWF-associated hand dermatitis.

*Fig. 5.* Clearing or improvement of hand dermatitis as a function of occupational changes. Still in job = the worker remained at the same workplace; no longer in job = the worker changed the workplace in the same company, chose a new profession, was unemployed or pensioned.

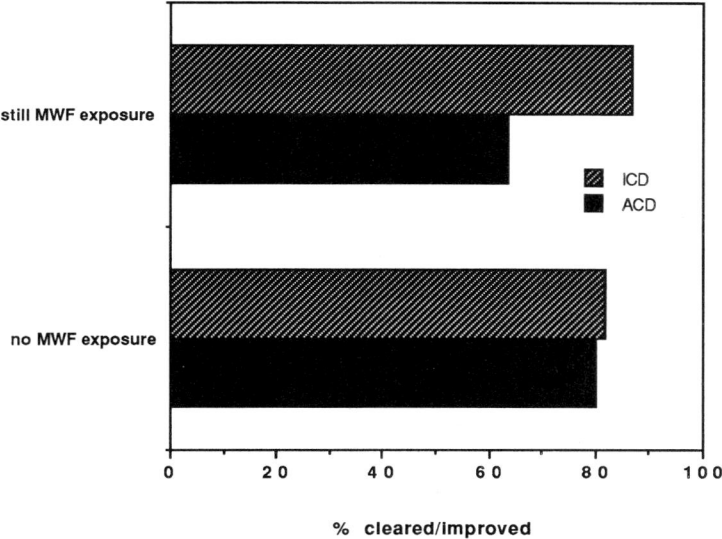

*Fig. 6.* Clearing or improvement of hand dermatitis as a function of continued MWF exposure.

Workers still exposed to MWF were asked about their changes in working habits concerning MWF. While more than 40% in ICD patients and 30% in ACD patients had not changed their habits, the majority of patients (90% in the ACD and 64% in the ICD group) had improved their personal protection (which includes the use of gloves and barrier creams). Technical improvements of MWF use such as replacement of irritating MWFs, more frequent changes and better control of MWF were mentioned only rarely (fig. 7).

### Discussion

The present study is a questionnaire follow-up study based on historical controls. From its design, it suffers from certain inherent weaknesses. Although the response rate (nearly 80%) was relatively high, there may be a bias towards an increased response rate from patients who were more skilled, more motivated, who felt better treated by the medical system or who had a better outcome. As is always the case in retrospective studies, the precision of the responses will depend on what length of time has passed since the questioned event thus increasing

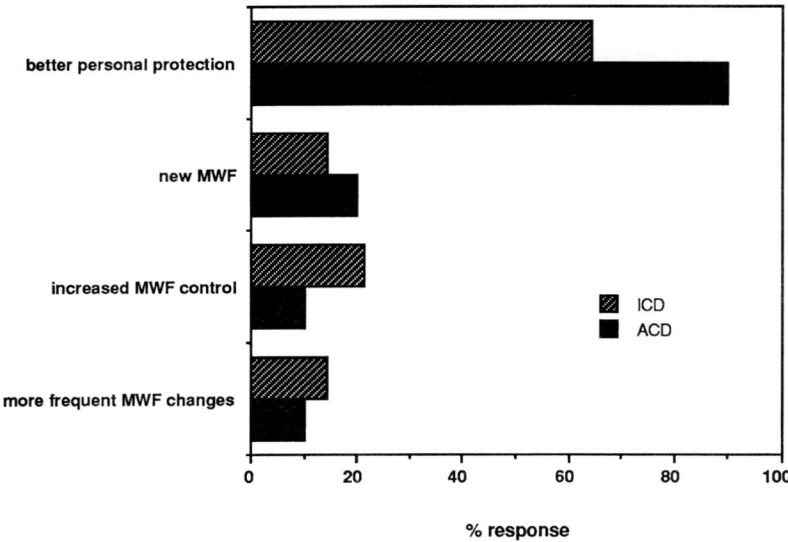

*Fig. 7.* Changes in working with MWFs in workers with continued exposure. More than one item could be mentioned.

stochastic error. The problem of diagnostic (mis-)classification was already mentioned in 'Methods'; since there is a considerable overlap between irritant and allergic dermatitis in MWF-associated hand dermatitis [9], we pragmatically distinguished between purely irritant and nonirritant cases that will be a collection of patients with irritant, allergic or combined reactions. Endogenous hand dermatitis was not considered in this study since the retrospective data on the diagnosis of atopy were not sufficient. Finally, the power of the present study suffers from the limited number of cases in both groups.

The authors are aware of three recent studies on the prognosis of MWF dermatitis. All of them are retrospective questionnaire studies like the present one.

Pryce et al. [10] sent questionnaires to 121 machine operators diagnosed over a 5-year period. Of the 100 responders, 78% of those with continued MWF exposure and 70% of those with discontinued exposure had not healed. This may come close to our results since we found a clearing rate of only 18% in the ACD and 39% in the ICD group. From these low clearing rates, we do not draw the conclusion that prognosis needs to be judged too unfavorably taking improvement into account. Unfortunately, the authors of the mentioned study did not report on improvement rates. If clearing cannot be achieved, improvement seems to be sufficient enough for many patients enabling them to cope with their disease and to continue work. It is noteworthy that in the study of Pryce et al. [10] no significant differences in clearing rates were observed between etiological groups of hand dermatitis quite parallel to our findings.

The questionnaire study of Grattan and Foulds [11] in 56 patients had a mean follow-up time of 20.3 months and a response rate of 71% (40 responses). No data were given on the percentage of irritant and allergic cases. Thirty (75%) patients improved, 6 (15%) remained stable and 4 (10%) worsened, which is again very similar to our findings.

Finally, in a recent review on occupational dermatitis by MWF, de Boer and Bruynzeel [12] mention a small questionnaire study in 10 patients tested 1–3 years before. Two of 5 patients with irritant dermatitis and 1 of 5 patients with allergic contact dermatitis had healed, irrespective of further MWF exposure.

There is a noteworthy agreement between all studies that the prognosis of hand dermatitis in metalworkers is doubtful as far as complete healing is concerned, but the outlook is better considering improvement. None of the studies considered therapy as a variable influencing the prognosis. This may be more a consequence of the retrospective design not allowing to evaluate different therapies than of a fatalistic approach. The prognosis of MWF dermatitis does not seem to be influenced by the disease mechanism, be it irritant or allergic. In other words, persistence of a chronic dermatitis with negative patch tests and despite avoiding MWF exposure should not be used as an argument in favor of endogenous hand dermatitis as is sometimes done by insurers denying a causal relationship between work and hand dermatitis, but this patient may just as well suffer from persistent irritant dermatitis. Since a change of job or discontinuation of MWF exposure does not seem to influence healing or improvement considerably, there are few arguments in favor of advising patients to a job change. Instead, the consultation should be focussed on improvement of skin protection and skin care. As our data show, the majority of patients had reacted to this advice and had improved their working habits, but there remained a subset of about 30% in ICD patients and 10% in ACD patients that had not changed their way of handling MWF. They may be a worthwile target for secondary prevention.

While these retrospective prognostic studies give a rather congruent picture, prospective studies are needed to overcome the inherent pitfalls in the design and to assess the effects of therapeutic interventions and secondary prevention in patients with MWF dermatitis.

## References

1    Baxmann F: Kühlschmiermittel-Dermatitis. Darstellung und Diskussion der dermatologischen und rechtlichen Problematik auf Grundlage einer Verlaufsstudie bei Metallarbeitern; Inauguraldiss, Medical Faculty of the University of Würzburg, 1993.

2    Jost M, Ruppen L: Berufskrankheitenstatistik der SUVA: Aktueller Stand und Trends. SUVA Arbeitsmed 1992;22:72–74.

3    Diepgen TL, Coenraads PJ: What can we learn from epidemiological studies on irritant contact dermatitis? In Elsner P, Maibach HI (eds): Irritant Dermatitis: New Clinical and Experimental Aspects. Basel, Karger, 1994.

4    deBoer EM, van Ketel WG, Bruynzeel DP: Dermatoses in metal workers. Irritant contact dermatitis. Contact Derm 1989;20:212–218.

5    Pryce DW, White J, English JS, Rycroft RJ: Soluble oil dermatitis: A review. J Soc Occup Med 1989;39:93–98.

6    deBoer EM, Scholten RJ, van Ketel WG, Bruynzeel DP: The irritancy of metalworking fluids: A laser Doppler flowmetry study. Contact Derm 1990;22:86–94.

7    Hogan DJ, Dannaker CJ, Maibach HI: Contact dermatitis: Prognosis, risk factors, and rehabilitation. Semin Dermatol 1990;9:233–246.

8    Hogan DJ, Dannaker CJ, Maibach HI: The prognosis of contact dermatitis. J Am Acad Dermatol 1990;23:300–307.

9    Grattan CE, English JS, Foulds IS, Rycroft RJ: Cutting fluid dermatitis. Contact Derm 1989;20: 372–376.

10   Pryce DW, Irvine D, English JS, Rycroft RJ: Soluble oil dermatitis: A follow-up study. Contact Derm 1989;21:28–35.

11   Grattan CE, Foulds IS: Outcome of investigation of cutting fluid dermatitis. Contact Derm 1989; 20:377–378.

12   deBoer EM, Bruynzeel DP: Occupational dermatitis by metalworking fluids; in Menné T, Maibach HI (eds): Hand Eczema. Boca Raton, CRC Press, 1994, pp 217–230.

Peter Elsner, MD, Department of Dermatology, University of Zurich,
Gloriastrasse 31, CH–8091 Zurich (Switzerland)

Elsner P, Maibach HI (eds): Irritant Dermatitis. New Clinical and Experimental Aspects.
Curr Probl Dermatol. Basel, Karger, 1995, vol 23, pp 87–94

..........................

# The Occlusive Effects of Protective Gloves on the Barrier Properties of the Stratum corneum[1]

*C.J. Graves, C. Edwards, R. Marks*

Department of Dermatology, University of Wales College of Medicine, Cardiff, UK

## Background

Industrial safety procedures aim to separate the operator from hazardous processes. This has been helped by the increased use of automated handling techniques in many food and industrial processes. However, human intervention is still required and the use of protective gloves and clothing is often necessary. This is because of the wide range of potentially damaging chemicals present in aqueous and organic solvents, and because of the hazards of mechanical injury. The use of protective gloves and clothing in medical and associated laboratory procedures is increasing.

The protection offered by gloves is usually occlusive by nature. This may be counter-productive, as it may alter the stratum corneum and reduce the protective barrier properties of the skin. Small amounts of material may enter the glove or other protective material and become trapped on the skin under occlusion for many hours, possibly leading to irritation, and more seriously to dermatitis or eczematous changes [1]. Much work has been reported on the allergic or irritant effects of the glove's constituent chemicals [2–4]. However, there is little mention in the literature of the effects on the barrier properties of the stratum corneum due simply to the occlusion caused by wearing impermeable gloves for extended periods [5]. In addition the hydrated and macerated stratum corneum is less mechanically protective. It may allow injury more easily as well as the enhanced penetration of toxic substances [5].

[1] Supported by a grant from the UK Health and Safety Executive Health Policy Division.

The aim of this study was to characterise the effects of occlusion by gloves on the stratum corneum in terms of its physical and functional properties and to investigate whether glove use has any detrimental effect.

## Methods

We chose to use glove patches in the first instance, because of the difficulties in persuading volunteers to wear occlusive gloves for long periods of time. The advantage of using patches was that we were able to use up to four sites per volunteer, per trial. A series of volunteer studies were carried out, looking at the effect of occlusion by patches of PVC glove material on the stratum corneum. We chose to use $3 \times 5$ cm patches of glove material held in place with Tegaderm transparent dressings. Sites on the volar aspect of the forearms were used (fig. 1). The general format for these trials was: on day 1 to occlude one site for 4 h, a second for 8 h and to use a third as an unoccluded control; and on day 2 to use a fourth site as a covered control for 8 h. On this fourth site we used an empty Tegaderm dressing (i.e. not containing a glove patch) as an active control.

To assess the effects of such occlusion on the stratum corneum, measurements were made of parameters relating to percorneal permeability, impairment of barrier function, skin surface roughness and skin surface compliance. These measurements were carried out before and at varying time points following the 4- or 8-hour periods of occlusion – to enable the time course effect of occlusion and recovery afterwards to be determined. Thus, we conducted a series of volunteer trials employing the four measurement techniques which are described below.

### Hyperaemic Response to Topical Hexyl Nicotinate/Percorneal Permeability

The change in percorneal permeability after occlusion of the forearm skin of 10 volunteers (3 M, 7 F, mean age 30, SD 9) was assessed. Percorneal permeability of the stratum corneum is a property for which there is no perfectly satisfactory in vivo measurement method. A relatively simple method would seem to offer a good estimate of permeability within clearly defined restrictions [6]. The method in question uses the vascular response to hexyl nicotinate penetration. The nicotinate induces hyperaemia which can be monitored using a laser Doppler flowmeter. A solution of 5 µl of 0.1 $M$ hexyl nicotinate in monopropylene glycol was applied and spread over an area not greater than $1 \times 1$ cm, for 1 min exactly using a spatula. Immediately following this any excess solution was wiped off with an absorbent pad and the laser Doppler flowmeter (PF1, Perimed) probe was placed on the centre of the site. Readings were recorded on a chart recorder for a maximum of half an hour, but more typically 15–20 min until maximum blood flow had clearly been observed. Parameters taken from the recordings were the time to onset of hyperaemia ($t_0$), and the peak value of the skin blood flow ($V_{max}$). The whole measurement procedure was first carried out on the unoccluded control site and then on each designated site following removal of the patch. On the site occluded for 8 h the measurement procedure was repeated the next day. The results have been expressed as a change in parameter as compared with the control measurement. Since readings have only been compared within subjects, differences in physiological reaction to the nicotinate between subjects are controlled for.

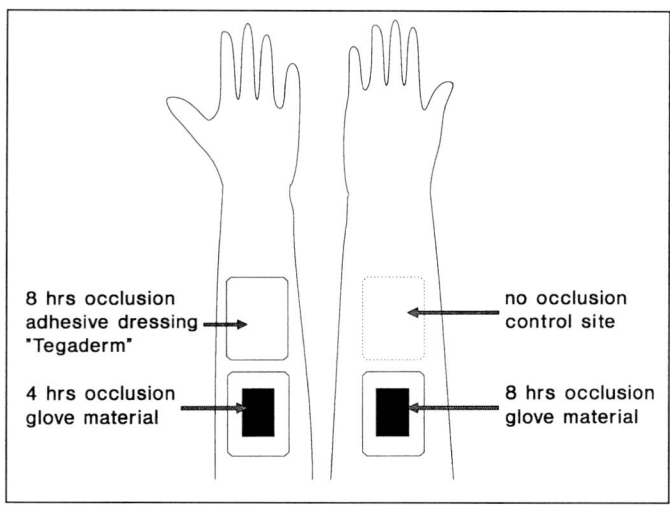

8 hrs occlusion
adhesive dressing →
"Tegaderm"

no occlusion ←
control site

4 hrs occlusion →
glove material

← 8 hrs occlusion
glove material

*Fig. 1.* Arrangement of glove patches.

### Transepidermal Water Loss

The effect of occlusion on the barrier function of the stratum corneum was assessed in 14 subjects (7 M, 7 F, mean age 29, SD 8) using measurements of transepidermal water loss (TEWL), using an instrument based on the design of Nilsson [7]. Each subject was rested for 10 min after preparation of the sites to be measured. The room temperature was maintained at 19–22 °C and humidity varied between 29 and 47%. Readings were taken on each occluded site 20 min after patch removal. On the site which had been occluded for 4 h the measurement was repeated 4 h after patch removal. On the site which had been occluded for 8 h, the measurement was repeated the following day. Two types of unoccluded control were used: first, the unoccluded control site was used and measurement of TEWL was taken in parallel with all measurements after removal of patches; second, measurement of TEWL was carried out on each designated site prior to application of the patch. As measurement of TEWL is sensitive to ambient factors – we were thus able to control for any site differences and any time point differences. In order to monitor in more detail the recovery of TEWL after glove patch occlusion for 4 h, a second group of 12 volunteers (5 M, 7 F, mean age 30, SD 8) was recruited for TEWL measurements to be carried out on the site prior to patch application and at 1, 2 and 3 h after its removal.

### Skin Surface Roughness

The effect of occlusion on skin surface roughness was assessed using 11 volunteers (5 M, 6 F, mean age 26, SD 4). Surface roughness was measured on each site at the time points described, using silicone rubber skin surface replicas and a stylus profilometer [8, 9]. At each site silicone dental impression material (Minerva Dental Ltd., Cardiff, UK) was applied to the skin, while the arms were in a resting position, so as to obtain a natural negative impression of the skin surface. Each impression obtained was mounted on a glass slide for later assessment. Measurements were made using a Hommeltester (Hommelwerke Ltd.), to assess the roughness of the skin impression surface, perpendicular to the longitudinal axis of the arm. The parameters used were: $R_a$ (microns), the arithmetic average value of all absolute distances of the roughness profile from the centre line within the measuring line of 4.8 mm; and $R_z$ (microns), the average maximum peak to valley height of 5 consecutive sampling lengths within the measuring length of 4.8 mm.

### Skin Surface Compliance

The deformability or compliance of the skin surface was assessed by measuring the change in skin surface roughness after a linear extension of 15% using a device specially constructed for the purpose [10]. The compliance technique was carried out in conjunction with the skin surface roughness measurements already described, on the same subjects at the same time. The compliance device encloses a 43 × 20 mm rectangular area of skin within a metal frame 3 mm thick. Following taking a resting impression of the skin surface, the compliance device was glued to the skin surface in a longitudinal position using cyanoacrylate adhesive. Only the two 20-mm sides were glued to the skin. The device was operated by moving one of these sides 7 mm, via a manual lever mechanism, thus stretching the skin by approximately 15%. Whilst the skin was in a stretched state a second skin surface impression was taken. Roughness measurements were again carried out perpendicular to the longitudinal direction, the direction of extension.

### Statistical Analysis

Nonparametric comparisons were made using the Wilcoxon rank sum test, with significance at 95%. Confidence interval analysis was performed using the difference of medians and the Wilcoxon nonparametric test.

## Results and Discussion

The results of the percorneal permeability measurements are shown in figure 2. After 4 hours' occlusion there was a median reduction in the time to onset of hyperaemia, to 59% of its preocclusion value (95% confidence interval 41–76%, p = 0.006 Wilcoxon test). After 8 hours' occlusion this was further reduced to 38% of its pre-occlusion value (95% confidence interval 22% to 53%, p = 0.006 Wilcoxon test). However, onset times returned to normal after an overnight rest period. This indicates that percorneal permeability is increased by the use of occlusive glove material, albeit only temporarily. The application of the Tegaderm control had no effect on percorneal permeability. (We have omitted the Tegaderm data from the following sections for the sake of clarity, however in most cases there was no significant or obvious effect.) There was no significant change in $V_{max}$ – indicating that individual physiological reactivity to hexyl nicotinate was well controlled for.

The effects of occlusion on the barrier function of the stratum corneum as measured by TEWL are shown in figure 3. There was a significant increase in

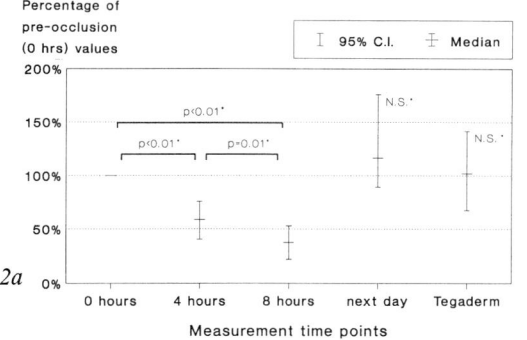

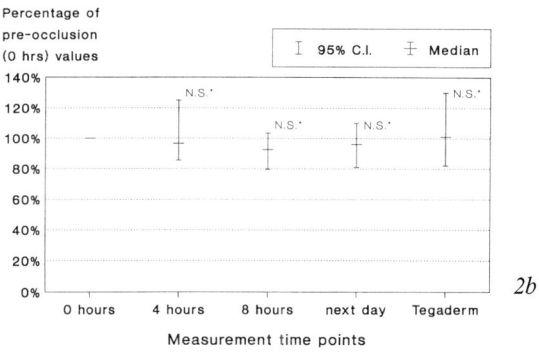

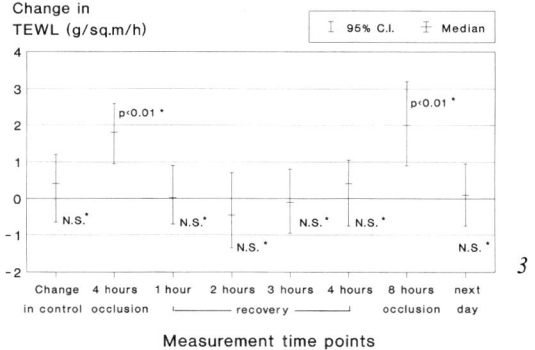

*Fig. 2. a* Change in 'time to onset' of nicoti-
nate-induced hyperaemia after occlusion by PVC
glove material. *b* Change in maximum nicotinate-
induced blood flow after occlusion of skin by PVC
glove material. * Wilcoxon test.

*Fig. 3.* Effect of occlusion by PVC glove material on TEWL. * Wilcoxon test re: pre-control.

TEWL after periods of both 4 and 8 h of occlusion. However, there was no significant difference between the 4- and 8-hour results. The median increases were 1.8
$g^{-2}$ $h^{-1}$ (95% confidence interval 0.95–2.6 $g^{-2}$ $h^{-1}$, p = 0.002 Wilcoxon test) and
2.0 $g^{-2}$ $h^{-1}$ (95% confidence interval 0.9–3.2 $g^{-2}$ $h^{-1}$, p = 0.004 Wilcoxon test),
respectively. These measurements were completed 20 min after removal of the
occlusive glove patches. Using the site that had been occluded for 4 h, measurements of TEWL were made after rest periods of 1, 2, 3 and 4 h. A further measurement was made on the site that had been occluded for 8 h, after an overnight rest
period. In each case, the TEWL readings had returned to their preocclusion values. These results suggest that there is temporary impairment of barrier function
after the use of occlusive glove material for just 4 h, although the influence of
residual surface water evaporation cannot entirely be discounted. They also suggest that doubling this period to 8 h does not produce entirely be discounted. They
also suggest that doubling this period to 8 h does not produce a proportional
increase in barrier impairment. What is not clear at this point is whether or not

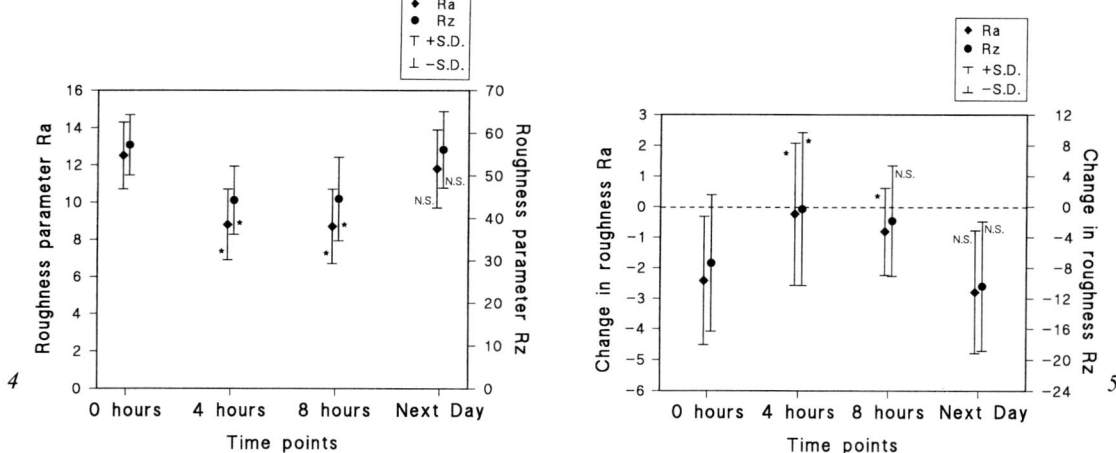

*Fig. 4.* Surface roughness of stratum corneum after occlusion by PVC glove patches. * p < 0.01 Wilcoxon text; N.S. = not significant.

*Fig. 5.* Change in compliance of stratum corneum after occlusion by PVC glove patches. * p < 0.05 Wilcoxon test; N.S. = not significant.

there would be a longer lasting chronic effect from repeated use of occlusive gloves.

The results of skin surface roughness measurements are presented in figure 4. After occlusion by periods of both 4 and 8 h, the skin surface roughness was significantly reduced in terms of the roughness parameters Ra and Rz. Our results indicate that the effect of occlusion on surface roughness reaches a threshold, before or at 4 hours' occlusion. After an overnight resting period, readings returned to preocclusion values. Such findings are consistent with increased hydration of the stratum corneum when the expansion of hydrated corneocytes acts to reduce the roughness profile.

The premise behind compliance measurements is that a more compliant and flexible stratum corneum will give an increase in roughness laterally when stretched longitudinally. The basis of this premise is that as the skin surface is extended in the longitudinal axis, the skin between the two stretch points will pull in from the sides (laterally). When the skin is compliant and flexible (low coefficient of elasticity) the skin surface 'rucks up', increasing the roughness profile laterally. If the skin is less compliant, i.e. more stiff (thus having a greater coefficient of elasticity), the surface is extended in the same way longitudinally but lateral tension counterbalances the force produced by the extension frame and thus no 'rucking' occurs. In fact, if the coefficient of elasticity is large enough then there will be a small decrease in the lateral roughness profile. The results of skin compliance measurements are presented in figure 5. After occlusion by periods of both 4 and 8 h, the compliance parameter (pre-stretch roughness-post-stretch roughness) was significantly greater than prior to occlusion. The explanation for this is that as the skin was extended longitudinally, the lateral forces produced were counterbalanced by the natural tension in the skin, but this was not as great as prior to occlusion because the coefficient of elasticity had been reduced by hydration. The coefficient of elasticity was sufficiently great prior to occlusion to give a stiff (less compliant) response. After occlusion, the response was that of less stiff more compliant skin. The question that arises from this is – does this result mean that the skin after occlusion is less mechanically protective? In order to answer this, comparison with measurements of stratum corneum point hardness after occlusion is needed. As with surface roughness measurements, our results indicate that the effect of occlusion on compliance reaches a threshold, before or at 4 hours' occlusion. Following an overnight resting period, the compliance measured had returned to the pre-occlusion value, again indicating that the effect may only be temporary.

In a further study reported elsewhere [11], we recruited volunteers who were willing to wear whole gloves for 6-hour periods for 2 days. A significant increase in TEWL was observed at all time points after glove use compared with before. What is particularly interesting is that even following an overnight recovery period TEWL readings were significantly elevated ($p < 0.05$). This was not observed in previous studies, and indicates that there may be a cumulative effect. To our knowledge, this has not been reported before. It would seem that the glove patches used were too small to simulate sufficiently the effects of heavy glove use. It is clear that a longer cumulative study is needed to test our findings.

## Conclusions

We have demonstrated at least a temporary impairment in the barrier function of the stratum corneum after glove patch use. Further data suggest that repeated occlusion by gloves may have a cumulative effect. Thus, the stratum corneum barrier could be impaired in the long term.

Clearly more needs to be done to follow up this work. There are a number of questions arising that are important from an occupational health and safety view:

(1) What level of occlusive glove use would promote a significant impairment of stratum corneum barrier function?
(2) How often and for how long would rest periods need to be to prevent a cumulative effect?

(3)  Is this reduction in barrier efficacy of sufficient magnitude to enhance the penetration of irritant substances present in the work place to an extent likely to increase the incidence of irritant contact dermatitis?

(4)  Is there a corresponding impairment in the mechanical competence of the stratum corneum rendering users more susceptible to the effects of mechanical trauma?

### References

1   van der Valk PG, Maibach HI: Post-application occlusion substantially increases the irritant response of the skin to repeated short-term sodium lauryl sulfate (SLS) exposure. Contact Derm 1989;21:335–338.
2   Estlander T, Jolanki R, Kanerva L: Dermatitis and urticaria from rubber and plastic gloves. Contact Derm 1986;14:20–25.
3   Wrangsjo K, Wahlberg J, Axelsson I: IgE-mediated allergy to natural rubber in 30 patients with contact urticaria. Contact Derm 1988;19:264–271.
4   Fisher A: Management of dermatitis due to surgical gloves. J Dermatol Surg Oncol 1985;11:628–631.
5   Shumnes E, Darby T: Contact dermatitis due to endotoxin in irradiated latex gloves. Contact Derm 1984;10:240–244.
6   Kohli R, Archer W, Li Wan Po A: Laser velocimetry for the non-invasive assessment of the percutaneous absorption of nicotinates. Int J Pharm 1987;36:91–98.
7   Nilsson GE: Measurement of water exchange through skin. Med Biol Eng Comput 1977;15:209–218.
8   Marks R, Pearse AD: Surfometry. A method of evaluating the internal structure of the stratum corneum. Br J Dermatol 1975;92:651–657.
9   Grove GL, Grove MJ: Objective methods for assessing skin surface topography non-invasively; in Leveque J (ed): Cutaneous Investigation in Health and Disease. New York, Marcel Dekker, 1990, pp 1–30.
10  Marks R, Edwards C, Black D: Non-invasive assessment of stratum corneum and function. Int J Cosmet Sci 1989;11:59–65.
11  Graves CJ, Edwards C, Marks R: The effects of protective occlusive gloves on stratum corneum barrier properties. Contact Derm, in press.

Mr. C.J. Graves, Department of Dermatology, University of Wales College of Medicine, Heath Park, Cardiff CF4 4XN (UK)

Elsner P, Maibach HI (eds): Irritant Dermatitis. New Clinical and Experimental Aspects.
Curr Probl Dermatol. Basel, Karger, 1995, vol 23, pp 95–103

..........................

# Human Barrier Formation and Reaction to Irritation

*Manigé Fartasch*

Department of Dermatology, University of Erlangen, Germany

The epidermal differentiation leads to the formation of the stratum corneum (SC). An important function of the SC is to serve as barrier and thus provide protection from the penetration of irritants and loss of water (transepidermal water loss, TEWL). Biophysical, morphological and biochemical data indicate that the SC form a continuous sheath of protein-enriched corneocytes embedded in an intercellular matrix enriched in nonpolar lipids, organized as lamellar lipid layers [1–5]. These intercellular lipid layers are thought to mediate transcutaneous water loss and play a vital role in the function of the barrier.

Irritation of the skin is accompanied by a complex array of epidermal and dermal metabolic responses. The role of the epidermal barrier and its repair seems to be crucial for the responses especially of the epidermal components. A growing body of data indicates that the occurring repair responses of the epidermis, which seemed to be regulated by the interplay of $Ca^{2+}$ and $K^+$ ions [6], involves the immediate secretion of preformed LB by outer stratum granulosum cells [7–10] and later the formation of new LB, the increase in epidermal lipid synthesis by the activation of regulatory enzymes [11] and the stimulation of DNA synthesis [12] leading to epidermal hyperplasia.

Recent studies give evidence that epicutaneously applied irritants like sodium dodecyl sulfate, acetone and other irritants or mechanical trauma (like SC tape stripping) leads to activation and expression of cytokines by the epidermal keratinocytes both shown on messenger RNA and protein level [13–16]. Additionally, adhesion molecules (like ICAM 1) are not only expressed by dermal components after irritation [17], but it has also been shown that there is an up-regulation of integrins by the keratinocytes [18].

The modulation and initiating effect of the keratinocytes is influenced by the reaction of the barrier and the pathway of the irritant into the nucleated parts of the epidermis. To date, there have been studies investigating the different structural changes of the nucleated parts of the epidermis induced by different chemical irritants [17, 19–24]. However, systematic investigations of the pathologic effects of a range of different irritants on the human horny layer (e.g. especially barrier structures) could not be performed, since with routine electron microscopy using double fixation with glutaraldehyde and osmium ($OsO_4$) the intercellular spaces of the SC appear empty, $OsO_4$ has not been effective in demonstrating the spatial organization of the epidermal lipids in SC. The use of ruthenium tetroxide ($RuO_4$) fixation [25–29] opened the possibility of detailed studies of the spatial organization of the intercellular lipids of the SC.

## Material and Methods

*Irritation with SDS.* Patch tests with Finn chambers on scanpore tape (Hermal, Reinbek, Germany) (for 24 h with 0.5% aqueous SDS and 1% SDS) on 9 human volunteers with no history of skin disease. Contralateral sites on the volar aspects of each forearm were treated. To perform TEWL measurements in the recovery phase, the experiments were performed simultaneously on both volar forearms and biopsies were only taken from the test sites of the left arm.

*Irritation with Acetone.* Patch test (see above) were performed for 1 (n = 2), 3 (n = 2), 5 (n = 3) and 12 h (n = 3).

*Controls.* For comparison, healthy skin from 5 patients without skin disease undergoing plastic surgery was obtained.

*Transepidermal Water Loss Measurements.* The quantitative measurements of TEWL were carried out using the Servo Med Evaporimeter (Servo Med, Sweden) and by the Tewameter (Courage and Kasaka, Cologne, Germany). The TEWL is calculated automatically and expressed in $g/m^2/h$. The reading was performed after the guidelines of Pinnagoda et al. [30] before irritation, 30 min after removal of patch tests.

*Morphologic Preparations.* Punch biopsies were divided into two parts: half of the biopsy was processed for routine electron microscopy: double fixation with 2.5% glutaraldehyde and osmium tetroxide ($OsO_4$). The other half was fixed in acrolein vapor and 2.5% glutaraldehyde for 2 h, rinsed in buffer for 3 h, postfixation with 0.5% ruthenium tetroxide ($RuO_4$, Polyscience, Warrington, Pa., USA) with 0.25% potassium ferrocyanide, ph 6.2, for 1 h in darkness at 4°C. Embedding: Spurr's resin.

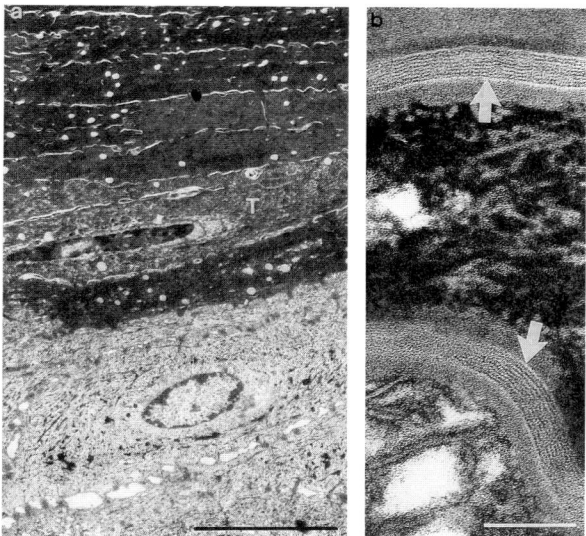

Fig. 1. *a* Occurrence of lipid droplets in the matrix of the corneocytes. Additionally transitional cell zone (T). RuO$_4$. × 19,400. *b* Regular lamellar arranged epidermal lipids (white arrows) in the upper regions of SC inspite of SDS exposure. RuO$_4$. × 121,000.

## Results

### SDS-Treated Skin

In the probands with 0.5% SDS-induced dry skin with scaly surface the epidermal lipids still showed lamellar arrangement (fig. 1b, white arrow). A one-layered transit-cell zone (sign of premature keratinization) and lipid droplets were found in the matrix of horny cells (fig. 1a). With 1% SDS the intercellular spaces showed focal widening with alterations in substructures of lipid layers with loss of parallel arrangement.

Additionally, disturbance of rearrangement of LB sheets into parallel lipid layers appeared in the stratum compactum which could be demonstrated by routine fixation of the 1% irritated tissue. The cornified envelope showed broadening as sign of structural alteration of the horny cell which could be demonstrated by routine fixation (OsO$_4$) of the 1% irritated tissue. In some areas of the epidermis paranuclear vacuoles appeared in the keratinocytes mostly indenting one side of the nucleus as typical sign of toxic damage [31] of vital epidermis. Light microscopy of the routinely fixed semithin section of the proband who showed clinically severe reaction revealed detergent-induced cell dissociation of SC [32].

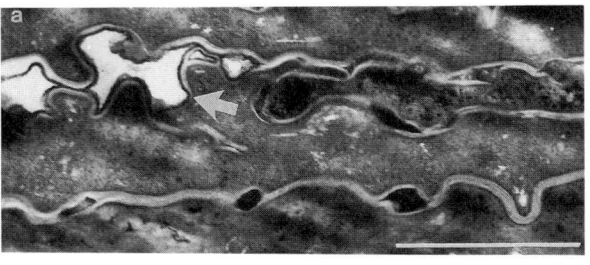

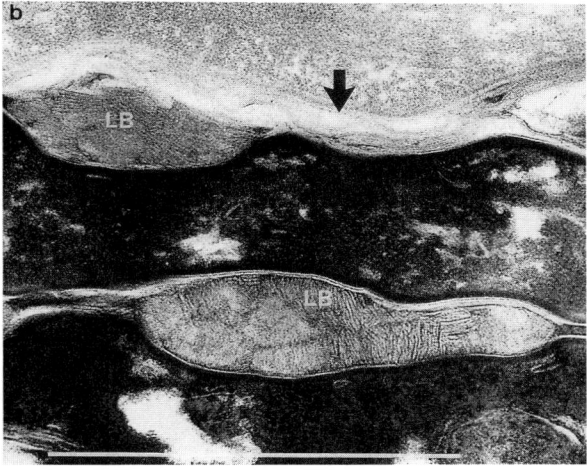

*Fig. 2. a* ICS showed partly lipid extracted regions (arrow). RuO$_4$. × 24,000. *b* The LB lipids seemed to be unaltered compared to the lipid bilayers (arrow) which showed loss of cohesion. RuO$_4$. × 54,000.

### Acetone-Treated Skin

After 1 and 3 h application of acetone no TEWL enhancement could be observed. From the morphological point of view the SC seemed to be unaltered. The alteration of barrier structure and the viable parts of the epidermis after 5 h occlusion were as follows: Partial expansion of the ICS (fig. 2a), loss of lamellar arrangement of the lipid structures. The lamellae were transformed into an electron-dense granular substance, which seemed to be partly extracted (fig. 2a, white arrow) and then are only partly filled with the granular material (fig. 2b). In some regions, the lamellar arrangement of lipid bilayers could still be recognized but there seemed to be a loss of cohesion between the lamellae (fig. 2b, arrow). In the lower parts of stratum corneum, the more polar LB sheets seem to be more resistant to acetone and looked structurally unaltered while the already transformed more unpolar lipids showed disruption.

## Discussion

Since chemically induced skin irritation is a major human health problem, there is considerable interest in understanding the effects of exposure of the human barrier of the skin (e.g. SC) to the various irritants.

Ultrastructural and biochemical studies on the intercellular compartment of SC have previously shown that the horny layer should not be considered to be uniform in functional barrier characteristics [32]. Recent studies clearly indicate that the different layers of the SC vary functionally, structurally and biochemically [33–39]. For example, in the lower parts of the SC (stratum compactum) tranformation ('lipid maturation') of hte initially secreted LB lipids (predominantly glucoceramides, cholesterol, and phospholipids) into a nonpolar mixture, enriched in ceramides, cholesterol, and the free fatty acids [4] is seen. It has been shown that in humans the processing of the LB lipids takes place within saccular compartments of the ICS. These compartments might offer a unique microenvironment and facilitate the transformation process of LB sheets by actively maintaining an optimal pH necessary for activating the 'lipid transformation' enzymes (β-glucocerebrosidase, phospholipase A, sphingomyelinase) [26–28]. The regular persistence of the LB sheets (polar precursor lipids) in the intercellular spaces of the lower parts of SC may also imply a different behavior to penetrating substances compared to the upper regions of the SC.

Previous studies have shown that structurally unrelated chemical irritants induce a different pattern of cellular damage to the nucleated parts of the epidermis when topically applied to human skin [17]. The aim of the study has been to characterize and perform systematic analysis of the intimate structural changes produced by two chemically different irritants which are typically used in irritancy studies.

The subjects were exposed to different concentrations of SDS, an anionic surfactant which has been widely used as a model detergent in irritancy studies, to induce a dry scaly skin condition. The dry skin was characterized by an enhanced TEWL as a sign of an impaired water permeability barrier of the horny layer. The events that lead to the dry and scaly skin with high TEWL are still being elucidated. The supposition that surfactants degrease the skin (e.g. selective depletion of the lipids from the ICS [39]) has persistence in the literature [40]. On the other hand, the induction of SC cell dissociation [41] (possibly by degradation of protein structures) or a direct interaction between the surfactant and keratin proteins are discussed. In our study, the tissue response to SDS was quite varied. Some probands only showing a slight dryness or scaliness of the SC, others exhibiting erythema with histologically manifested epidermal cell necrosis and parakeratosis, hyperplasia [42, 43] and the occurrence of vacuoles and lipid droplets [22]. A predominant feature of cutaneous responses to the detergent was parakeratosis or the formation of a nuclei retaining SC. Such an appearance may have a number of causes, including increased epidermal cell proliferation, accelerated keratinization, or direct cytotoxic injury [44]. Interestingly, in the upper regions of SC still regular lamellar arrangements of lipids were seen after exposure to 0.5% SDS. This corroborates biochemical studies which have shown that in experimentally in vivo-induced scaly skin the amount of ceramides did not differ from the concentration in healthy control skin and only 4–7% removal of lipids had been found [45]. On the other hand, Fulmer and Kramer [46] showed that the lipid composition of skin was altered. Our findings show that these changes are not likely to reflect the immediate consequences of surfactant exposure. It seems to be due to an action of SDS upon lipogenesis since the surfactant (in low concentration) did not seem to alter the existing lipid structure, but instead the synthesis of new lipids: (1) by the direct alteration of the LB secretory system and disturbance of the formation of LB sheets, which in consequence leads to abnormal ceramide biosynthesis, and (2) indirectly via inducing a disturbed differentiation by alterating the viable parts of the epidermis. All these observed changes might explain the long-lasting TEWL enhancement since normalization of barrier function occurred earliest after 10–12 days. Other reasons for the early subclinical in vivo effect of SDS, especially the induction of dryness, roughness and scaliness, could be the direct interaction with the protein structures inducing cell dissociation [41] and the desquamation of the upper layers [47]. This is intensified by the already described occurrence of parakeratosis which may be due to an increased epidermal cell proliferation or an accelerated SC turnover [17, 48, 49]. Thus, on the whole, factors other than lipid removal are likely to be eliciting the flaking [45].

Topical acetone treatment is believed to induce disruption of the barrier by the selective removal of SC lipids, accompanied by minimal cytotoxicity [6, 8, 10]. In our probands, we could not detect any morphological changes of the skin after 1 and 3 h application of acetone. Increase of TEWL could only be detected after more than 5 h of exposure.

Our studies on human skin have shown that the structural alterations of the horny layer with acetone were quite different as seen in SDS-treated skin. Previous ultrastructural observations with conventional $OsO_4$ staining on human skin after irritation with acetone [50] and our studies showed predominantly nuclear changes of the stratum granulosum cells in the viable parts of the epidermis with no alteration in the basal layers.

Most of the recent ultrastructural studies have focussed on acetone-treated hairless murine skin [8, 50–52]. In human skin, the penetration of acetone showed that the LB sheets derived intercellular bilayers seemed to be more sensitive to the irritant.

Such results do support the theories that the individual irritants themselves are not only exerting rather specific morphological effects on the cellular components of the skin, particular the epidermal keratinocyte, but also inducing distinct morphological alterations of the different components of the SC.

## References

1   Elias PM: Epidermal lipids, barrier function, and desquamation. J Invest Dermatol 1983;80:44–49.
2   Elias PM, Goerke J, Friend DS: Mammalian epidermal barrier layer lipids: Composition and influence on structure. J Invest Dermatol 1977;69:535–546.
3   Landmann L: Epidermal permeability barrier: Transformation of lamellar granule disks into intercellular sheets by a membrane-fusion process. A freeze-facture study. J Invest Dermatol 1986;87:202–209.
4   Elias PM, Menon GK: Structural and lipid biochemical correlates of the epidermal permeability barrier. Adv Lipid Res 1991;24:1–26.
5   Elias PM, Menon GK, Holleran WM, Ghadially R, Williams ML, Feingold KR: Epidermal permeability barrier homeostasis: Normal mechanisms and pathophysiology; in Dahl MV, Lynch PJ (eds): Current Opinion in Dermatology. London, Current Science, 1994.
6   Lee CH, Kawasaki Y, Maibach H: Effect of surfactant mixtures on irritant contact dermatitis potential in man: Sodium lauroyl glutamate and sodium lauryl sulphate. Contact Derm 1994;30:205–209.
7   Wilgram GF, Krawczyk WS, Cole PL: Sunburn effect on keratinosomes. Arch Dermatol 1970;101:505–510.
8   Menon GK, Feingold KR, Elias PM: Lamellar body secretory response to barrier disruption. J Invest Dermatol 1992;98:279–289.
9   Menon GK, Elias PM, Feingold KR: Integrity of the permeability barrier is crucial for maintenance of the epidermal calcium gradient. Br J Dermatol 1994;130:139–147.
10  Menon GK, Price LF, Bommannan B, Elias PM, Feingold KR: Selective obliteration of the epidermal calcium gradient leads to enhanced lamellar body secretion. J Invest Dermatol 1994;102:789–795.
11  Grubauer G, Feingold KR, Harris RR, Elias PM: Lipid content and lipid type as determinants of the epidermal permeability barrier. J Lipid Res 1989;30:89–96.
12  Proksch E, Feingold KR, Qiang M, et al: Barrier function regulates epidermal DNA snythesis. J Clin Invest 1991;87:1668–1673.
13  Wood LC, Jackson SM, Elias PM, et al: Cutaneous barrier perturbation stimulates cytokine production in the epidermis of mice. J Clin Invest 1992;90:482–487.
14  Nickoloff BJ, Naidu Y: Perturbation of epidermal barrier function correlates with initiation of cytokine cascade in human skin. J Am Acad Dermatol 1994;30:535–546.
15  Tsai JC, Feingold KR, Crumrine D, Wood LC, Grunfeld C, Elias PM: Permeability barrier disruption alters the localization and expression of TNFalpha protein in the epidermis. Arch Dermatol Res 1994;286:242–248.
16  Wilmer JL, Burleson FG, Kayama F, Kanno J, Luster MI: Cytokine induction in human epidermal keratinocytes exposed to contact irritants and its relation to chemical-induced inflammation in mouse skin. J Invest Dermatol 1994;102:915–922.
17  Willis CM, Stephens CJM, Wilkinson JD: Selective expression of immune-associated surface antigens by keratinocytes in irritant cintact dermatitis. J Invest Dermatol 1991;96:505–511.
18  von den Driesch P, Fartasch M, Ponec M: Expression of integrin receptors and ICAM-1 in the in vitro reconstructed epidermis: Effect of sodium lauryl sulphate. Arch Dermatol Res 1994;in press.
19  Tovell PWA, Weaver AC, Hope J, Sprott WE: The action of sodium lauryl sulphate on rat skin: An ultrastructural study. Br J Dermatol 1974;90:501–506.

20   Mahmoud G, Lachapelle JM, Van Neste D: Histological assessment of skin damage by irritants: Its possible use in the evaluation of a barrier cream. Contact Derm 1984;11:179–185.

21   Hamami I, Marks R: Structural determinants of the response of the skin to chemical irritants. Contact Derm 1988;18:71–75.

22   Willis CM, Stephens CJM, Wilkinson JD: Epidermal damage induced by irritants in man: A light and electron microscopic study. J Invest Dermatol 1989;93:695–699.

23   Willis CM, Stephens CJM, Wilkinson JD: Differential effects of structurally unrelated chemical irritants on the density of proliferating keratinocytes in 48 h patch test reactions. J Invest Dermatol 1992;99:449–453.

24   Willis CM, Stephens CJM, Wilkinson JD: Differential patterns of epidermal leukocyte infiltration in patch test reactions to structurally unrelated chemical irritants. J Invest Dermatol 1993;101: 364–370.

25   Madison KC, Swartzendruber DC, Wertz PW, Downing DT: Presence of intact intercellular lipid lamellae in the upper layers of the stratum corneum. J Invest Dermatol 1987;88:714–718.

26   Fartasch M, Bassukas ID, Diepgen TL: Disturbed extruding mechanism of lamellar bodies in dry non-eczematous skin of atopics. Br J Dermatol 1992;127:221–227.

27   Fartasch M, Bassukas ID, Diepgen TL: Structural relationship between epidermal lipid lamellae, lamellar bodies and desmosomes in human epidermis: An ultrastructural study. Br J Dermatol 1993;128:1–9.

28   Fartasch M, Ponec M: Improved barrier structure formation in air-exposed human keratinocyte culture systems. J Invest Dermatol 1994;102:366–374.

29   Hou SYE, Rehfeld SJ, Plachy WZ: X-ray diffraction and electron paramagnetic resonance spectroscopy of mammalian stratum corneum lipid domains. Adv Lipid Res 1991;24:141–171.

30   Pinnagoda J, Tupker RA, Agner T, Serup J: Guidelines for transepidermal water loss (TEWL) measurement. Contact Derm 1990;22:164–178.

31   Nakamura M, Rikimaru T, Yano T, Moore KG, Pula PJ, Schofield BH, Dannanberg AM: Full thickness human skin explants for testing toxicity of topically applied chemicals. J Invest Dermatol 1990;95:325–332.

32   Blank IH: Cutaneous barriers. J Invest Dermatol 1965;45:249–256.

33   Scheuplein RJ: Percutaneous absorption after twenty-five years: or 'old wine in new wineskins'. J Invest Dermatol 1976;67:31–38.

34   Bowser PA, White RJ: Isolation, barrier properties and lipid analysis of stratum compactum, a discrete region of the stratum corneum. Br J Dermatol 1985;112:1–14.

35   Elias PM, Feingold KR: Lipid-related barriers and gradients in the epidermis. Ann NY Acad Sci 1988;548:4–13.

36   van der Valk PGM, Maibach HI: A functional study of the skin barrier to evaporative water loss by means of repeated cellophane-tape stripping. Clin Exp Dermatol 1990;15:180–182.

37   Bommannan D, Potts RO, Guy RH: Examination of stratum corneum barrier function in vivo by infared spectroscopy. J Invest Dermatol 1990;95:403–408.

38   Potts PO, Francoeur ML: The influence of stratum corneum morphology on water permeability. J Invest Dermatol 1991;96:495–499.

39   Proksch E, Feingold KR, Elias PM: Localisation and regulation of epidermal 3-hydroxy-3-methyl-glutaryl-coenzyme. A reductase activity by barrier requirements. Biochim Biophys Acta 1991;1083: 71–79.

40   Imokawa G, Akasaki S, Minematsu Y, Kawai M: Importance of intercellular lipids in water-retention properties of the stratum corneum: induction and recovery study of surfactant dry skin. Arch Dermatol Res 1989;281:45–51.

41   Egelrud T, Lundstrom A: The dependence of detergent-induced cell dissociation in non-palmoplantar stratum corneum on endogenous proteolysis. J Invest Dermatol 1990;45:456–457.

42   Prottey C, Hartop PJ: Changes in glycerolipid metabolism in rat epidermis following exaggerated washing with soap solutions. J Invest Dermatol 1973;61:168–179.

43   Lindberg M, Forslind B, Sagström S, Roomans GM: Elemental changes in guinea pig epidermis at repeated exposure to sodium lauryl sulfate. Acta Derm Venereol (Stockh) 1992;72:428–431.

44   Willis CM, Stephens JM, Wilkinson JD: Selective expression of immune-associated surface antigens by keratinocytes in irritant contact dermatitis. J Invest Dermatol 1991;96:505–511.

45  Froebe CL, Simion FA, Rhein LD, Cagan RH, Kligman A: Stratum corneum lipid removal by surfactants: Relation to in vivo irritation. Dermatologica 1990;181:277–283.
46  Fulmer AW, Kramer GJ: Stratum corneum lipid abnormalities in surfactant-induced dry scaly skin. J Invest Dermaol 1986;86:598–602.
47  Denda M, Koyama J, Namba R, Horii I: Stratum corneum lipid morphology and transepidermal water loss in normal skin and surfactant-induced scaly skin. Arch Dermatol Res 1994;286:41–46.
48  Baker H, Kligman AM: Technique for estimating turnover time of human stratum corneum. Arch Dermatol 1967;95:408–411.
49  Suzuki Y, Nomura J, Koyama J, Horii I: The role of proteases in stratum corneum: Involvement in stratum corneum desquamation. Arch Dermatol Res 1994;286:249–253.
50  Lupulescu AP, Birmingham DJ, Pinkus H: An electron microscopic study of human epidermis after acetone and kerosene administration. J Invest Dermatol 1973;60:33–45.
51  Ghadially R, Halkier-Sorensen L, Elias PM: Effects of petrolatum on stratum corneum structure and function. J Am Acad Dermatol 1992;26:387–396.
52  Man M-Q, Feingold KR, Elias PM: Exogenous lipids influence permeability barrier recovery in acetone-treated murine skin. Arch Dermatol 1993;129:728–738.

Manigé Fartasch, MD, Department of Dermatology, University of Erlangen,
D–91054 Erlangen (Germany)

Elsner P, Maibach HI (eds): Irritant Dermatitis. New Clinical and Experimental Aspects.
Curr Probl Dermatol. Basel, Karger, 1995, vol 23, pp 104–107

...........................

# Physiologic Response of Chronically Inflamed and Accommodated Human Skin

*Robert L. Rietschel*

Department of Dermatology, Ochsner Clinic and Alton Ochsner Medical
Foundation, New Orleans, La., USA

The down-regulation of inflammation due to irritants applied to human skin has been termed accommodation or 'hardening' [1–3]. To learn more about this phenomenon, 50 human subjects were recruited to apply 20% maleic acid in a vehicle of 20% propylene glycol, 50% ethanol, and 30% water to one forearm daily for a period of 6 weeks (pH 1.4) [4]. These subjects were then observed for clinical 'hardening' or accommodation. This particular irritant was chosen for its ability to produce an even and tolerable degree of inflammation in most subjects.

Between the fourth and sixth weeks of application, 17 subjects developed an acute vesicular dermatitis deemed to be the acquisition of allergic contact dermatitis to the irritant. These subjects were dropped from further study. At the end of the 6-week period of treatment, a 4-hour closed patch test to 20% maleic acid was performed on both the normal and treated forearm to assess the status of the subject. The results in table 1 show that subjects could be grouped as to their reactivity based on the clinical picture as well as the patch test reaction. Only 5 subjects were deemed to have accommodated. The remainder had varying degrees of inflammation or hyperirritable skin.

To determine whether the repeated application of maleic acid had led to protection against other unrelated irritants, a challenge of 4 hours' duration under a closed patch test was undertaken with 17% benzalkonium chloride (pH 4.52) and 5% sodium lauryl sulfate (pH 8.94). The results are shown in table 2. Reactivity to other irritants was lower in the accommodated than in the inflamed group. However, the accommodated forearm skin was not as resistant to irritation as the untreated normal skin on the contralateral forearm. Only 1 of the 32 subjects reacted to 17% benzalkonium chloride under a 4-hour closed patch test on the healthy normal forearm.

Table 1. Determination of clinical status

| Clinical status of treated skin | Reactivity of closed patch test on treated and normal forearms (4-hour) | Determined status |
|---|---|---|
| Inflamed (n = 9) | treated > normal | hyperirritated |
| Inflamed (n = 8) | treated = normal | inflamed |
| Inflamed (n = 11) | treated < normal | inflamed |
| Normal (n = 5) | virtually no reactivity on treated skin | accommodated |

Table 2. Reactivity of maleic acid treated skin to unrelated irritants*

| Status | Subjects | Irritants | | % reacting |
|---|---|---|---|---|
| | | 17% BZK | 5% SLS | |
| Accommodated | 5 | 1 | 0 | 20 |
| Inflamed | 18 | 5 | 1 | 33 |
| Hyperirritated | 9 | 6 | 0 | 67 |

BZK = Benzalkonium chloride; SLS = sodium lauryl sulfate; * 4-hour closed patch test reactions.

To assess whether inflammation was broadly and nonspecifically blunted in accommodated skin, recall antigens were tested. These included streptokinase-streptodonase, *Candida,* and urshiol. These antigens produced reactions of virtually equal intensity on both accommodated and unaccommodated skin alike, indicating that immunologically driven inflammation was not broadly and nonspecifically blunted in the accommodated group.

To assess whether the accommodated skin could acquire new delayed hypersensitivity allergens normally, a challenge with 2,000 μg of dinitrochlorobenzene (DNCB) was used as an allergen-induction dose. Three weeks later, elicitation with 100, 50 and 5 μg of DNCB was performed on both the normal skin and the accommodated forearm skin. All 5 of the accommodated subjects sensitized normally. The subjects who had persistent inflammation were likewise tested for the acquisition of new allergy, but their results were very different. Of this group, 4 received the DNCB induction dose of 2,000 μg on both their normal forearm and their inflamed forearm, while 7 received the DNCB induction dose on their inflamed arm only. When challenged 3 weeks later with the 100, 50 and 5 μg

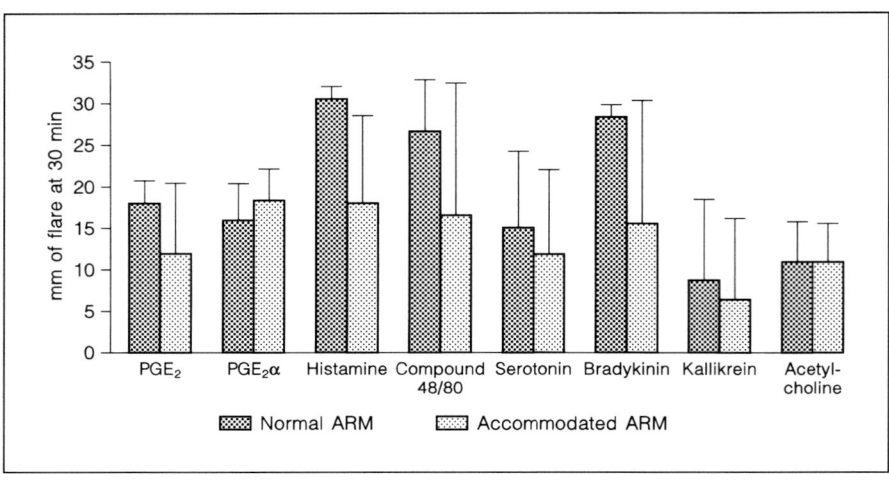

*Fig. 1.* Physiological differences between normal and accommodated skin in well-accommodated subjects.

patch tests on *normal* skin, only 2 of the 4 subjects who were given an induction dose on both forearms were found to be elicitable and sensitive, and 4 of the 7 who received the induction dose on the accommodated (inflamed) arm were deemed to be sensitive. Importantly, these results are in contrast to the 5 well-accommodated subjects who all reacted on their normal forearm. When the *inflamed* skin was used as the test area to determine whether induction of DNCB sensitization occurred, at the three dosages used for elicitation in the 11 inflamed subjects, only 1 of 4 subjects who were sensitized on both forearms was still deemed as not sensitive, and 3 of 7 who received the induction dose on their inflamed forearm only were still considered nonsensitive.

To learn more about the status of accommodated skin, mediators of inflammation were studied. The following substances were injected intradermally and read for wheal and flare reactions at 30 and 60 min: prostaglandin $E_2$ 0.001 mg/ml, prostaglandin $F_{2\alpha}$ 0.1 mg/ml, histamine 0.01 mg/ml, compound 48/80 0.01 mg/ml, serotonin 0.01 mg/ml, bradykinin 0.01 mg/ml, kallikrein 4 U/ml, and acetylcholine 1:1,000.

When comparing the normal forearm of the accommodated subjects with the accommodated forearm, there was significantly less wheal to bradykinin at 30 and 60 min and significantly less flare to histamine at 30 and 60 min in the accommodated forearm ($p < 0.05$). When comparing the *normal* skin of accommodated subjects with the *normal* skin of subjects who had persistent inflammation, the accommodated group showed more histamine flare and more acetylcholine flare. To accurately read the degree of inflammation on the forearm that showed persistent inflammation was difficult, thus allowing for a reliable comparison between the two groups in the normal skin only.

The comparison between normal and accommodated skin in the well-accommodated subjects is illustrated in figure 1, which demonstrates that the flare phenomenon is altered. This finding is the most significant in this comparison and would seem to suggest that the axon reflex-mediated events are blunted in accommodation. This blunting would lead to less vasodilatation and could have a more general effect on the response to other irritants. Since neurons would be involved in this process, a receptor fatigue phenomenon could be a mechanism for the changes observed, which would explain why similar changes were seen whether endogenous histamine was released by compound 48/80 or whether exogenous histamine was administered by the intradermal test. In both cases, the axon reflex flare was blunted.

When the normal skin responses of the subjects who had persistent inflammation but *did not* sensitize to DNCB were compared with the normal skin responses of those with persistent inflammation who *did* sensitize to DNCB, a striking significant reduction in the prostaglandin $E_2$ response was seen in the subjects who did not sensitize. This result would seem to suggest that the presence of inflammation had created an inhibitor of prostaglandin $E_2$ that could possibly circulate systemically and suppress the induction phase of sensitization – an unexpected finding in the study that may deserve further investigation.

### Acknowledgement

This work was accomplished with the technical assistance of Joyce Klemm of Emory University, Atlanta, Ga., USA. Supported by National Institute of Occupational Safety and Health grant R01-OHO1124.

### References

1  Mitchell JS: Hardening in allergic contact dermatitis and immunologic tolerance. A review of the literature with special reference to hyposensitization induced by topically applied agents. Trans St Johns Hosp Derm Soc 1969;55:141–159.
2  Skog E: The influence of preexposure of alkyl benzene sulfonate detergent, soap and acetone on primary irritant and allergic eczematous reactions. Acta Derm Venerol 1958;38:1–14.
3  McOsker DE, Beck LW: Characteristics of accommodated (hardened) skin. J Invest Dermatol 1967;48:372–383.
4  Britz MB, Maibach HI: Human cutaneous vulvar reactivity to irritants. Contact Dermatitis 1979;5: 375–377.

Robert Rietschel, MD, Ochsner Clinic, 1516 Jefferson Highway, New Orleans, LA 70121 (USA)

Elsner P, Maibach HI (eds): Irritant Dermatitis. New Clinical and Experimental Aspects.
Curr Probl Dermatol. Basel, Karger, 1995, vol 23, pp 108–113

..........................

# An Immunohistochemical Study of Contact Irritant and Contact Allergic Patch Tests

*A. Verheyen*[a], *L. Matthieu*[a], *J. Lambert*[a], *E. Van Marck*[b], *P. Dockx*[a]

Departments of
[a] Dermatology and
[b] Pathology, University of Antwerp, Belgium

There are still many controversies concerning the pathogenesis and the pathogenetic pathways in allergic and irritant contact dermatitis. Allergic and irritant contact dermatitis show a remarkable similarity with respect to clinical appearance, histology and immunohistochemistry. Our study was designed to compare the role of the keratinocyte in contact allergic and contact irritant dermatitis. The expression of different surface antigens on keratinocytes by means of patch tests in allergic and irritant contact dermatitis were done to examine the role of the keratinocyte in the pathogenesis of contact allergic and irritant dermatitis.

A retrospective study of the standard allergic patch test series showed a high percentage of positive (+) and dubious reactions for cocamidopropylbetaine (1% in water). Could there be an irritant factor?

A prospective study was set up to compare the dubious patch tests of cocamidopropylbetaine immunohistochemically with manifest allergic and irritant patch tests, in order to make a distinction between an allergic or irritant mechanism.

## Methods

### Patients

We studied a group of patients who where patch tested with the standard or the cosmetic batteries (Trolab), when we suspected a contact allergic eczema. A group of volunteers were tested with 1% aqueous sodium lauryl sulfate (SLS), to elicit an irritant contact reaction.

All the test products were applied under occlusion in small Finn Chambers® on the back. After 48 h, the patch tests were removed and the intensity of the response was visually scored following the standard codes. Nine patients showed a positive reaction (code +, ++) after 48 h for different allergens: nickel, cobalt, fragrance mix, paraphenylenediamine, form-aldehyde, chloracetamide. Twelve patients had a +/– reaction for cocamidopropylbetaine (1% in water).

Ten positive (redness and edema) contact irritant reactions due to SLS were selected after 48 h.

*Materials*
Punch biopsies (2 mm) were then taken from each selected test site. The biopsy was snap frozen in liquid nitrogen and stored at – 70 °C.

*Staining of the Frozen Sections.* Cryostat sections (6 μm) were placed on poly-*L*-lysine-coated glass slides, air dried, fixed in acetone for 10 min and washed in phosphate-buffered saline (PBS). The sections were then incubated in a moist chamber at room temperature for 30 min or 1 h, depending on the monoclonal antigen used, with monoclonal antibodies diluted in PBS + 1% bovine serum albumin.

The panel of mouse monoclonal antibodies was composed of: anti-ICAM-1 (CD54): dilution 1/100, incubation 1 h, Boehringer Mannheim Biochemica, Mannheim, Germany; OKM5 (CD36): dilution 1/200, incubation 30 min, Orthomune, Raritan, N.J., USA; anti-HLA-DR: undiluted, Clonab Biotest, Dreieich, Germany.

After washing, the bound monoclonal antibodies were labelled using the PAP tech-nique. The peroxidase enzyme was visualised by diaminobenzidine. The sections were counterstained with hematoxylin. Negative controls were obtained by omitting the primary monoclonal antibody.

*Antigen expression and cell numbers* in the infiltrates were assessed semiquantitatively. The extent of antigen expression by different cell types and the number of cells in the infil-trates in the biopsies were graded as absent (0), weak/rare (+), moderate/few (++), strong/numerous (+++), and very strong/very numerous (++++) on sections randomly assessed by two observers (A.V. and P.D.).

## Results

### ICAM-1 Expression

Examination of contact allergic tests and tests with cocamidopropylbetaine revealed areas of keratinocyte ICAM-1 expression in, respectively, 55 and 50% of the samples. In the irritant-treated skin only 1 of 10 cases showed expression for ICAM-1 on epidermal keratinocytes. The staining was the heaviest in the basal layer of the epidermis (fig. 1). ICAM-1 was also expressed by endothelial cells and a proportion of dermal mononuclear cells of the infiltrate in the three groups.

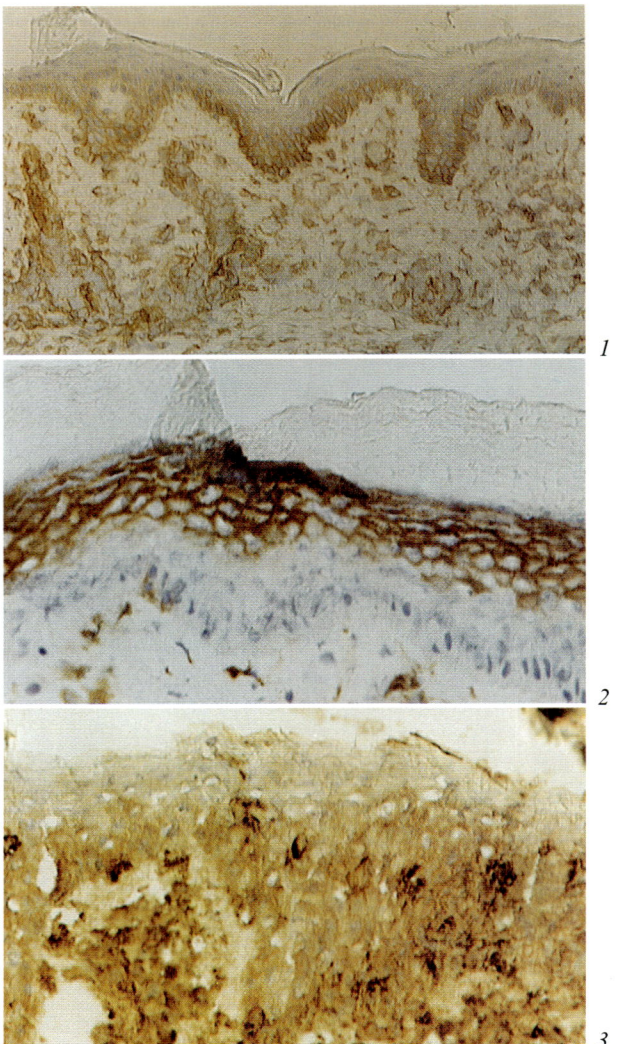

*Fig. 1.* ICAM-1 expression on the keratinocytes in the basal layer of the epidermis in a test with cocamidopropylbetaine.

*Fig. 2.* OKM5 expression on the keratinocytes of the stratum granulosum in a test with cocamidopropylbetaine.

*Fig. 3.* Immunolabeling of a section from a contact irritant reaction showing an area in which keratinocytes have been induced to express HLA-DR.

*Table 1.* Results for the expression of ICAM-1,HLA-DR, OKM5 by epidermal keratinocytes in contact allergic, irritant tests and in the tests with cocamidopropylbetaine

| Expression of antigens on keratinocytes | Contact allergic tests (n = 9) | | Contact irritative tests (n = 10) | | Cocamidopropylbetaine tests (n = 12) | |
|---|---|---|---|---|---|---|
| | n | % | n | % | n | % |
| ICAM-1 | 5/9 | 55 | 1/10 | 10 | 6/12 | 50 |
| OKM5 | 4/9 | 44 | 6/10 | 60 | 6/12 | 50 |
| HLA-DR | 2/9 | 22 | 2/10 | 20 | 6/12 | 50 |

*OKM5 Expression*

The positivity of OKM5 was expressed on the keratinocytes of the stratum granulosum (fig. 2). Our results showed in the three groups the same degree of expression: 44% in contact allergic tests, 60% in the contact irritant tests and 50% in the tests with cocamidopropylbetaine. The findings in the dermis were also very similar for the three groups. In all the cases there was a positive perivascular infiltrate and positive mononuclear cells in the interstitial spaces.

*HLA-DR Expression*

Foci of positive keratinocytes were observed in 2 cases in contact allergic tests, 2 cases in contact irritant tests and 6 cases in the group of cocamidopropylbetaine. In all the cases there were increased numbers of HLA-DR-positive mononuclear cells in the dermis and in a few cases also in the epidermis, in comparison with normal skin (fig. 3). In the areas where HLA-DR labeling was seen, T lymphocytes were generally present within the same region of the epidermis.

In the groups of allergic tests and the tests with cocamidopropylbetaine, there was a correlation between ICAM-1 and HLA-DR expression and the exocytosis of T lymphocytes. These correlations wer not found in the irritant tests. The combined results for the expression of ICAM-1, HLA-DR and OKM5 by keratinocytes in contact allergic-contact irritant tests and in the tests with cocamidopropylbetaine are given in table 1.

**Discussion**

There was a clear difference in ICAM-1 expression between contact allergic and contact irritative patch tests. Our data confirm the hypothesis of Veylsgaard et al. [1] that ICAM-1 plays a role in the specific immune response by facilitating the antigen presentation or lymphocytic infiltration. On the other hand, our findings contradict the publication of other investigators, e.g. Willis et al. [2], who found a high percentage of ICAM-1 in tests with SLS.

Interpretation of the immunolabeling patterns with OKM5 proved to be difficult. Our results showed the same expression for OKM5 in the three groups. The OKM5-positive keratinocyte was thought to play an important role in the cutaneous immune reaction [3]. It was also stated that in positive intracutaneous tests for delayed-type hypersensitivity epidermal keratinocytes displayed OKM5 antigens [4].

A positive OKM5 expression was found most frequently in the contact irritant tests. According to the literature there may possibly be a connection between OKM5 expression in the stratum granulosum and the proliferative state of the epidermis [2].

Expression of HLA-DR by keratinocytes was absent from many reactions and only limited in others in contact allergic and contact irritant tests, confirming the results of several previous studies [5, 6], but contradicting that of Gawkrodger et al. [7].

Concerning the doubtful reactions for cocamidopropylbetaine, ICAM-1 expression correlates best with the results of contact allergic tests. These findings do not correlate with results of noninvasive methods in the evaluation of cocamidopropylbetaine. Vilaplana et al. [8] concluded that the skin response to cocamidopropylbetaine can be described as an irritant contact dermatitis but not as an allergic contact dermatitis. The immunohistochemical data of the cocamidopropylbetaine tests are as dubious as the clinical ones to make a differential diagnosis between an allergic or irritant reaction.

Finally, we think that the results of our investigation bring forward arguments for the hypothesis that there is an overlap in the pathogenesis of allergic and irritative contact dermatitis.

### References

1  Vejlsgaard GL, Ralfkiaer E, Avnstorp C, Czajkwoski M, Marlin S, Rothlein R: Kinetics and characterization of intercellular adhesion molecule-1 (ICAM-1) expression on keratinocytes in various inflammatory skin lesions and malignant cutaneous lymphomas. J Am Acad Dermatol 1989;20: 782–790.
2  Willis C, Stephens CJM, Wilkinson D: Selective expression of immune-associated surface antigens by keratinocytes in irritant contact dermatitis. J Invest Dermatol 1991;96:505–511.
3  Soyer HP, Smolle J, Kerl H: Distribution patterns of the OKM5 antigen in normal and diseased human epidermis. J Cutan Pathol 1989;16:60–65.
4  Hunyadi J, Simon M Jr: Expression of OKM5 antigen on human keratinocytes in vitro upon stimulation with gamma-interferon. Acta Derm Venereol (Stockh) 1986;66:527–530.

5    Scheynius A, Fischer T: Phenotypic difference between allergic and irritant patch test reactions in man. Contact Derm 1986;14:297–302.
6    Suitters AJ, Lampert IA: Expression of Ia antigen on epidermal keratinocytes is a consequence of cellular immunity. Br J Exp Pathol 1982;63:207–212.
7    Gawkrodger DJ, Carr MM, McVittie E, Guy K, Hunter JAA: Keratinocyte expression of MHC class II antigens in allergic sensitization and challenge reactions and in irritant contact dermatitis. J Invest Dermatol 1987;88:11–16.
8    Vilaplana J, Mascaro JM, Trullas C, Coll J, Romaguera C, Zemba C, Pelejero C: Human irritant response to different qualities and concentrations of cocamidopropylbetaines: A possible model of paradoxical irritant response. Contact Derm 1992;26:289–294.

Ann Verheyen, MD, Department of Dermatology, University Hospital Antwerp,
Wilrijkstraat 10, B–2650 Edegem-Antwerp (Belgium)

Elsner P, Maibach HI (eds): Irritant Dermatitis. New Clinical and Experimental Aspects.
Curr Probl Dermatol. Basel, Karger, 1995, vol 23, pp 114–120

# Modulation of Integrins on Epidermal Keratinocytes in vivo and on in vitro Reconstructed Epidermis

*P. von den Driesch*[a], *U. Kämmerer*[b], *M. Ponec*[c], *A. Hüner*[a], *M. Fartasch*[a, 1]

Departments of
[a] Dermatology and
[b] Molecular Cardiology, University of Erlangen-Nürnberg, Erlangen, Germany;
[c] Department of Dermatology, University of Leiden, The Netherlands

Barrier perturbation and irritation of the epidermis leads to a cascade of subsequent events. Activation and increased proliferation of lesional keratinocytes [1], induction and release of proinflammatory cytokines [1, 2], as well as induction of inflammation-associated adhesion molecules on either keratinocytes [1, 3, 4] or dermal vessels [1, 4] have been described.

Integrins are comprised of noncovalently associated heterodimers and are involved in a variety of cell-cell and cell-matrix adhesion interactions [5]. Basal epidermal keratinocytes continuously express four different heterodimers, namely $\alpha2\beta1$, $\alpha3\beta1$, $\alpha6\beta4$, and $\alpha v\beta5$. In inflammatory dermatoses [6–8] as well as during wound healing process [9, 10] upregulation and suprabasal expression of these integrins has been described. For the screening of the irritant potential of a given substance in vivo or in vitro, parameters are needed for the objectivation of the cytotoxic effect. Therefore, the present study was aimed to clarify whether irritation with sodium dodecyl sulfate (SDS) causes changes of epidermal integrin expression either in vivo or in vitro and whether this can be suggested as a possible sensitive parameter for the objectivation of irritation. The results were compared to the already described [3] induction of ICAM-1 on keratinocytes.

[1] The authors would like to thank Mrs. C. Wagner, Mrs. J. Kempenaar and Mrs. S. Weiler for their excellent technical assistance.

*Table 1.* Monoclonal antibodies used

| Clone | CD | Specificity | Source |
|-------|------|-------------|--------|
| A1-A5 | 29 | β1 chain | PG A.E.G. van den Borme, Amsterdam, The Netherlands |
| 10G11 | 49b | α2 chain | PG A.E.G. van den Borme, Amsterdam, The Netherlands |
| J-143 | 49c | α3 chain | PG C.E. Klein, Würzburg, Germany |
| IOP49e | 49e | α5 chain | Immunotech, Marseille, France |
| GOH3 | 49f | α6 chain | PG A. Sonnenberg, Amsterdam, The Netherlands |
| LV-230 | 51 | αv chain | PG C.E. Klein, Würzburg, Germany |
| 3E1 | 104 | β4 chain | Biomol, Hamburg, Germany |
| B5-IA9 | –/– | β5 chain | PG M. Hemler, Boston, Mass., USA |
| RR 1/1 | 54 | ICAM-1 | PG T.S. Springer, Boston, Mass., USA |

PG = Personal gift.

## Material and Methods

50 µl of SDS (1 % in water, water as control) was applied on the back of healthy volunteers using an occlusive Finn chamber test on scanpore tape. Using the evaporimeter EP-1 (Servomed, Stockholm), a threefold increase of mean transepidermal water loss was achieved after 24 h. After 1, 2, 3, 4, 16, and 24 h biopsies were obtained after written informed consent. Biopsies were then stored at $-70\,°C$ until use for immunohistochemical procedures.

In vitro, air-exposed epidermis reconstructed on de-epidermized dermis (RE-DED) was subjected to SDS (1 %, water as control) for 24 h as described [11, 12].

For immunohistochemical stainings 4-µm sections were stained with appropriate dilutions of the first monoclonal antibody (table 1) using either the alkaline phosphatase anti-alkaline phosphatase [13] or a three-step biotin-streptavidin-fluorescein method [4].

For semiquantitative reverse PCR epidermis of 4-hour biopsies was separated after microwave heating. After RNA extraction with guanidinium isothiocyanate [15], 1 µg of extracted RNA was subjected to a reverse transcriptase using the 3′ β1 integrin primer outlined below for 60 min at $42\,°C$. cDNA was then amplified with standard reagents [16] during 15–40 cycles of PCR in a PREM III thermocycler (LEP Scientific) with denaturation for 45 s at $94\,°C$, annealing for 45 s at $55\,°C$ and primer extension for 45 s at $72\,°C$. Reagent grade primers were designed based on the previously published [Ignatius et al.: J Cell Biol 1990;111:709–720] sequence for the human β1 integrin chain (5′-ACAGTGAAGACATG-GATGCTTA-3′ and 3′-TTGCTCCAGTACCAAGTACAAC-5′). After amplification, the product of each reaction was subjected to electrophoresis through a 1.5 % agarose gel and stained with ethidium bromide.

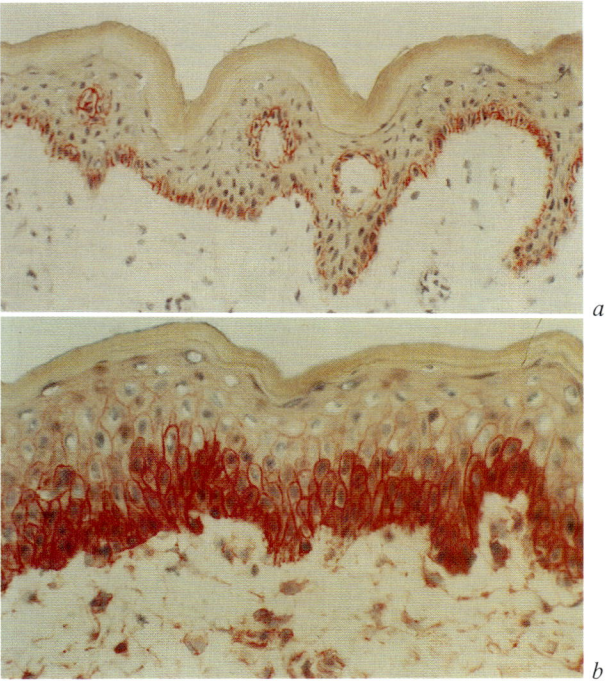

*Fig. 1.* Expression of integrins in normal skin: basal expression of β1 integrin chain in unirritated skin (*a*) and suprabasal up-regulation after application of SDS 1% for 16 h (*b*).

## Results

### Normal Epidermis

In normal epidermis all investigated integrins showed an expression on basal keratinocytes as described [17]. Only αv and β5 integrin chains exhibited a faint suprabasal expression. For α6 and β4 the characteristic [14] basal polarisation was demonstrable. No ICAM-1 was present in normal epidermis.

### Irritated Epidermis in vivo

No difference was found between control and 1-, 2- and 3-hour biopsies. In the 4-hour biopsies, a faint up-regulation and suprabasal expression of all investigated integrin chains was demonstrable. Exposure of epidermis to SDS for 16 (fig. 1a, b) and 24 h caused a prominent up-regulation of all integrin chains. This was most prominent for the αv and β5 chains whereas α2 showed the most moderate modulations. ICAM-1 was demonstrable after 4 h of irritation and became

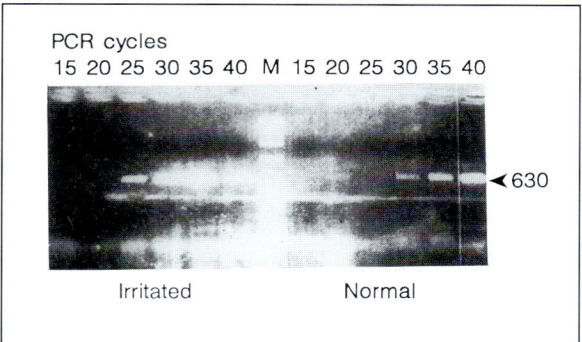

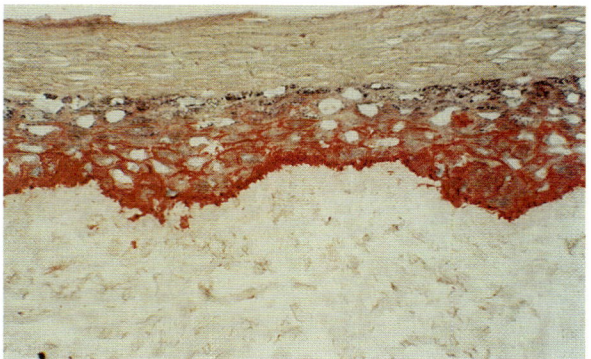

*Fig. 2.* Demonstration of β1 integrin mRNA after 4 h of application using a reverse PCR technique. In unirritated epidermis the 630-bp β1 integrin chain band was demonstrable after 30 cycles, in irritated epidermis after 25 cycles of the PCR.

*Fig. 3.* Integrin expression in reconstructed epidermis on de-epidermized dermis (RE-DED). Abnormal basal polarisation and suprabasal up-regulation of β1 integrin chain in unirritated RE-DED.

further up-regulated after 16 and 24 h. Semiquantitative rPCR revealed up-regulated epidermal mRNA for the β1 integrin chain after 4 h. The differences to the control biopsies were demonstrable between 25 and 35 cycles of the rPCR (fig. 2).

### Reconstructed Epidermis on Deepidermized Dermis (RE-DED)

Three divergent findings were revealed for the integrin expression in RE-DED (fig. 3). First, a continuous suprabasal expression was present for all investigated integrin chains except α2. Second, an abnormal polarisation towards the basement membrane, in healthy skin only demonstrable for α6 and β4, was found

---

in all investigated integrin chains. Third, an expression of α5, not present in vivo, along the basement membrane and on basal keratinocytes was demonstrable. No ICAM-1 expression was found in unirritated RE-DED.

*Irritated RE-DED*

Application of SDS 1% on air-exposed RE-DED for 24 h caused only a limited further up-regulation of all integrin chains investigated. No ICAM-1 expression was inducible.

**Discussion**

The present study was aimed to investigate the influence of contact irritant dermatitis on the expression of epidermal integrins in vivo and in vitro and to clarify whether this modulation represents a parameter for the objectivation of the cytotoxic effect of SDS. In vivo, first up-regulation of all integrins investigated was demonstrable even after 4 h. This parallels the induction of ICAM-1 suggesting that similar pathways may cause both adhesion molecule modulations. After 16 and 24 h of SDS exposure integrin up-regulation became more prominent except α2, which remained to be expressed more basally. At the transcriptional level, we found an induction of β1 integrin chain mRNA after 4 h, suggesting that integrin modulation is regulated on the transcriptional level and not only caused by an increased mobilisation of intracellular 93-kD integrin precursors [18]. In epidermis reconstructed in vitro, however, the baseline expression of integrins was elevated and immunohistochemically only limited up-regulation was demonstrable after irritation. This may be due to the fact that keratinocytes in RE-DED have features of activation such as expression of keratins 6 and 16 and suprabasal induction of involucrin and transglutaminase [19, 20]. Therefore, immunohistochemical integrin stainings are not sensitive enough for the evaluation of cytotoxicity in vitro. Additionally, no ICAM-1 was inducible in vitro supporting the view that T cells, which are not present in the RE-DED system, are the major cause of epidermal ICAM-1 induction [21, 22].

In vivo, however, modulation of integrins may be used as a parameter for irritation on both the protein and the mRNA level. Further studies are now needed to clarify whether this is true for all irritational substances and can be used for screening purposes.

The functional role of epidermal integrins is not completely elucidated so far. Immunolocalisation data and experimental results evidence their role (i) in keratinocyte adhesion to the basement membrane [23]; (ii) in keratinocyte differentiation and activation [18, 24–26]; (iii) in cell-cell adhesion between keratinocytes [27–31], and (iv) in keratinocyte migration on extracellular matrix proteins [32].

# References

1   Nickoloff BJ, Naidu Y: Perturbation of epidermal barrier function correlates with initiation of cytokine cascade in human skin. J Am Acad Dermatol 1994;30:535–546.
2   Tsai JC, Feingold KR, Crumrine D, Wood LC, Grunfeld C, Elias PM: Permeability barrier disruption alters the localization and expression of TNF alpha/protein in the epidermis. Arch Dermatol Res 1994;286:242–248.
3   Willis CM, Stephens CJM, Wilkinson JD: Selective expression of immune-associated surface antigens by keratinocytes in irritant contact dermatitis. J Invest Dermatol 1991;96:505–511.
4   Brasch J, Burgard J, Sterry W: Common pathogenetic pathways in allergic and irritant contact dermatitis. J Invest Dermatol 1992;98:166–170.
5   Hynes RO: Integrins: Versatility, modulation, and signaling in cell adhesion. Cell 1992;69:11–25.
6   Boehncke W, Kellner I, Konter U, Sterry W: Differential expression of adhesion molecules on infiltrating cells in inflammatory dermatoses. J Am Acad Dermatol 1992;26:907–913.
7   von den Driesch P, Simon M Jr: Cellular adhesion antigen modulation in purpura pigmentosa chronica. J Am Acad Dermatol 1994;30:193–200.
8   von den Driesch P, Gruschwitz M, Hornstein OP, Sterry W: Adhesion molecule modulation in Sweet's syndrome compared to erythema multiforme. Eur J Dermatol 1993;3:393–397.
9   Cavani A, Zambruno G, Marconi A, Manca V, Marchetti M, Gianetti A: Distinctive integrin expression in the newly forming epidermis during wound healing in humans. J Invest Dermatol 1993;1101:600–604.
10  Juhasz I, Murphy GF, Yan HC, Herlyn M, Albelda SM: Regulation of extracellular matrix proteins and integrin cell substratum adhesion receptors on epithelium during cutaneous human wound healing in vivo. Am J Pathol 1993;143:1458–1469.
11  Ponec M, Kempenaar J: The use of human skin recombinants as an in vitro model for testing the irritation potential of cutaneous irritants. Skin Pharmacol 1994, in press.
12  von den Driesch P, Fartasch M, Hüner A, Ponec M: Expression of integrin receptors and ICAM-1 on keratinocytes in vivo and in the in vitro reconstructed epidermis: Effect of sodium lauryl sulphate. Arch Dermatol Res 1994, in press.
13  Cordell JL, Falini B, Erber WN, Ghosh AK, Addulaziz Z, MacDonald S, Pulford KAF, Stein H, Mason DY: Immunoencymatic labelling of monoclonal antibodies using immune complexes of alkaline phosphatase and monoclonal anti-alkaline phosphatase (APAAP complexes). J Histochem Cytochem 1984;32:219–229.
14  Kanitakis J, Zambruno G, Vassileva S, Giannetti A, Thivolet J: Alpha-6 (CD 49f) integrin expression in genetic and acquired bullous skin diseases. J Cutan Pathol 1992;19:376–384.
15  Wilmer JL, Burleson FG, Kayama F, Kanno J, Luster MI: Cytokine induction in human epidermal keratinocytes exposed to contact irritants and its relation to chemical-induced inflammation in mouse skin. J Invest Dermatol 1994;102:915–922.
16  Kämmerer U, Kunkel B, Korn K: A nested PCR for specific detection and rapid identification of human enteroviruses. J Clin Microbiol 1994;32:285–291.
17  Konter U, Kellner I, Klein E, Kaufmann R, Mielke V, Sterry W: Adhesion molecule mapping in normal human skin. Arch Dermatol Res 1989;281:454–462.
18  Kim LT, Ishihara S, Lee CC, Akiyama SK, Yamada KM, Grinnell F: Altered glycosylation and cell surface expression of beta 1 integrin receptors during keratinocyte activation. J Cell Sci 1992;103:743–753.
19  Heenen M, De Graef C, Parent D, De Dobbeler G, Galand P: Renewal and differentiation of keratinocytes cultured on dead de-epidermalized dermis. Cell Prolif 1992;25:311–319.
20  Ponec M: Reconstruction of human epidermis on de-epidermized dermis: expression of differentiation-specific protein markers and lipid composition. Toxicol In Vitro 1991;5:597–606.
21  Griffiths CEM, Voorhees JJ, Nickoloff BJ: Characterisation of intercellular adhesion molecule-1 and HLA-DR expression in normal and inflamed skin: Modulation by recombinant gamma interferon and tumor necrosis factor. J Am Acad Dermatol 1989;20:617–629.
22  Gatto H, Viac J, Charveron M, Schmitt D: Study of immune-associated antigens (IL-1 and ICAM-1) in normal human keratinocytes treated by sodium lauryl sulphate. Arch Dermatol Res 1992;284:186–188.

23   Sonnenberg A, Calafat J, Janssen H, Daams H, van der Raaij-Helmer LMH, Falcioni R, Kennel SJ, Aplin JD, Baker J, Loizidou M, Garrod D: Integrin alpha 6/β4 complex is located in hemidesmosomes, suggesting a major role in epidermal cell-basement membrane adhesion. J Cell Biol 1991; 113:907–917.

24   Adams JC, Watt FM: Expression of β1, β3, β4, and β5 integrins by human epidermal keratinocytes and non-differentiating keratinocytes. J Cell Biol 1991;115:829–841.

25   Hertle MD, Kubler MD, Leigh IM, Watt FM: Aberrant integrin expression during epidermal wound healing and psoriatic epidermis. J Clin Invest 1992;89:1892–1901.

26   Carter WG, Wayner EA, Bouchard TS, Kaur P: The role of integrins alpha2-β1 and alpha3-β1 in cell-cell and cell-substrate adhesion of human epidermal keratinocytes. J Cell Biol 1990;110:1387–1404.

27   Larjava H: Expression of β1 integrins in normal human keratinocytes. Am J Med Sci 1991;301: 63–68.

28   Kaufmann R, Frösch D, Westphal C, Weber L, Klein CE: Integrin VLA-3: ultrastructural localization at cell-cell contact sites of human cell cultures. J Cell Biol 1989;109:1807–1815.

29   Symington BE, Takada Y, Carter WG: Interaction of integrins alpha 3 beta 1 and alpha 2 beta 1: Potential role in keratinocyte intercellular adhesion. J Cell Biol 1993;120:523–535.

30   Larjava H, Lyons JG, Salo T, Makela M, Koivisto L, Birkedal Hansen H, Akiyama SK, Yamada KM, Heino J: Anti-integrin antibodies induce type IV collagenase expression in keratinocytes. J Cell Physiol 1993;157:190–200.

31   Tenchini ML, Adams JC, Gilbert C, Steel J, Hudson DL, Malcovati M, Watt FM: Evidence against a major role of integrins in calcium-dependent intercellular adhesion of epidermal keratinocytes. Cell Adh Commun 1993;1:55–66.

32   Scharffetter-Kochanek K, Klein CE, Heinen G, Mauch C, Schaefer T, Adelmann-Grill BC, Goerz G, Fusenig NE, Krieg T, Plewig G: Migration of a human keratinocyte cell line (HACAT) to interstitial collagen type I is mediated by the α3β1-integrin receptor. J Invest Dermatol 1992;98:3–11.

Priv.-Doz. Dr. P. von den Driesch, Department of Dermatology, Hartmannstrasse 14,
D–91052 Erlangen (Germany)

Elsner P, Maibach HI (eds): Irritant Dermatitis. New Clinical and Experimental Aspects.
Curr Probl Dermatol. Basel, Karger, 1995, vol 23, pp 121–130

..........................

# Human in vivo Microdialysis Technique Can Be Used to Measure Cytokines in Contact Reactions

*C. Anderson*[a], *C. Svensson*[a], *F. Sjögren*[a], *T. Andersson*[a], *K. Wårdell*[b]

Departments of
[a] Dermatology and
[b] Biomedical Engineering, University Hospital, Linköping, Sweden

Microdialysis is a bioanalytical sampling technique, which can be used to measure endogenous and exogenous substances in vivo in a target tissue [1, 2]. For several years we have used a method for cutaneous microdialysis in which the dialysis membrane at the tip of a steel-shafted probe can be placed in the skin via a guide, 2–3 cm from the point of insertion, at a depth of around 1 mm [3, 4]. The standard dialysis membrane is permeable to substances with a molecular weight up to 20 kD. We have studied 2 main areas – percutaneous absorption [3] (fig. 1) and cutaneous inflammation using histamine as the model inflammatory mediator [4]. We found variable levels of histamine in the period after probe insertion, presumably due to biological variation in tissue histamine levels and responses, which returned to resting levels after 40–60 min 'equilibration' [4]. Using a new development of laser Doppler flowmetry, laser Doppler perfusion imaging (LDI) [5], we have also shown that the axon reflex flare caused by probe insertion subsides over the same period of time [6]. We have studied the histamine response after provocation with compound 48/80 and in physical urticaria (fig. 2). Microdialysis technique, a method of increasing clinical interest [7], can also be performed using 'sewn in' microdialysis tubing [8, 9].

Just as the morphological patterns of irritant contact dermatitis vary, so can the pathogenetic mechanisms [10]. Detailed study of these mechanisms will give basic knowledge which will hopefully result in improved treatment regimes for

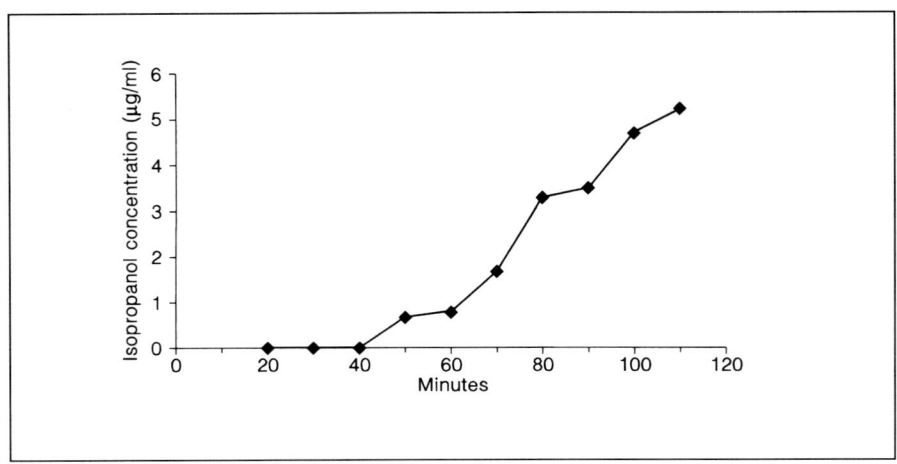

*Fig. 1.* Percutaneous absorption of 70% isopropanol. Isopropanol was applied in excess to the skin of the ventral forearm as described in Anderson et al. [3]. Pump speed 3 µl/min, sampling interval 10 min, probe depth 1.1 mm, application at 40 min. Analysis by gas chromatography using a flame ionisation detector (Vadstena Kemanalys AB, Vadstena, Sweden).

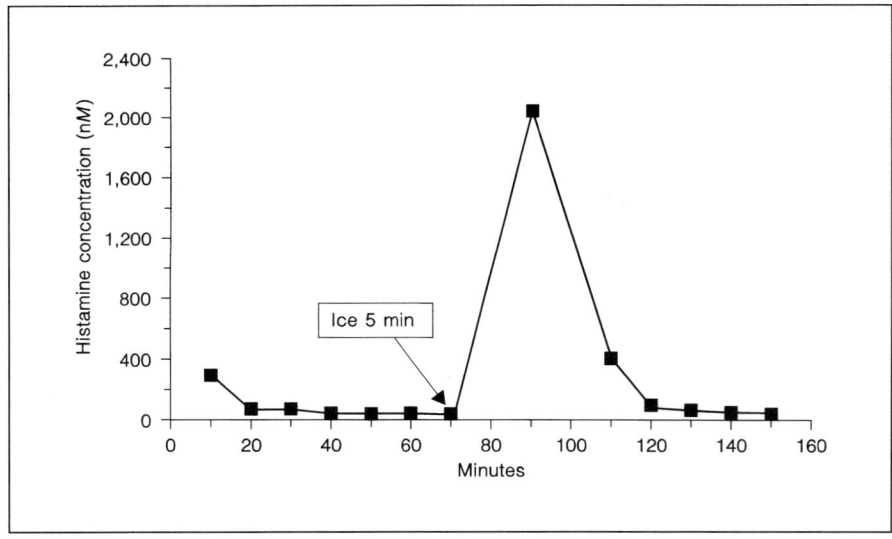

*Fig. 2.* Histamine response measured by HPLC after ice provocation in a patient with cold urticaria. Pump speed 5 µl/min, sampling interval 10 min, probe depth 1.0 mm.

contact dermatitis. Our ability to observe the chronology of the skin physiological aspects of contact reactions has improved greatly over recent years, with methods such as evaporimetry and laser Doppler flowmetry. Methods for following the pharmacology of cutaneous inflammation have, however, not enjoyed the same improvement. The main traditional methods, tissue analysis after biopsy or tissue fluid analysis after suction blisters preclude, by their destructive nature, subsequent chronological study of an individual reaction and thus represent the endpoint for any attempt at multiple parameter assessment of reactions. We have previously reported [International Symposium on Irritant Contact Dermatitis, Groningen, 1991] use of a cutaneous microdialysis technique to study our model inflammatory mediator, histamine, in contact reactions to sodium lauryl sulfate (SLS). To be really useful, cutaneous microdialysis must also manage the measurement of molecules of larger molecular weight such as cytokines. To achieve this, a more permeable membrane, the ultrafiltration membrane (a membrane permeable to molecules up to 100 kD), must be used. This is technically more difficult at several levels than microdialysis with the standard membrane.

The aim of the present pilot study was to investigate whether cutaneous microdialysis using an ultrafiltration membrane would make possible the measurement of cytokines in contact reactions. Two cytokines (IL-6 and IL-2R) were selected for study based chiefly on ubiquity and the availability of analysis methods for small aliquotes. IL-6 is produced by many cells including keratinocytes, and has many effects. It can be detected in small amounts in around 40% of normal sera, and is known to be present at relatively high levels in a range of inflammatory conditions. IL-2R was chosen because it might indicate T-lymphocyte involvement. An additional aim was to use LDI to illustrate the possibility of multiple parameter assessment during cutaneous microdialysis and to observe possible effects of probe insertion on the subsequent course of observed reactions.

## Materials and Methods

*Subjects*
Volunteers and patients were examined after giving their informed consent. When nickel reactions were provoked, nickel allergy had been present for several years. The study was approved by the Regional Ethics Committee for Human Research, Linköping University Hospital.

*Microdialysis*
The essentials for the performance of microdialysis are a probe (containing the dialysis membrane), a pump and a collection system.

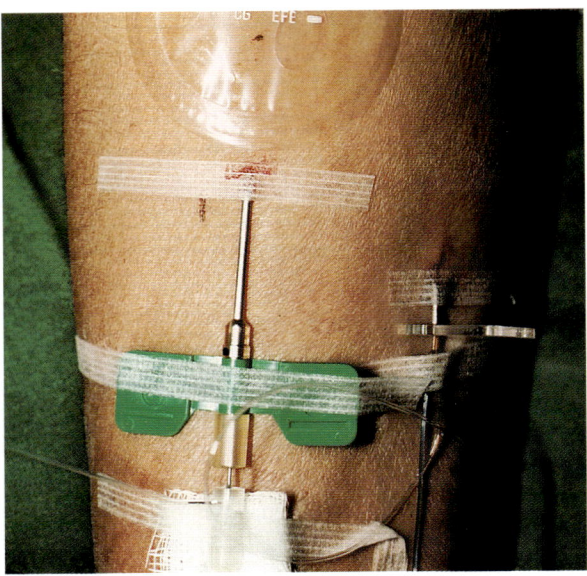

*Fig. 3.* Experiment in which the two probe types (CMA/10 and CMA/20) have been placed in the skin of the ventral forearm for study of percutaneous absorption (CMA/10 to the left) and IL-6 in early contact reaction to nickel (CMA/20 to the right).

*Probes.* The standard microdialysis probe used (CMA/10, CMA/Microdialysis Research AB, Sweden) has a shaft length of 70 mm and a membrane length of 10 mm. The steel shaft of the probe has an outer diameter of 0.64 mm. The diameter at the tip of the probe, where the polyamide dialysis membrane is located, is 0.5 mm. The molecular weight cut off point for the standard membrane is 20 kD [3, 4]. For the present ultrafiltration cutaneous microdialysis experiments, a flexible, shorter probe (CMA/20) has been used (ultimately longer probes will be used). Figure 3 shows the 2 types of probe in place.

*Probe Insertion.* A site for insertion of the microdialysis probe on the ventral forearm was chosen so as to avoid obvious veins. The point of insertion was anesthetized with a 3-mm weal of local anesthetic (mepivacaine 5 mg/ml), injected intradermally. For insertion of the standard (CMA/10) probe, an apheresis needle (Terumo 1.60 × 32 mm or Baxter 1.60 × 25 mm) was inserted as a guide, subcutaneously for the first 1.5–2 cm and then intradermally for the last centimeter in order to position the dialysis probe membrane as superficially as possible. The probe was inserted through the guide, which was then withdrawn and taped in position. Thus, the dialysis membrane at the tip of the probe was 2.5–3 cm from the point of insertion. The procedure of insertion for the ultrafiltration probes (CMA/20) was similar but the guide was a shorter custom-built guide. Because of the shorter distance from the point of insertion to the measurement area, local anesthesia was not used in the cytokine experiments. The probes were perfused with physiological (Ringer's) solution at room temperature.

*Pump and Sample Collector.* The perfusate was pumped by a CMA/100 microinjection pump. For sample collection, an automatic collector (CMA/140 microfraction collector) synchronized with the pump was used.

### Localisation of Probes

In the standard experimental procedure, ultrasound measurement (Dermascan A, Sonotron AB, Sweden) [11] allowed the estimation of the actual probe depth as well as the skin thickness (epidermis + dermis) near the probe tip. Probes are consistently placed at a depth of around 1 mm. In the cytokine experiments using shorter guides, estimation of probe depth was not performed.

### Laser Doppler Perfusion Imaging

LDI uses a low-power He-Ne laser to scan the tissue. At each measurement site the laser beam penetrates the tissue to a depth of a few hundred micrometers [5]. Moving blood cells give rise to Doppler shifts in the backscattered light. These Doppler components are detected and processed to generate an output signal (within the range 0–10 V) which is linearly proportional to tissue blood perfusion. A map of an area 64 × 64 measurement points large is produced. Smaller areas can be studied as required. A color-coded image is displayed on a monitor. All data are stored for subsequent data analysis and statistical calculations. The resolution in the system (distance between measurement points) can be set to 'high' or 'low'. With a measurement distance of 16 cm between the scanner head and the measurement area, and the resolution set to 'high', the actual area mapped is approximately 16 cm$^2$.

### Analysis of Cytokines

Levels of IL-6 were measured by an EASIA kit (Medgenix Diagnostic Fleures, Belgium). Soluble IL-2 receptor was measured by an ELISA procedure, Predicta IL-2 receptor kit (Genzyme, Cambridge, Mass., USA). These kits include an antibody-coated microtiter plate and standard solutions. After following the manufacturer's assay procedure, the tests were analyzed on a Dynatech MR 5000 microplate reader at 450 nm. The concentrations of the two cytokines are given in pg/ml.

### Test Reactions

24-hour reactions to nickel sulfate 5% in petrolatum or 5% sodium lauryl sulfate (SLS) were applied with the standard Finn chamber technique to sites on the ventral forearm which had been selected as suitable for cutaneous microdialysis. Occlusion was removed approximately 20 min prior to probe insertion.

### Experimental Design

The experiments were performed in two series. In the first series IL-6 was examined at 4 test sites (one normal skin, two 24-hour reactions to 5% SLS and one 24-hour reaction to nickel) in 3 individuals. The perfusion speed was 1 µl/min and 30-µl aliquots were collected. In the second series of experiments, 5% SLS and 5% nickel reactions were induced in 2 nickel-allergic patients. In 1 patient normal skin was also examined. The probes were in place for 2–3 h. Samples were stored frozen until analysis. LDI was performed on the contact reactions prior to probe insertion, at intervals when the probe was in place, after probe removal and approximately 24 h after probe removal.

---

*Table 1.* Results series 1: IL-6 levels (pg/ml) in normal skin, 24-hour reaction to SLS 5%, 24 hour reaction to nickel sulfate 5% in 3 patients

| Normal skin | SLS 5% | | Nickel 5% |
|---|---|---|---|
| subject 1 | subject 2 | subject 3 | subject 3 |
| 18 | 25 | 34 | 39 |
| 17 | 18 | 23 | 29 |
| 14 | 19 | 27 | 43 |
| 15 | 13 | 52 | 33 |
| 14 | | 74 | 35 |
| 17 | | 67 | 32 |

Pump speed 1 μl/min, sample size 30 μl. Values are shown in chronological order after probe insertion. IL-6 could be detected in all samples. The experimental method does not correct for several variables and thus does not show absolute levels. Valid comparison between different test sites cannot be performed.

## Results

We have found the cutaneous microdialysis procedure to be well tolerated by subjects. Probe insertion through anesthetized skin is painless but tunnelation to the probe tip area can cause slight pain, which is usually well tolerated. Apart from occasional ecchymoses, we have not encountered side effects.

In the first series of experiments, IL-6 was detected at the four sites in the 3 subjects examined (table 1). Since a number of variables have not been corrected for, no valid comparison between sites of the results obtained can be performed.

In the second series of experiments, the attempt to use smaller aliquots (in order to allow estimation of 2 cytokines at the same time) pushed the analysis method to and beyond its limit. At several sites cytokines were not detectable (table 2). At other sites levels were low, near the lower end of the standard curve. Both IL-6 and IL-2 soluble receptor could, however, be detected at some point. Again no comparison between sites is valid.

LDI is not only noninvasive but also nontouch and thus easily used during cutaneous microdialysis experiments. Figure 4a, b shows the LDI maps before, during and after the cutaneous microdialysis in the SLS and nickel reactions, respectively. There was a slight decrease in perfusion while the probe was in place for the SLS 5% reaction.

*Table 2.* Results series 2: results of IL-6 and IL-2R estimations (pg/ml) in 2 patients

| Subject 1 | | | | Subject 2 | | | | | |
|---|---|---|---|---|---|---|---|---|---|
| nickel 5% 24 h | | SLS 5% 24 h | | normal | | nickel 5% | | SLS 5% | |
| IL-6 | IL-2R | IL-6 | IL-2R | IL-6 | IL-2R | IL-6 | IL-2R | IL-6 | IL-2R |
| 6 | n.d. | <5 | 190 | <5 | n.d. | <5 | n.d. | <5 | 248 |
| 80 | | <5 | | <5 | n.d. | <5 | n.d. | <5 | 203 |
| | | | | <5 | | 40 | | | |

Sample size is smaller (15 µl) than in series 1 in an attempt to perform more estimations within the experimental time frame. Pump speed 1 µl/min. Values for both cytokines were consistently near or below the lower detection limit for the analysis method.

n.d. = Not detected.

## Discussion

While it must be emphasized that no conclusions as to the roles of specific cytokines can be drawn from the present study, the results indicate that cutaneous microdialysis using the ultrafiltration membrane can be used to study secretory cytokine levels in contact reactions (and in other cutaneous inflammation).

Our experience after the insertion of hundreds of cutaneous microdialysis probes is that the procedure is well tolerated by patients. We have encountered no serious side effects. Using the standard microdialysis membrane we have studied the percutaneous absorption of organic solvents and topical drugs and histamine secretion after provocation with compound 48/80 and in physical urticarias [to be publ.] Histamine is, of course, but one of a large number of inflammatory mediators which participate in cutaneous inflammation from its induction to its resolution. Microdialysis technique utilizing the standard membrane has already moved from the experimental to the practical clinical area [12, 13]. Use of the ultrafiltration membrane presents some additional practical problems but the methods promise motivation for tackling them. If the microdialysis technique is to gain general dermatological use, standardized manufactured probes will be a necessity, for which reason we have not used the less-expensive 'sewn in' microdialysis technique. In the near future newer probes, portable pumps and other modifications will further enhance the possibilities which human microdialysis technique offers.

The present paper demonstrates that cutaneous microdialysis does not preclude the parallel use of other methods of assessment in the study of the chronology of acute inflammatory cutaneous reactions. Although much remains to be

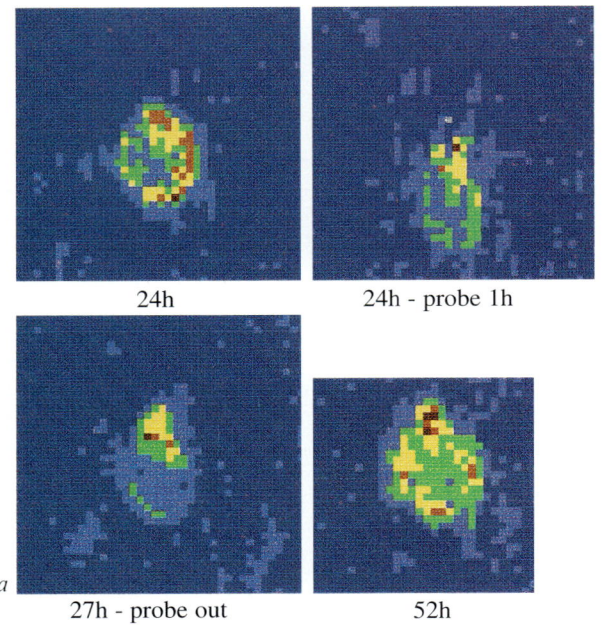

24h                    24h - probe 1h

27h - probe out            52h

*a*

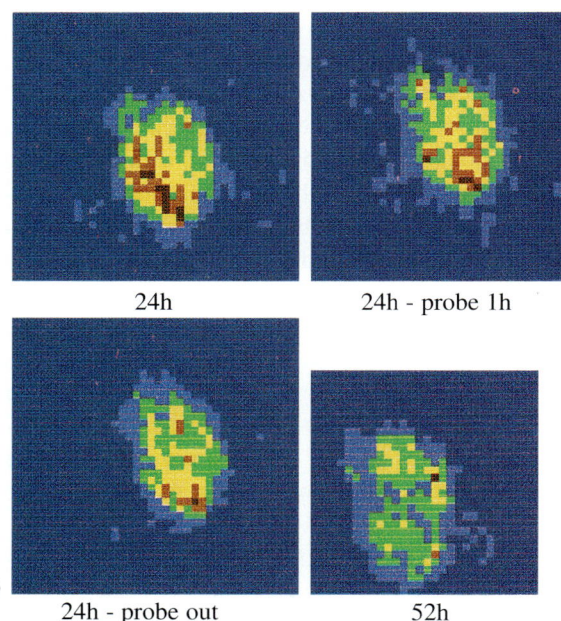

24h                    24h - probe 1h

24h - probe out            52h

*b*

*Fig. 4.* LDI maps in 24-hour reaction to SLS 5% *(a)* and 24-hour reaction to Ni 5% *(b)*. The color scale is corrected to the maximal reaction. Increasingly bright colours (from blue to red) show increasing skin perfusion. Maps at 24 h (20 min after patch removal), 1 h after probe insertion, immediately after probe removal, and 24 h later (52 h) are shown.

investigated, it seems that the presence of the probe has little effect on the course of the cutaneous reaction. LDI can be expected to find an important place in the multiple parameter assessment of cutaneous reactions. Cutaneous microdialysis though not 'noninvasive' may well ultimately find its place amongst the physiological noninvasive methods, adding chronological pharmacology to the investigative tools available to dermatologists.

## Conclusion

Cutaneous microdialysis is a relatively new clinical method for the measurement of endogenous and exogenous substances in the human dermis. The method involves placement of a probe in the skin at a depth of about 1 mm, a procedure which is well tolerated by patients. An important advantage of the method is that cutaneous inflammatory reactions under study seem to be only slightly affected by the technique. Multiple parameter assessment of reactions such as contact reactions can be performed at the same time as chronological pharmacological information is obtained. The present paper investigates, in pilot studies, the use of an ultrafiltration membrane, which is permeable to molecules up to 100 kD, for the measurement of cytokines in contact reactions. Although no conclusions can be drawn on the significance of the results obtained, the study demonstrates that cytokines can be detected in human skin in vivo using cutaneous microdialysis technique.

## Acknowledgements

We are grateful for fruitful collaboration with Magnus Molander and Jan Kehr. These studies have been performed with financial assistance from the Swedish Work Environment Fund, the Swedish Association against Asthma and Allergy and the Edvard Welander Foundation.

## References

1   Ungerstedt U: Measurement of neurotransmitter release by intracranial dialysis; in Marsden CA (ed): Measurement of Neurotransmitter Release in vivo. Bath, Wiley, 1984, 81–105.
2   Ungerstedt U: Microdialysis – principles and applications for studies in animals and man. J Intern Med 1991;230:365–373.
3   Anderson C, Andersson T, Molander M: Ethanol absorption across human skin measured by in vivo microdialysis technique. Acta Derm Venereol (Stockh) 1991;71:389–393.
4   Anderson C, Andersson T, Andersson RGG: In vivo microdialysis estimation of histamine in human skin. Skin Pharmacol 1992;5:177–183.

5   Wårdell K, Jakobsson A, Nilsson GE: Laser Doppler perfusion imaging by dynamic light scattering. IEEE Trans Biomed Eng 1993:40:309–316.
6   Anderson C, Andersson T, Wårdell K: Changes in skin circulation after insertion of a microdialysis probe visualized by laser Doppler perfusion imaging. J Invest Dermatol 1994:102:807–811.
7   Microdialysis (editorial): Lancet, 1992;339:1326–1327.
8   Petersen LJ, Stahl Skov P, Bindslev-Jensen C, Søndergaard J: Histamine release in immediate-type hypersensitivity reactions in intact human skin measured by microdialysis. Allergy 1992;47:635–637.
9   Petersen LJ, Kristensen JK, Bulow J: Microdialysis of the interstitial water space in human skin in vivo: Quantitative measurement of cutaneous glucose concentrations. J Invest Dermatol 1992:99:357–360.
10  Anderson C: The spectrum of non-allergic contact reactions: An experimental view. Contact Derm 1990;23:226–229.
11  Alexander H, Miller DL: Determining skin thickness with pulsed ultrasound. J Invest Dermatol 1979;72:17–19.
12  Bolinder J, Hagström E, Ungerstedt U, Arner P: Microdialysis of subcutaneous adipose tissue in vivo for continuous glucose monitoring in man. Scand J Clin Lab Invest 1989;49:465–474.
13  Arner P, Bolinder J: Microdialysis of adipose tissue. J Intern Med 1991;230:381–386.

Dr. Chris Anderson, Department of Dermatology, University Hospital,
S–581 85 Linköping (Sweden)

Elsner P, Maibach HI (eds): Irritant Dermatitis. New Clinical and Experimental Aspects.
Curr Probl Dermatol. Basel, Karger, 1995, vol 23, pp 131–143

# The Spectrum of Irritancy and Application of Bioengineering Techniques

*Jørgen Serup*

Bioengineering and Skin Research Laboratory, Department of Dermatology,
Bispebjerg Hospital, University of Copenhagen, and Department of Dermatological
Research, Leo Pharmaceutical Products, Ballerup, Denmark

The last decennium was an exceptional period of sophistication of our insight into irritants and irritancy. The introduction of bioengineering techniques for noninvasive characterization of pathophysiological skin phenomena on an objective basis is one major reason for this progress, which was associated with increased understanding of the clinical manifestations of irritancy. The focus is no more simply on the irritant chemical but also on factors related to the individual, the environment and clinical and experimental exposure situations. We can today benefit from the continuous work of many dermatology research laboratories.

It is the purpose of the present synopsis to present an up-to-date status of definitions of irritants and irritancy taking the achievements due to the bioengineering techniques into consideration. A distinction of irritants into corrosive and noncorrosive irritants is, moreover, discussed and highlighted.

Detailed and basic informations in the field are already covered by different monographs and review papers [1–5].

## Definition of an Irritant Substance

Irritant literally means to incite, to provoke, etc. (Latin: irritare), which does not cover the use of the word in medical and especially dermatological contexts.

Discussion of irritants should always start with the reading of the dissertation of Björnberg [6]. In 1940, Sulzberger stated that a primary irritant is 'an agent capable of producing gross reactions without the necessity of a previous sensitization having taken place'. In 1957, Schwartz and co-workers stated that a primary irritant is 'an agent which will cause dermatitis by direct action on normal skin at the site of contact if it is permitted to act in sufficiently intensity or quantity for sufficient length of time'. In 1967, Kligman and Wooding [7] stated that 'primary irritants are substances which damage skin by direct cytotoxic action. This is in contrast to contact allergens which incite inflammation by indirect immunologic reactions involving changes in the whole organism'. In 1951, Shelanski [8] stated that 'a primary irritant is one which is toxic at the first contact and a secondary irritant is one producing a reaction only after repeated applications'. The Kligman and Wooding paper, which is very much up to date, also expresses that 'the potency range of irritants is very great. Some may irreversibly damage the skin after a moments contact; others produce mild and transient injury only after many days. It seems artificial to countenance special terms for these differences, as they simply represent a continuous spectrum of toxicity. We simply use the old fashion terms strong and weak in designation irritating capacity in words'. Moreover, it is expressed that 'it is urgent to abandon the conventional thinking in terms of irritants versus non-irritants. A sounder biological view is to regard all substances as potential irritants. The question then becomes how irritating. This raises the problem of measuring degrees of irritation and highlights the necessity of developing quantitative methods'.

Irritants are specific or characteristic in the sense that different chemicals may elicit clinically different reactions due to different skin penetration profiles and different modes of irritant actions directly on the exposed tissue.

Thus, it can be summarized that there is no clear definition of what an irritant substance is, and language is pauvre. Skin reactivity is not simply determined by the chemical but strongly influenced by the concentration applied, the kind of exposure and factors related to the individual. However, irritant reactions must be non-immunological. Previous distinctions such as primary versus secondary irritants, direct versus indirect irritants and relative versus absolute irritants may be obsolete.


### Definition of an Allergenic Substance

Allergy literally means a state or condition of different or other reactivity (Greek: allos, other). The allergy patch test was introduced in 1896 by Jadassohn, and allergy patch testing was further developed in 1911 by Bloch and in the thirties by Bonnevie, and by Sulzberger and Wise [4]. Thus, allergic reactions and allergic sensitization including testing of contact allergy was known in dermatology at least for a century. It is popular in modern textbooks simply to state that contact allergy is defined as a type IV allergic reaction of the skin, however, there is no general, validated and direct method to verify that a type IV allergic skin reaction definitely took place in clinical cases where allergic sensitization to a substance is claimed. It is well known that histology and estimation of mediators and markers of inflammatory reaction in positive patch test reactions have generally failed to distinguish irritant from allergic reactions. There is no real gold standard of contact allergy. The diagnosis of contact allergy is de facto based on a careful patient history, dermatological examination and patch testing in order to possibly find the allergenic substance. The known key elements in the diagnosis is the possibility of previous sensitization, actual exposure and skin reactivity at a very low concentration. Allergy patch testing is a fine art of dermatology, and not a validated test with high precision and an acceptable predictive value of a positive test and a negative test. Patch testing is, obviously, subject to bias. The distinction toward irritant reactions belongs to the borderline of the performance of the test. Allergy testing of irritant substances and substances with pharmacologicals actions as exemplified by the corticosteroids never found an optimal solution.

Allergic reactions are considered highly specific with respect to the elicitation of reactions (the afferent loop) while the inflammatory response (the efferent loop) is considered in common with the note that responses may sometimes be modified due to different skin penetration profiles of allergens and simultaneous irritant effect of the substance or application system.

Allergy patch test is a pragmatic red spot test but a useful practical tool in the hands of masters of clinical dermatology. The fundamentals about allergy patch testing and definition of allergenic substances should be considered whenever reactions are characterized by bioengineering methods and especially if clinical relevance is in question. There is no universally true test of cutaneous allergy.


### Overlap Problems between Irritants, Allergens and Therapeutics

The overlap problem of substances which at the same time are irritants and allergens is mentioned above. Some irritants show edge phenomenon, but others with benzalconium chloride as the well-known example can entirely mimic allergic reactions with papules, vesicles, etc. The bioengineering methods cannot solve the problem of distinguishing such reactions with the same phenomenology.

Several topical therapeutics are local irritants. All-trans retinoic acid, benzoyl peroxide, azelaic acid, dithranol, vitamin D and vitamin D analogues are known examples (fig. 1–5). It is uncertain whether the mechanism of therapeutic actions and the mechanism of irritancy is the same. Note that the irritancy response to all-trans retinoic acid is special with subepidermal edema, flattening of surface markings and noncorrosivity. Is irritation with flattening and antiwrinkling effect the same? Keratolytics may be considered special irritants which

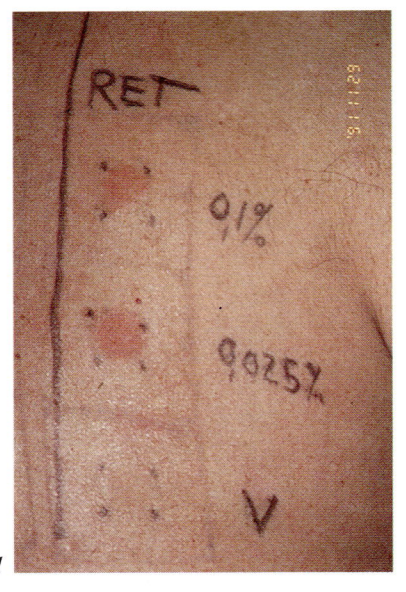

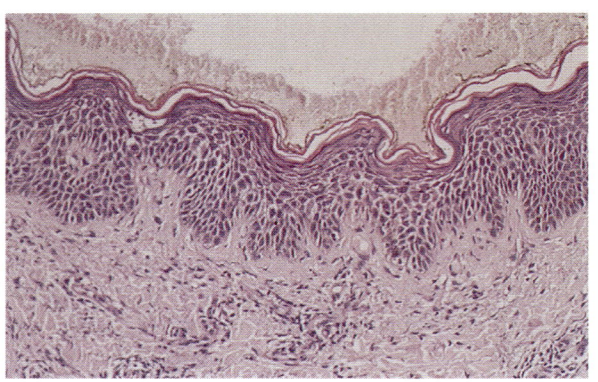

Fig. 1. Patch test (48 h) with all-trans retinoic acid 0.1 and 0.025% in ethanol, reading day 72.

Fig. 2. Punch biopsy from a patch test reaction to all-trans retinoic acid 0.1% showing acanthosis, spongiosis and cellular infiltration of the papular dermis indicating pharmaceutical action and/or irritant effect (?).

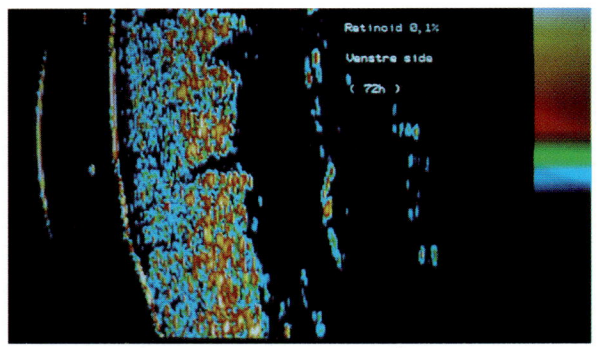

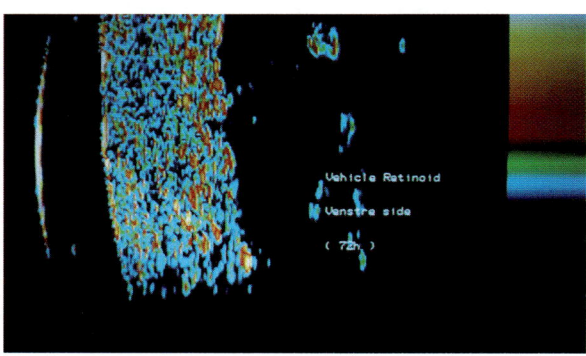

Fig. 3. 20-MHz ultrasound scanning (Dermascan C®, Cortex, DK) of a patch test reaction to all-trans retinoic acid 0.1% (a) showing a distinct echolucent band in the papular dermis representing a uniform layer of inflammatory edema immediately under the epidermis, which is seen as a vertical line to the left. The dermis block is oriented vertically and seen with strong echoes (yellow/red) and weaker echoes (blue) with a sharp interface toward the black and echolucent subcutaneous fat to the right. The curved line to the left is a plastic membrane of the measuring system. Vehicle exposed skin is also shown (b).

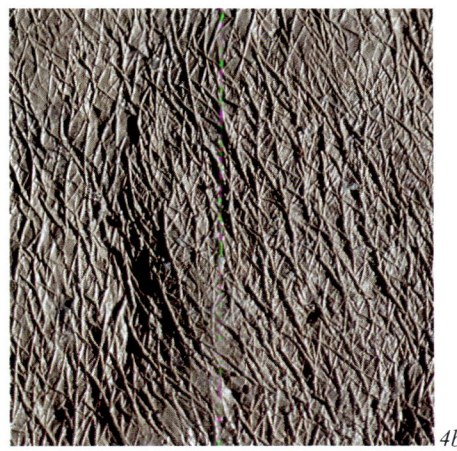

4a

4b

*Fig. 4.* Silicon rubber replica of a patch test reaction to all-trans retinoic acid 0.1% (*a*) and regional control (*b*). Note the flattening of the skin markings of tested skin and the edematous impression ring of the test chamber. The reaction is entirely nonpapular.

*Fig. 5.* Laser profilometry (UBM GmbH, Ettlingen, Germany) of replica obtained from positive patch test reactions to all-trans retinoic acid (RET) with flattening and replica from a positive reaction to SLS with increase of skin markings, with normal skin as a control (CTL).

5

induce nonerythematous irritation. Caustic chemicals for the treatment of warts and clavi are ultimate irritants. Peeling agents used for cosmetic purposes are clearly special irritants. 5-Fluoruracil is used directly as a cytotoxic agent. To improve hair growth patients may be treated with dinitrochlorobenzene or other substances to create a iatrogenic allergic sensitization. Thus, there are many examples of irritants and allergens primarily appreciated as therapeutics and not directly as hazardous chemicals simply because they are used with some useful purposes. The definition of irritant and allergenic substances is not a simple and logic choice on a purely biological basis between two separate boxes but the fact whether the chemical is considered harmful or useful strongly influences distinctions. We can with the bioengineering techniques quantify and characterize objective phenomena, but we cannot of course measure whether substances are harmful or useful in practical use.

## Irritancy in a Clinical Perspective – Irritancy Situations Rather than Irritant Substances

Taking into account that irritation or irritant dermatitis is now understood to be a complex biologic event, with a diverse pathophysiology, natural history and clinical appearance and depending on multiple external and internal factors the distinction of irritant contact dermatitis into different clinical types was proposed as follows [9]: (1) acute irritant dermatitis; (2) irritant reaction; (3) delayed acute irritant dermatitis; (4) cumulative irritant contact dermatitis; (5) traumatic irritant dermatitis; (6) pustular and acneiform dermatitis, and (7) nonerythematous irritation.

Readers may find the details in the textbook publication [9]. The list illustrates the diversity of clinical manifestations. The list can hardly be complete. Definition of types also includes data about onset and prognoses. It may in the practical clinical situation be difficult to characterize individual cases. It is interesting and important that nonerythematous irritation with very minor scaling and maybe microcracking and microfissuring is included with an overlap toward xerosis and ichthyosis. The word 'syndrome' is now and then used in relation to the clinical types of irritancy but this use of the word is inappropriate since 'syndrome' normally implies that the condition is constitutional and involves different organ systems.

From the point of view of bioengineering methods the range of clinical types of irritancy illustrates that evaluation with bioengineering techniques cannot simply be based on one selected method or a limited number of methods which can be used to assess inflammation. Methods should also include techniques designed to quantify desquamation and the scaling process.

### Noninvasive Methods for Irritancy Testing, Their Scope and Relevance

The bioengineering methods measure separate features of the cutaneous response independently, i.e. vasodilatation (laser Doppler flow), redness (colorimetry), water barrier damage (evaporimetry), edema formation (ultrasound thickness and image analysis) and different alterations of the epidermal surface.

Thus, the methods are narrow and focused and suited for in-depth study in contrast to the overall clinical reading. The methods are on the other hand not highly focused based on minute samples or fragments of mechanism with obscure clinical relevance.

The selection of bioengineering techniques normally used for the characterization of irritant skin reactions was previously reviewed [5, 10]. A handbook of noninvasive methods with key information on the complete range of bioengineering techniques is just published [11]. The clinically relevant elements of skin irritation is disturbance of the desquamation process resulting in scaling or hyperkeratosis, i.e. epidermal events, and an inflammatory response with vasodilation and redness in combination with extravasation of water and edema formation, i.e. essentially events taking place in dermis. These are the clinically relevant features to be measured. We cannot directly see the water molecule, the direct hydration of the skin or the water barrier function, and these are only of theoretical interest and indirect clinical relevance. For the study of epidermal changes measurement of electrical conductance, capacitance and different replica techniques have become popular. For the study of the inflammatory component laser Doppler flowmetry and colorimetry have become widely used, and for the study of edema formation high frequency ultrasound is used either for measurement of skin thickening or as a basis for advanced image analysis [12, 13]. It is impossible to give a detailed description of the devices, but readers are referred to the sources of information mentioned above, and to the guidelines on transepidermal water loss measurement and laser Doppler flowmetry published by the European Society of Contact Dermatitis [14, 15]. Skin color and visual erythema is primarily dependent on the volume of blood under a given area of skin and not directly on the blood flow or flux. Colorimetry and fluxmetry are therefore supplementary [16, 17]. Colorimeters are recommended as a substitute of the human eye, but laser Doppler flowmetry may be more sensitive for the detection of minor inflammation with faint erythema only. Laser Doppler imaging of the horizontal distribution of blood flow was recently introduced [18]. This is a major step forward in the objective evaluation of patch test reactions.

There is an important distinction between *monoparametric* and *multiparametric assessment* of experimental reactions using bioengineering techniques. It is common that a research laboratory only has one piece of equipment, for example an evaporimeter, and for that practical reason this equipment is used in a study of different irritants irrespective their nature and irrespective the fact that the water barrier function as such has no direct clinical relevance. It will work well if detergents are studied but it will be insensitive and inappropriate if for example the irritancy due to all-trans retinoic acid or nonanoic acid is studied because the latter are noncorrosive. Thus, when applying bioengineering technique it is rational to divide the research project into two phases:

(1) Characterization of the *profile of irritancy* due to the irritant using a multiparametric assessment with measurement of epidermal changes, water barrier function, redness and edema.

(2) *Further evaluation* of irritants with a known profile for selected or dedicated study based on one main feature of the irritant reaction and *monoinstrumental assessment* or monitoring.

Nomination of one (or two) especially relevant method will save resources and allow scale up of activities.

It is in most laboratories a routine to *backup measurements with a clinical evaluation* based on a defined rating scale even if this reference may be less reproducible and less sensitive. A defined main purpose already in the study protocol with a nominated main parameter supported by sample size calculations prior to the study can clarify the question of which is statistically more valid, results of bioengineering measurements or clinical grading.


### Definition of Corrosive and Noncorrosive Irritants and Their Distinctions

As a result of the many bioengineering studies of a broad spectrum of irritants a major distinction between corrosive and noncorrosive irritants has appeared. *Corrosive irritants can be defined as irritants which in experimentally provoked weak reactions (redness only, faint redness) or subclinical reactions induce impairment of the water barrier function and increase of transepidermal water loss while noncorrosive irritants can be defined as irritancy of low degree but with no increase of transepidermal water loss.* The detergents with sodium lauryl sulfate (SLS) as the standard example are clearly corrosive [10, 19]. Nonanoic acid is an example of a noncorrosive irritant [20]. Therapeutics like all-trans retinoic acid, vitamin D and vitamin D analogues and dithranol are noncorrosive.

Corrosive irritants characteristically show a linear dose/irritation (inflammatory reaction) curve with ulceration if the concentration is sufficiently high. In high concentrations the insult is cytotoxic damage resulting in a chemical burn. Noncorrosive irritants may have a linear dose/irritation curve at lower concentration levels and reach a plateau of maximum irritant reaction at high concentrations with no direct chemical burn at the application site even after exposure of very high concentrations. Noncorrosive irritants may be less hazardous than corrosive irritants and act indirectly, and gradually result in disintegration of epidermal functions and structures resulting in dermatitis clinically or they may specifically insult selected dermal structures such as the vasculature and the pilosebaceous unit.

The distinction can only be made accurately in weak or faint clinical reactions since moderate and strong inflammatory reactions of the skin after 1–2 days will present secondary impairment of the water barrier function [21]. Advanced inflammation is one common soup with hundreds of general ingredients and a poor chance to find the subtle trigger.

Therapeutics with irritant adverse effect typically belong to the noncorrosive type.

Light exposure falls into the group of noncorrosive irritation. It is quite surprising that sunburn reaction of the skin in the early phase is dominated by redness with no significant increase of transepidermal water loss although the light on its way to the vasculature in the dermis passed the epidermal level of the location of the water barrier.

### Bioengineering, Timing of Reactions and Choice of Instrumentation

Acute irritant dermatitis as represented by the patch test reaction is uncommon in the dermatological outpatient clinic. What we see is typically the result of multiple exposures and a response modified by defence and repair mechanisms in a very complicated way with interactions between a large number of mediators and tissue factors impossible to overview or put on some simple formula (fig. 6, 7). If the manifestations of dermatitis gradually increase following repeated exposures we talk of 'cumulative irritant dermatitis' [22]. Thus, manifestations are no longer determined simply by the irritant substance but also by the exposure practices and the reactivity or sensitivity of the individual. 'Hardening' describes a situation with adaptation and recovery in spite of continued exposures.

In the patch test situation, a single exposure with induction of an acute irritant reaction typically using 24 or 48 h patch test application is used. After removal of patches clinical symptoms are scored, i.e. redness, edema and papules, and vesicles, and bullae representing consequences of edema. At the 7-day reading scaling or epidermal changes will often be more prominent. If patches are reapplied on the same site the reactivity may have changed due to the previous reactions which have left some memory in the skin [23]. Thus, there are three phases, i.e. an induction phase, a phase with clinical symptoms and ongoing repair and a subclinical phase with further repair and a memory of the proceeding trauma. The choice of bioinstrumentation in the different phases differs as a consequence of

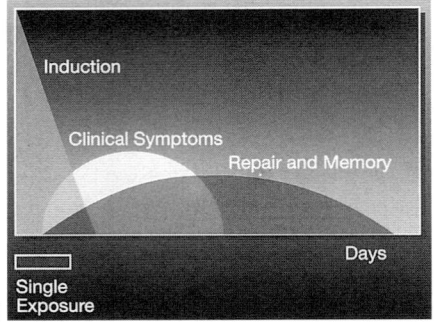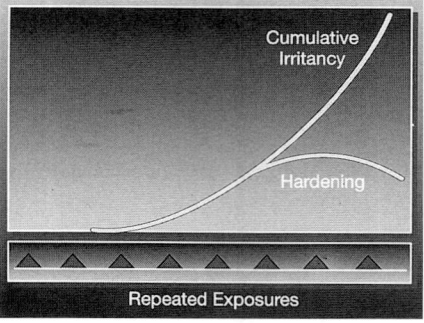

*Fig. 6.* Schematic presentation of single exposure to an irritant with an induction phase where the irritant penetrates the skin, reaches the targets of action and induce clinical symptoms of dermatitis with vasodilation and redness, extravasation of water and edema may be observed as papules, vesicles or bullae and finally disturbance of the epidermis with scaling. Defence and repair mechanisms already starts few hours after induction has started and proceeds a period after clinical symptoms have declined or disappeared, ending with a memory and change of threshold for a period following clinical healing.

*Fig. 7.* Schematic presentation of repeated exposures to an irritant with the development of cumulative irritant contact dermatitis or hardening depending on defence and repair mechanisms of the individual.

the change of phenomenology of the reactions over time. In the induction phase and the phase with clinical symptoms the signs of inflammation are of major clinical relevance, and transepidermal water loss relevant for the distinction between corrosive and noncorrosive irritation while epidermal features and measurement of electrical parameters, scaling and evaluation of the desquamation process become more relevant later. Wilhelm et al. [24] described the change in phenomenology over time in a recent publication. Paye et al. [25] recently evaluated the desquamation rate by dansyl chloride labelling. The choice of instrumentation relative to type of irritant is presented in table 1. The table has to be adopted to the researchers actual experiment. As stressed out previously, *transepidermal water loss has not a direct clinical relevance* but in the study of corrosive irritants the method is very sensitive and very useful but results need be backed up with some other measure of relevance. It is demonstrated in another chapter in this book [Fullerton and Serup] on the laser Doppler imaging method that TEWL measurements overestimate reactions elicited by SLS as compared to visual reading, colorimetry and blood flow measurements by the Doppler technique, indicating a relatively too high sensitivity of the SLS/transepidermal water loss experiment.

*Table 1.* Irritant reactions and choice of bioinstrumentation

*Table 1.* Irritant reactions and choice of bioinstrumentation

---

*Induction phase*

Transepidermal water loss (TEWL) to evaluate *corrosive* or *noncorrosive* mode of irritation

---

*Acute irritation phase*

a  Erythematous irritation
   Laser Doppler flowmetry/imaging
   Colorimetry a.m. CIE or erythema index
   High-frequency ultrasound, skin thickness/image analysis

b  Nonerythematous irritation
   Electrical conductance/capacitance
   Desquamation rate, dansyl labelling

---

*Late irritation phase including repair and memory*

a  Erythematous irritation
   Electrical conductance/capacitance
   Laser Doppler flowmetry/imaging
   Colorimetry a.m. CIE or erythema index
   High frequency ultrasound, skin thickness/image analysis

b  Nonerythematous irritation
   Electrical conductance/capacitance
   Desquamation rate, Dansyl labelling

---

The table primarily includes methods of direct clinical relevance, i.e. laser Doppler flowmetry and colorimetry as a measure of redness, high frequency ultrasound as a measure of edema of more advanced reactions, conductance, capacitance and Dansyl chloride labelling as measures of scaling, skin surface dryness and the desquamation process. Other methods such as replica and tape stripping techniques may be used for this purpose as well.

Methods are indicated in situations where they have a major relevance and may serve for quantitation. Transepidermal water loss measurement may be a useful quantitative tool in different phases although the measure is indirect does not directly quantify the clinical signs of dermatitis.

## The Future of Bioengineering Methods for the Evaluation of Irritant Reactions of the Skin

The future is bright. A large percentage of research papers in the field includes the methods. They are used more and more professionally and guidelines and information on proper use of methods are available [14, 15, 26].

It is proposed in the present review paper to categorize irritants into two groups, i.e. corrosive and noncorrosive irritants. The distinction may have important implications for the dose/irritation figures, potential hazards and practical use of the substances. A patch test/transepidermal water loss experiment is the key element in this distinction. It need be followed by further evaluations to document clinical relevance.

## References

1 Frosch P: Hautirritation und empfindliche Haut. Grosse scripta 7. Berlin, Grosse, 1985.
2 Marzulli FN, Maibach HI: Dermato-toxicology, ed 4. New York, Hemisphere, 1991.
3 Jackson EM, Goldner R: Irritant Contact Dermatitis. New York, Marcel Dekker, 1990.
4 Cronin E: Contact Dermatitis. Edinburgh, Churchill-Livingstone, 1980.
5 Serup J: Noninvasive techniques for quantification of contact dermatitis; in Rycroft RJG, Menné T, Frosch PH, Benezra C (eds): Textbook of Contact Dermatitis. Berlin, Springer, 1992, pp 323–338.
6 Björnberg A: Skin reactions to primary irritants in patients with hand eczema. An investigation with matched controls. Göteborg, Oscar Isacsons Tryckeri AB, 1968.
7 Kligman AM, Wooding WM: A method for the measurement and evaluation of irritants on human skin. J Invest Dermatol 1967;49:78–94.
8 Shelanski HA: Experiences with and considerations of the human patch test method. J Soc Cosmet Chem 1951;11:324–337.
9 Bason M, Lammintausta K, Maibach HI: Irritant dermatitis (irritation); in Marzulli FN, Maibach HI (eds): Dermato-toxicology, ed 4. Hemisphere, New York, 1991, pp 223–253.
10 Agner T: Noninvasive measuring methods for the investigation of irritant patch test reactions. A study of patients with hand eczema, atopic dermatitis and controls. Acta Derm Venereol (Stockh) 1992;suppl 173:1–26.
11 Serup J, Jemec GBE: Handbook of Noninvasive Methods and the Skin. Boca Raton, CRC Press, 1995.
12 Serup J: Ten years of experience with high-frequency ultrasound examination of the skin: Development and refinement of technique and equipments; in Altmeyer P, et al (eds): Ultrasound in Dermatology. Berlin, Springer 1992, pp 41–54.
13 Seidenari S, di Nardo A: B-scanning evaluation of irritant reactions with binary transformation and image analysis. Acta Derm Venereol (Stockh) 1992;suppl 175:9–13.
14 Pinnagoda J, Tupker RA, Agner T, Serup J: Guidelines for transepidermal water loss (TEWL) measurement. A report from the standardization group of the European Society of Contact Dermatitis. Contact Derm 1990;22:164–178.
15 Bircher A, de Boer E, Agner T, Wahlberg JE, Serup J: Guidelines for measurement of cutaneous blood flow by laser Doppler flowmetry. A report from the standardization group of the European Society of Contact Dermatitis. Contact Derm 1994;30:65–72.
16 Serup J, Agner T: Colorimetric quantification of erythema – comparison of two colorimeters (Lange Micro Color and Minolta Chroma Meter CR–200] with a clinical scoring scheme and laser Doppler flowmetry. Clin Exp Dermatol 1990;15:267–272.
17 Takiwaki H, Serup J: Variation in color and blood flow of the forearm skin during orthostatic maneuver. Skin Pharmacol 1994;7:226–230.
18 Wårdell K: Laser Doppler perfusion imaging. Methodology and skin applications. Thesis 329, Linköping University, 1994.
19 van der Valk PGM: Water vapour loss measurements on human skin. Thesis, State University Hospital, Gröningen, 1983.
20 Agner T, Serup J: Skin reactions to irritants assessed by non-invasive bioengineering methods. Contact Derm 1989;20:352–359.
21 Serup J, Staberg B: Differentiation of allergic and irritant reactions by transepidermal water loss. Contact Derm 1987;16:129–132.

22    Malten KE: Thoughts on irritant contact dermatitis. Contact Derm1981;7:238–247.
23    Widmer J, Elsner P, Burg G: Skin irritant reactivity following experimental cumulative irritant contact dermatitis. Contact Derm 1994;30:35–39.
24    Wilhelm KP, Freitag G, Wolff HH: Surfactant-induced irritation and skin repair. Evaluation of the acute human irritation model by noninvasive techniques. J Am Acad Dermatol 1994;30:944–949.
25    Paye M, Simion FA, Pierard GE: Dansyl chloride labelling of stratum corneum: Its rapid extraction from skin can predict skin irritation due to surfactants and cleansing products. Contact Derm 1994; 30:91–96.
26    Serup J: Bioengineering and the skin: From standard error to standard operating procedure. Acta Derm Venereol (Stockh) 1994; suppl 185:5–8.

Jørgen Serup, MD, PhD, Associate Professor of Dermatology, Leo Pharmaceutical Products, Department of Dermatological Research, Industriparken 55, DK–2750 Ballerup (Denmark)

Elsner P, Maibach HI (eds): Irritant Dermatitis. New Clinical and Experimental Aspects.
Curr Probl Dermatol. Basel, Karger, 1995, vol 23, pp 144–151

# Irritant Dermatitis: Experimental Aspects

*Klaus-P. Wilhelm*

Department of Dermatology, Medical University of Lübeck, and
proDERM, Institute for Applied Dermatological Research, Hamburg, Germany

It is 12 years ago that Albert Kligman [1] wrote: 'We are scandalously igno-
rant about irritation.' Although our understanding of irritant dermatitis is still
limited, especially if compared with the tremendously increased understanding of
the mechanisms involved in allergic contact dermatitis down to the molecular
mechanisms, considerable progress has been made in recent years to fight this
ignorance [2–5]. The increased interest in irritant dermatitis among researchers
worldwide and the resulting increased knowledge about this disease can at least
partly be ascribed to the availability of new instrumental techniques for noninvasive measurements of skin function, the so-called bioengineering instrumentation. Hence, Kligman's [1] demand from one decade ago has been fulfilled: 'Most
important of all, we must invent methodologies that permit us to measure the
difference in irritant potential among various substances. Quantification is an
urgent requirement.'

## Irritant Dermatitis Syndrome

Skin irritation is not at all a uniform reaction but a potpourri of a great number of morphologically and biochemically diverse reactions (table 1). In order to
reflect this diversity the term irritant skin syndrome has been proposed [2]. In
experimental models, however, as many variables as possible are standardized, in
order to obtain a uniform response that can be quantified and measured biophysi-
cally and biochemically. Therefore, few model irritants have been utilized in
experimental studies. Hence, most of the studies on skin irritation in the last
5 years have been conducted with the nonionic surfactant sodium lauryl sulfate

*Table 1.* Irritant dermatitis syndrome

| Irritant | Type of reaction |
|---|---|
| Detergents | scaling, erythema, edema* |
| Anorganic solvents | scaling, erythema, edema* |
| Acids, alkalis | necrosis |
| Croton oil | pustules |
| Cantharidin | subepidermal blister |
| Nicotinic acid | erythema (edema)* |
| Lactic acid | stinging |

* Morphology depends upon dose, application time and time of observation.

(SLS). This is a *legitimate* approach. However, one needs to know that a study with SLS as the irritant provides information only about SLS irritation. Results might be different for other surfactants, more different for cationic and anionic surfactants and even more different for thousands of irritants that are no surfactants.

### Variability

Many variables may influence the irritant response. The type and severity of skin irritation depend not only on the chemical nature of the irritant but also on the individual susceptibility of the subject and on the exposure conditions, as summarized in table 2. This article critically reviews some recent experimental studies. Some selected examples rather than an in-depth review will illustrate the complexity of experimental skin irritation, i.e. some parameters listed in table 2 will not always influence the irritant response as anticipated.

Gender may serve as a first example. Although it is common belief among laymen and experts that women have a more delicate integument than men, most studies have failed to provide the experimental evidence [6–8].

The influence of age on the irritant response cannot be elucidated by one simple statement either. Epidemiological data suggest a lower incidence of irritant contact dermatitis with increasing age [9]. The majority of recent studies investigating the influence of aging on the cutaneous permeability barriers by means of transepidermal water loss (TEWL) measurements, percutaneous penetration, and skin irritation models give strong evidence that the permeability bar-

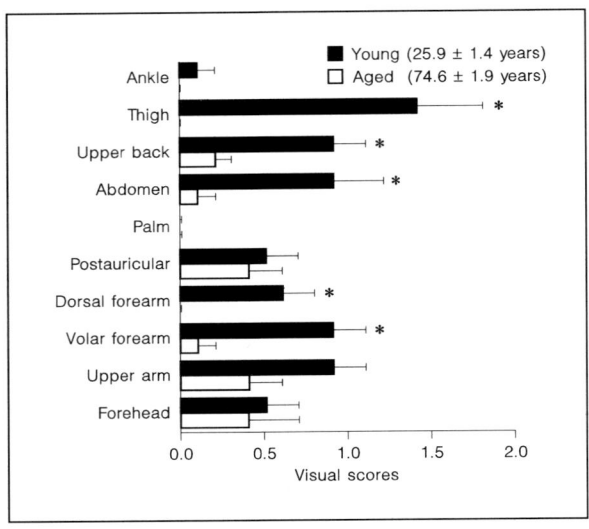

*Fig. 1.* Age dependence of proclivity to SLS irritation. Shown are visual erythema scores after exposure to 0.25% SLS for 24 h. On most anatomic sites less erythema was induced by SLS in the group of elderly individuals. Means ± SEM; n = 7–8. * Statistically significant difference between the age groups (p ≤ 0.05). Drawn according to the data provided in Cua et al. [7].

*Table 2.* Endogenous and exogenous factors influencing the irritant skin response

| Endogenous factors | Exogenous factors |
|---|---|
| Age | Concentration and amount of irritant |
| Gender | Irritation potential of the irritant |
| Race, pigmentation | Solvents |
| UV light sensitivity | Application area and time |
| Genetic predisposition, especially atopy | Ambient temperature and relative humidity |
| Skin diseases | Season |
| Predamaged skin | Time of observation |
| Anatomic site | Parameter of evaluation |
| Hormonal cycle | |

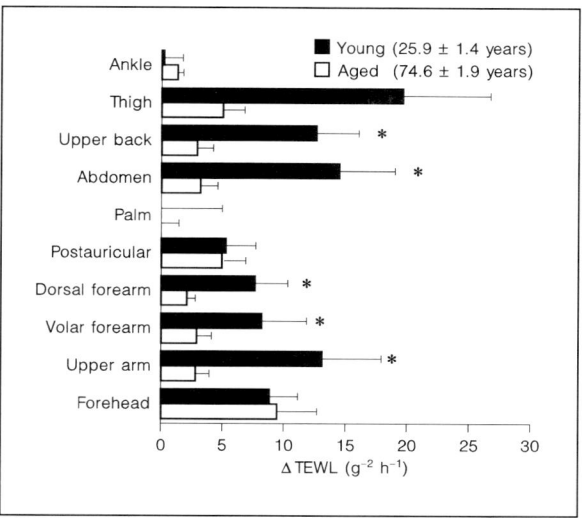

*Fig. 2.* Age dependence of proclivity to SLS irritation. Shown are increases in transepidermal water loss ($\Delta$ TEWL = TEWL$_{SLS}$ – TEWL$_{control}$) after exposure to 0.25% SLS for 24 h. On most anatomic sites SLS induced lower increases in TEWL in elderly individuals. Means ± SEM; n = 7–8. * Statistically significant difference between the age groups (p ≤ 0.05). Drawn according to the data provided in Cua et al. [7].

rier is not comprised in the elderly population [7, 10–13]. In contrast, experimental studies confirmed a decreased sensitivity to cutaneous irritants with increasing age. Thus, we have recently investigated the severity of the irritant response after application of SLS in two age groups on 11 anatomic sites [8]. Marked differences in the skin responses were noticed between the various regions (fig. 1, 2). We demonstrated that aged individuals had a significantly decreased irritant response on 5 of 11 anatomic sites. In our study, the severity of SLS-induced skin irritation was quantified by visual scores and by TEWL measurements (fig. 1, 2). Interestingly, aged individuals failed to demonstrate erythematous reactions at some anatomic sites completely, e.g. on the thigh and on the forearm (fig. 1). TEWL measurements, however, demonstrated that despite the lack of a visual reaction there was indeed significant barrier damage present in the aged group (fig. 2).

Limited information is available about the influence of the hormonal cycle on skin irritation. However, a recent study by Agner et al. [14] suggests a greater susceptibility to SLS at the beginning of the menstrual cycle as compared to the middle of the cycle (table 3).

In addition to the endogenous factors discussed above many exogenous variables will influence the irritant response of the skin to a compound (table 2). The concentration and amount of the irritant as well as its irritation potential will obviously determine the severity of the irritant reaction.

In addition, the solvent or the base in which the irritant is applied will influence the irritant response. The solvents, e.g. propylene glycol or ethanol, may be irritants and hence augment the irritant response [15, 16]. Or the solvents might not themselves be strongly irritating but might enhance the percutaneous penetration of the irritant and thereby increase the irritant response. Likewise, additional chemicals in the solution might influence the irritant response.

That additional surfactants in a solution with a constant dose of SLS do not always increase, but may in fact decrease the irritation potential of SLS was shown by Rhein et al. [17]. In their study, the addition of a mild surfactant, alkyl-7-ethoxy sulfate, to a constant dose of SLS resulted in a reduction of the human irritation response as assessed in a 21-day cumulative irritation test (fig. 3). The explanation for this result was unclear, but the authors suggested an alteration in the micellar solution properties of the SLS upon addition of the alkyl-7-ethoxy sulfate [17]. Hence, this experiment by Rhein et al. [17] demonstrates firstly that adding of one irritant to a second irritant will not necessarily increase its irritation potential. Secondly, the search for other effective co-surfactants that minimize the irritating surfactant in cleansing systems should be encouraged. Because of the diversity of the irritant syndrome the evaluation of the irritation potential of diverse chemical irritants depends significantly on the feature (erythema vs. TEWL vs. skin hydration) measured. Thus, we recently demonstrated in an acute irritation model [18] that the cationic DTAB and the anionic SLS produced similar degrees of erythema (table 4). In contrast, damage

*Table 3.* Menstrual cycle and irritant reactivity [adapted from ref. 14]

| | Females, SLS-exposed skin | | |
| --- | --- | --- | --- |
| | day 1 | days 9–11 | p value |
| TEWL | 15.8 (12.1–20.0) | 13.7 (11.0–15.9) | <0.05 |
| Blood flow, arbitrary units | 24 (14–40) | 17 (9–36) | n.s. |
| Skin thickness, mm | 1.23 (1.18–1.34) | 1.15 (1.04–1.25) | <0.05 |

n.s. = No statistical significance.

Median values and 25/75 percentiles of TEWL, superficial blood flow, and skin thickness as measured of skin of women after exposure to 0.5% SLS for 24 h at day 1 and days 9–11 of the menstrual cycle, respectively.

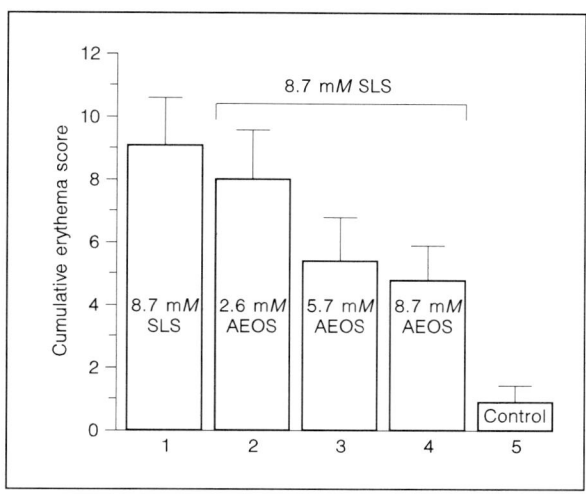

*Fig. 3.* 21-day cumulative irritation results for SLS and SLS/AEOS mixtures, respectively. The addition of alkyl-7-ethoxy sulfate (AEOS) to a constant dose of SLS resulted in a reduction of the human irritation response. Drawn according to the data provided in Rhein et al. [18].

*Table 4.* Comparison of the maximum irritant responses in an acute irritation model [adapted from ref. 18]

| Parameter | SLS | DTAB |
|---|---|---|
| SCR (a*) | $11.8 \pm 2.8$ | $11.1 \pm 2.2$ (n.s.) |
| TEWL, $g^{-2} h^{-1}$ | $35.6 \pm 17.6$ | $19.9 \pm 5.9$* |
| CAP, IU | $30.1 \pm 13.5$ | $47.0 \pm 11.6$ (n.s.) |

Given are the maximum responses for two surfactants with a high irritation potential in an acute irritation model: anionic SLS and cationic dodecyl trimethyl ammonium bromide (DTAB).

SCR = Skin color reflectance measurements, parameter a* (erythema); TEWL = transepidermal water loss for evaluation of barrier function; CAP = electrical capacitance (instrumental units) for quantification of stratum corneum hydration.

n.s. = Statistically not significant ($p > 0.05$); * statistically significant difference ($p \leq 0.05$).

Means $\pm$ SD; n = 11 volunteers.

of the SC was significantly more pronounced with SLS. This was evidenced by both the increase in TEWL and the decrease in SC hydration in the following repair phase (table 4).

## Conclusion

Skin irritation is a very complex disease entity the diversity of which is reflected by the term 'irritant skin syndrome'. Many exogenous variables determine how and when this yet to be tamed inflammatory disease will present itself. Although considerable progress has been made since Albert Kligman's complaint about our 'scandalous ignorance about skin irritation', our knowledge about this disease is still limited to the irritant action of a limited number of model irritants.

## References

1  Kligman AM: Assessment of mild irritants; in Frost P, Horwitz SN (eds): Principles of Cosmetics for the Dermatologist. London, Mosby, 1982, pp 265–274.
2  Bason M, Lammintausta K, Maibach HI: Irritant dermatitis (irritation); in Marzulli FN, Maibach HI (eds): Dermatotoxicology, ed 4. New York, Hemisphere, 1991, pp 223–254.
3  Frosch PJ: Irritant contact dermatitis; in Frosch PJ, Dooms-Goossens A, Lachapelle JM, Rycroft RFG, Scheper RJ (eds): Current Topics in Contact Dermatitis. Berlin, Springer, 1990, pp 385–398.
4  Andersen KE, Benezra C, Burrows D, Camarasa J, Dooms-Goossens A, Ducombs G, et al: Contact dermatitis. A review. Contact Derm 1987;16:55–78.
5  Wilhelm KP, Cua AB, Wolff HH, Maibach HI: Surfactant-induced stratum corneum hydration in vivo: Prediction of the irritation potential of anionic surfactants. J Invest Dermatol 1993;101:310–315.
6  Lammintausta K, Maibach HI, Wilson D: Irritant reactivity in males and females. Contact Derm 1987;17:276–280.
7  Cua AB, Wilhelm KP, Maibach HI: Cutaneous sodium lauryl sulfate irritation potential: Age and regional variability. Br J Dermatol 1990;123:607–613.
8  Björnberg A: Skin reactions to primary irritants in men and women. Acta Derm Venereol (Stockh) 1975;55:191–194.
9  Malten KE, Fregert S, Bandmann HJ, Calnan CD, Cronin E, Hjorth N, et al: Occupational dermatitis in five European dermatological departments. Berufsdermatosen 1971;19:1–13.
10  Wilhelm KP, Maibach HI: The effect of aging on the barrier function of human skin evaluated by in vivo transepidermal water loss measurements; in Frosch PJ, Kligman AM (eds): Noninvasive Methods for the Quantification of Skin Functions. Heidelberg, Springer, 1993, pp 181–189.
11  Roskos KV, Maibach HI, Guy RH: The effect of aging on percutaneous absorption in man. J Pharmacokinet Biopharm 1989;17:617–630.
12  Tagami H: Functional characteristics of aged skin. 1. Percutaneous absorption. Acta Dermatol (Kyoto) 1971/1972;66/67:19–21.
13  Lévêque JL: Measurement of transepidermal water loss; in Lévêque JL (ed): Cutaneous Investigation in Health and Disease: Noninvasive Methods and Instrumentation. New York, Dekker, 1989, pp 135–152.

14  Agner T, Damm P, Skouby SO: Menstrual cycle and skin reactivity. J Am Acad Dermatol 1991;24: 566–570.
15  Abrams K, Harvell JD, Shirner D, Wertz P, Maibach H, Maibach HI, et al: Effect of organic solvents on in vitro human skin water barrier function. J Invest Dermatol 1993;101:609–613.
16  Berardesca E, Andersen PH, Bjerring P, Maibach HI: Erythema induced by organic solvents: In vivo evaluation of oxygenized and deoxygenized haemoglobin by reflectance spectroscopy. Contact Derm 1992;27:8–11.
17  Rhein LD, Simion FA, Hill RL, Cagan RH, Mattai J, Maibach HI: Human cutaneous response to a mixed surfactant system: Role of solution phenomena in controlling surfactant irritation. Dermatologica 1990;180:18–23.
18  Wilhelm KP, Freitag G, Wolff HH: Surfactant-induced skin irritation and skin repair: Evaluation of the acute human irritation model by noninvasive techniques. J Am Acad Dermatol 1994;30: 944–949.

Dr. med. Klaus.-P. Wilhelm, proDERM, Institute for Applied Dermatological Research,
Industriestrasse 1, D–22869 Schenefeld/Hamburg (Germany)

Elsner P, Maibach HI (eds): Irritant Dermatitis. New Clinical and Experimental Aspects.
Curr Probl Dermatol. Basel, Karger, 1995, vol 23, pp 152–158

..........................

# Transepidermal Water Loss Measurements in Patch Test Assessment: The Need for Standardisation

*Vera Rogiers*

Department of Toxicology, Vrije Universiteit Brussel, Belgium

Tenside-containing cosmetics repeatedly brought in contact with the skin may not only remove a part of the natural protective hydrolipidic layer but may also cause interferences with the lamellar sheets and lamellar bodies [1]. Skin irritation can be assessed in vivo using transepidermal water loss (TEWL) measurements [2, 4].

A serious drawback, however, is the observation that several factors may influence TEWL assessment. Factors related to the volunteers include age [3, 5–10], sex and race [3, 7–9], anatomic region [4, 6–9, 11–13], diurnal rhythm [7], skin temperature [7, 11, 14], skin damage and disease [4, 11, 14, 15].

Other factors are related to the 'measurement environment' and 'apparatus'. They consist of air circulation, relative humidity, direct light, seasonal variations, ambient temperature, and probe pressure and a protective jacket, respectively [7].

Consequently, when the irritative properties of tenside-containing cosmetics are evaluated by TEWL measurements, it becomes essential to work under rigorously standardized conditions. The aim of this study was to develop such standardized conditions to be used after patch testing.

## Materials and Methods

*Volunteers*
Caucasian females 22–29 years, with normal skin.

*Apparatus*
A TEWA-meter® TM200 (Courage and Khazaka, Germany) and an Evaporimeter® EP1 (Servomed, Sweden), both calibrated, have been used for TEWL assessment. Skin temperature was measured via a contact thermometer HI 9063 (Hanna Instruments, USA) and

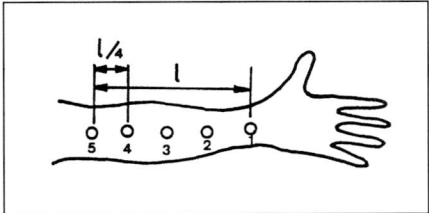

*Fig. 1.* Measuring sites on the volar forearm.

stratum corneum hydration by a Corneometer CM® 820 (Courage and Khazaka, Germany).

*Materials*

Patches consisted of glass fiber pre-filters (Sartorius, Germany), saturated with $50 \mu l/cm^2$ of either a tenside solution or deionised water (control). Patches with tenside solution were applied on the left volar forearm, and controls on identical sites on the right forearm. Both were fixed with hypoallergic Leukofix® (Beiersdorf, Germany). Occlusion was obtained by using tin foil (Mellita, Belgium) fixed with hypo-allergic Fixomull-stretch® (Beiersdorf, Germany) and Bandafix® (International Medical, Belgium).

The tenside solution used was a 1% (g/v) solution of sodium lauryl sulfate (SLS) in deionised water (Texapon® K12, 90.5% active, Henkel, Germany).

*Methods*

Volunteers were resting in an air-conditioned room (constant temperature of 20°C, relative humidity 35–45%) for at least 15 min. TEWL measurements were carried out on the left volar forearm (fig. 1).

Skin temperature was measured and the probe of the TEWA-meter was warmed up to that particular temperature on a place of the skin which was not to be used as a measurement site.

For patch testing, occlusion was maintained for 24 h, after which the skin was carefully rinsed with water and patted dry. TEWL measurements were carried out 24 h after exposure to the air.

## Results and Discussion

*Temperature of the Measuring Probe*

This particular factor has not yet been investigated in detail due to the fact that the probe temperature is unknown when an Evaporimeter is used. Only the TEWA-meter displays it on the screen. The TEWL of a number of subjects (n = 10) was assessed as a function of the probe temperature (fig. 2). The TEWL increases continuously and significantly until the probe has reached skin temperature, $30.9 \pm 1.0°C$ (n = 10). From this point onwards the TEWL remains con-

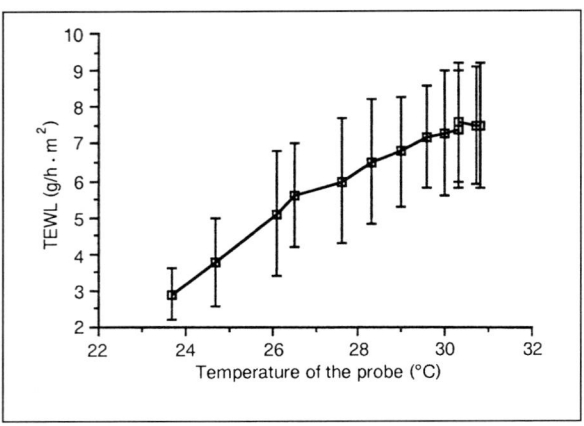

*Fig. 2.* Effect of probe temperature on TEWL (n = 10).

stant. Hence, only at this point can stable TEWL measurements be performed. This specific skin temperature is usually reached after an 11- to 12-min warm-up of the probe. In practice then, it is only from that moment onwards that the actual TEWL measurement may be carried out.

*Shielding Box*

The results show that the TEWA-meter is not sensitive to air circulation. Breathing in the direction of the probe or opening and closing doors at a distance of two meters from the measurement probe exerted no significant effect (n = 11). Only major changes in air circulation caused by the use of hair driers or fans significantly affected the measurements. In the shielding box a TEWL (in $g/m^2 \cdot h$) of 5.6 ± 1.3 and 5.9 ± 1.3 was measured (n = 11); out of the box 2.8 ± 1.0 and 2.0 ± 0.9 were found, respectively (paired Student's t test, $p < 0.05$).

*Location of the Measurement Site on the Body*

TEWL was measured on different sites on the left and right forearms, on the palm of the hand, in the fold of the arm and on the forehead (fig. 3). It seems that only the forearm is a good measurement site whereas the forehead and palm are not due to their high coefficients of variation. Corresponding places on the right and left forearm have the same TEWL, which is of practical relevance for the testing of various topical preparations.

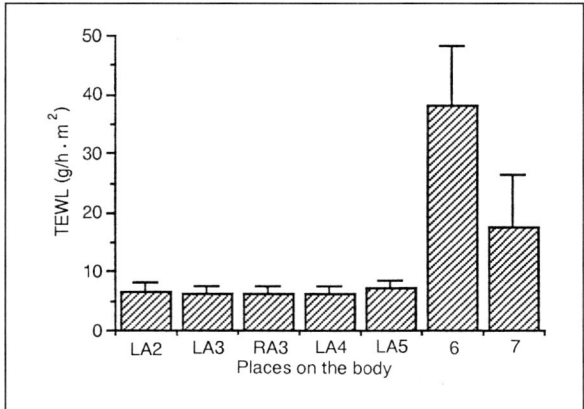

*Fig. 3.* Effect of the measurement site on TEWL measurements (n = 21). Sites 2, 3, 4 and 5 correspond with those indicated in figure 1. LA = Left forearm; RA = right forearm. Sites 6 and 7 represent the center of the palm and 1 cm above the nose on the forehead, respectively.

### Temperature of the Environment

Water baths of, respectively, 5 and 42 °C were used. Subjects' forearms were alternatively immersed in one of these for 1 min and patted dry before measuring TEWL, skin temperature and the hydration state. The two bathing sessions were administered at sufficiently large time intervals to exclude possible interference (table 1).

At the start of the experiments no significant differences were observed in skin temperature, TEWL and skin hydration. Bathing at 5 and 42 °C, however, had a significant effect: fluctuations in the temperature affected the hydration state of the skin and the TEWL. It is important, therefore, to keep the environmental temperature constant.

### Humidity of the Environment

The effect of humidity was studied by working alternatively inside and outside a water bath. The temperature of this bath approximated skin temperature as closely as possible (table 2). The results suggest that the TEWL and degree of hydration increase significantly as the environmental humidity increases.

### Intra- and Intergroup Variability

To determine the effect of the diurnal rhythm the skin of the left forearm of 21 females was cleaned at 9 a.m. with a hypo-allergic lotion. TEWL was assessed

*Table 1.* Effect of environmental temperature on TEWL, skin temperature and stratum corneum hydration (n = 20)

| Measured factors | | Before bathing | After bathing |
|---|---|---|---|
| TEWL, $g/m^2 \cdot h$ | 5°C | $6.6 \pm 1.3$ NS | $36.6 \pm 6.6$ * |
| | 42°C | $7.1 \pm 1.6$ | $40.9 \pm 9.5$ |
| Skin temperature, °C | 5°C | $30.7 \pm 1.0$ NS | $21.8 \pm 1.6$ * |
| | 42°C | $31.0 \pm 0.9$ | $34.0 \pm 0.8$ |
| Hydration (relative units) | 5°C | $54.6 \pm 5.6$ NS | $95.1 \pm 7.3$ * |
| | 42°C | $54.5 \pm 4.2$ | $81.2 \pm 5.7$ |

Paired Student's t tests have been applied to compare the measurements made at 5 and 42°C. NS = Not significant; * $p < 0.05$. Mean values ± SD.

*Table 2.* Effect of environmental humidity on TEWL, probe temperature and skin hydration (n = 20)

| Measured factors | Before bathing | After bathing |
|---|---|---|
| TEWL, $g/m^2 \cdot h$ | $6.6 \pm 1.2$ | $41.7 \pm 8.9$* |
| Skin temperature, °C | $30.1 \pm 0.8$ | $30.3 \pm 0.9$ NS |
| Hydration (relative units) | $54.6 \pm 5.6$ | $87.3 \pm 11.6$* |

Paired Student's t tests show the comparison of values from before and after bathing. NS = Not significant; * $p < 0.05$. Mean values ± SD.

2, 5 and 7 h thereafter. TEWL values were also measured on 4 consecutive days (= 1 week) and on the same day of 4 consecutive weeks (= 1 month) and no significant differences could be observed. When all the measurements done were taken together the intra- and intergroup variations were $10.8 \pm 3.5$ and $20.1 \pm 2.7$, respectively.

### TEWL Assessment by TEWA-meter or Evaporimeter

When blank patches were applied on both volar forearms on identical sites, no significant differences in TEWL values were found when both apparatuses were used (ANOVA test, $p > 0.05$) (table 3). The values obtained with the Evaporimeter, however, were significantly lower than those measured with the TEWA-meter (ANOVA test, $p < 0.001$). When patches with 1% SLS were applied on the

*Table 3.* TEWL values after patch testing measured by a TEWA-meter and an Evaporimeter (n = 13)

| | TEWL by TEWA-meter g/m²·h | | TEWL by Evaporimeter g/m²·h | |
|---|---|---|---|---|
| | LA | RA | LA | RA |
| Experiment 1 | 8.1 ± 1.2 (15) | 8.0 ± 1.1 (14) | 4.6 ± 1.0 (22) | 4.9 ± 1.1 (22) |
| Experiment 2 | 30.3 ± 11.6 (38) | 8.1 ± 1.4 (17) | 18.9 ± 8.5 (45) | 4.6 ± 1.2 (26) |

In the first experiment, blanks were patch tested on the left (LA) and right (RA) volar forearms. In the second experiment, blanks and a 1% SLS solution were applied on RA and LA, respectively. The coefficient of variation (in %) is shown in parentheses.

left forearm and blanks on identical sites of the right forearm, the blank-corrected TEWL values for the SLS solutions measured with the Evaporimeter were lower than those obtained with the TEWA-meter. Analysis of variance revealed a significant interaction (apparatus-test solution) pointing to the fact that for higher TEWL values (than the blank), the difference between the two apparatuses became significant ($p < 0.05$). The intergroup coefficient of variation was significantly higher for the Evaporimeter than for the TEWA-meter (F test, $p < 0.001$). An explanation may probably be found in the difficulties experienced in the standardisation of the former apparatus. Indeed, it was not possible to warm up the probe to the exact skin temperature since the probe temperature is not displayed on the screen.

## Conclusion

When the irritative properties of tenside-containing cosmetics are evaluated in vivo by patch testing, objective scoring is possible by TEWL measurements. Reliable measurements are obtained on the condition that all factors are controlled for and standardized conditions are rigorously applied. Attention should be paid to a constant temperature of 20°C and a constant relative humidity lower than 50% for the air-conditioned room. The volunteers should be resting for at least 15 min in that room before measurements are carried out. The most important factor is the probe temperature. The probe must be warmed up to skin temperature before any TEWL measurement is carried out. Consequently, the TEWA-meter is preferred to the Evaporimeter since the latter does not show the probe temperature on the monitor and is more difficult to standardize. In addition, it is safer to use a nonocclusive shielding box to avoid excessive air turbulence. Measurements should be carried out on identical anatomic sites for all the subjects involved. The volar forearm is a good measurement site and corresponding places on the right and left forearms exhibit the same TEWL. Diurnal rhythm seems not to affect TEWL measurements. When all these factors are taken into account the coefficient of variation for TEWL measurements at the level of the individuals involved is situated around 10% whereas the intergroup variation is around 15–20%.

## References

1 Fartasch M: Human barrier formation and reaction to irritation. Abstr Proc 2nd Int Symp Irritant Contact Dermatitis, Zurich, April 14–16, 1994.
2 Tupker RA, Pinnagoda J, Nater JP: The transient and cumulative effect of sodium lauryl sulphate on the epidermal barrier assessed by transepidermal water loss: Inter-individual variation. Acta Derm Venerol (Stockh) 1990;70:1–5.
3 Wilson DR, Maibach HI: Transepidermal water loss: A review; in Lévêque JL (ed): Cutaneous Investigation in Health and Disease. Non-Invasive Methods and Instrumentation. New York, Marcel Dekker, 1988, pp 113–130.
4 Lévêque JL: Measurement of transepidermal water loss; in Lévêque JL (ed): Cutaneous Investigation in Health and Disease. Non-Invasive Methods and Instrumentation. New York, Marcel Dekker, 1988, pp 135–151.
5 Thune P, Nilsen T, Handstad K, Gustavsen T, Lövig DH: The water barrier function of the skin in relation to the water content of stratum corneum, pH and skin lipids. Acta Derm Venerol (Stockh) 1988;68:277–283.
6 Cua AB, Wilhelm KP, Maibach HI: Cutaneous sodium lauryl sulphate irritation potential: Age and regional variability. Br J Dermatol 1990;123:607–613.
7 Pinnagoda J, Tupker RA, Agner T, Serup J: Guidelines for transepidermal water loss (TEWL) measurement. Contact Derm 1990;22:164–178.
8 Cua AB, Wilhelm KP, Maibach HI: Frictional properties of human skin: Relation to age, sex and anatomical region, stratum corneum hydration and transepidermal water loss. Br J Dermatol 1990; 123:473–479.
9 Wilhelm KP, Cua AB, Maibach HI: Skin aging. Arch Dermatol 1991;127:1806–1809.
10 Elsner P, Wilhelm D, Maibach HI: Effect of low-concentration sodium lauryl sulfate on human vulvar and forearm related differences. J Reprod Med 1991;36:77–81.
11 Potts RO: Stratum corneum hydration: Experimental techniques and interpretations of results. J Soc Cosmet Chem 1986;37:9–33.
12 Rougier A, Lotte C, Corcuff P, Maibach HI: Relationship between skin permeability and corneocyte size according to anatomic site, age and sex in man. J Soc Cosmet Chem 1988;39:15–26.
13 Van der Valk PGM, Maibach HI: Potential for irritation increases from wrist to the cubital fossa. Br J Dermatol 1989;121:709–712.
14 Grice KA, Bettley FR: Skin water loss and accidental hypothermia in psoriasis, ichthyosis, and erythroderma. Br Med J 1967;iv:195–198.
15 Lavrijsen APM, Oestmann E, Hermans J, Boddé HE, Vermeer BJ, Ponec M: Barrier function parameters in various keratinization disorders: Transepidermal water loss and vascular response to hexyl nicotinate. J Dermatol 1993;129:547–554.

Prof. Dr. Vera Rogiers, Department of Toxicology, Vrije Universiteit Brussel,
Laarbeeklaan 103, B–1090 Brussels (Belgium)

Elsner P, Maibach HI (eds): Irritant Dermatitis. New Clinical and Experimental Aspects.
Curr Probl Dermatol. Basel, Karger, 1995, vol 23, pp 159–168

..............................

# Laser Doppler Image Scanning for Assessment of Skin Irritation

*Ann Fullerton, Jørgen Serup*

Department of Dermatological Research, Leo Pharmaceutical Products Ltd.,
Ballerup, Denmark

Vasodilatation with increased blood flow is an essential part of an irritant inflammatory response and cutaneous blood flow measurement can therefore be used as an objective assessment of skin reactions as a supplement to the subjective clinical scoring of erythema and palpable edema.

Laser Doppler velocimetry (LDV) has been widely used as a noninvasive method for measurement of vascular changes accompanying cutaneous inflammation. In conventional LDV instruments the laser light is transmitted to and from the tissue by optical fibres positioned over the site of interest. The perfusion within an approximately 1 mm$^3$ tissue volume at the tip of the probe is recorded continuously and no information is attained about how the perfusion varies over the skin surface regardless that tissue perfusion frequently shows a substantial spatial variation resulting in significant differences in perfusion values even at adjacent sites.

Laser Doppler image scanning (LDI) is a new technigue for mapping cutaneous blood flow [1, 2]. The method employs a two-dimensional horizontal scanning of the flow of a specific tissue and makes it possible to visualize the spatial variation of the perfusion. The method is not only noninvasive but also nontouch. The latter reduces the risk of influence on skin perfusion mechanically. A scanning procedure with a maximum image format of 4,096 measurement points makes it possible to map the perfusion in an area up to 12 × 12 cm.

The usefulness of the LDI technique to investigate skin irritation after application of two different irritant substances, sodium lauryl sulfate (SLS) and all-trans retinoic acid was evaluated in a hairless guinea pig model.

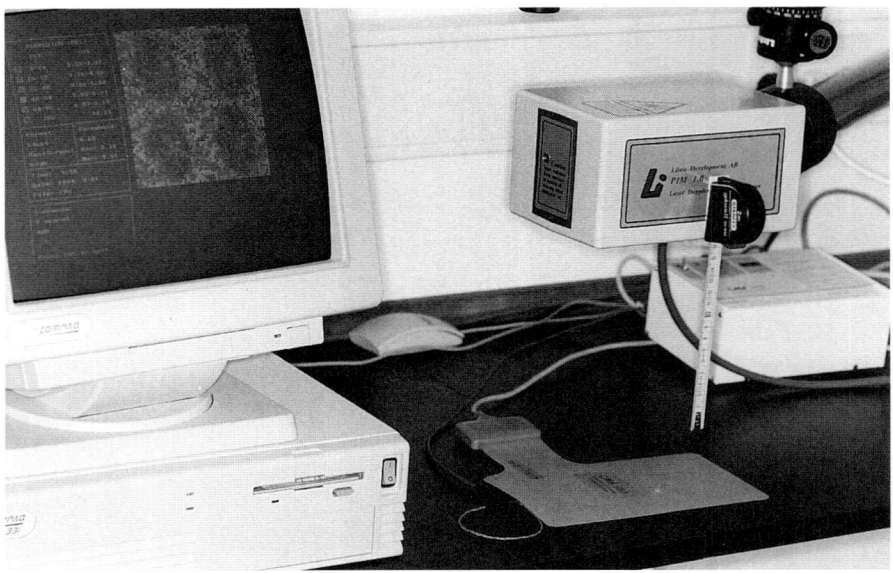

*Fig. 1.* The LDI system (Lisca Laser Doppler perfusion Imager®).

## Methods

### Animals

Hairless (hr/hr) guinea pigs (Charles River Laboratories) each weighing about 400–450 g were used in the study. The guinea pigs were placed in macrolon cages with free access to a commercial rodent diet and tap water. A 12-hour light/12-hour dark cycle was maintained. Temperature and humidity in the animal room was recorded by a thermohygrograph throughout the study.

### Test Preparations

SLS, (99% purity, Sigma) 2, 1 and 0.5% in water. All-trans retinoic acid (Sigma) 0.05, 0.01, 0.002% in ethanol with 0.01% BHT added as an antioxidant.

### Patch Test Procedure

Each group of animals was patch tested with SLS solutions and placebo vehicle or with all-trans retinoic acid solutions and placebo vehicle, respectively. Two test sites were marked on each flank of the animal. In order to remove the influence of an effect caused by location of treatment, the treatments were randomly assigned to locations in an as much as possible balanced way using for each group of 4 animals a Latin square.

Patch test application was performed using 60 µl of test preparation applied on a filter disc fitted into a large Finn Chamber® (diameter 12 mm) and placed on the skin. Test chambers were removed at day 1 (SLS) or at day 2 (all-trans retinoic acid). Evaluation of test sites was performed 2 h after removal of the patches.

---

*Test Evaluation*

Test evaluation was based on a clinical scoring of the degree of erythema and edema and objective measurement of transepidermal water loss (TEWL) and cutaneous blood flow using the laser Doppler image scanning technique. Test evaluations were done blinded.

*Clinical Scoring*

Clinical scoring of erythema and edema of test sites was done according to Draize et al. [8]. Test reactions were assessed visually and by palpation.

Evaluation of skin reactions

| | |
|---|---|
| *Erythema and eschar formation* | |
| No erythema | 0 |
| Very slight erythema (barely perceptible) | 1 |
| Well defined erythema | 2 |
| Moderate to severe erythema | 3 |
| Severe erythema (beet redness) to slight eschar formation | 4 |
| Total possible erythema score | 4 |
| *Edema formation* | |
| No edema | 0 |
| Very slight edema (barely perceptible) | 1 |
| Slight edema (edge of area well defined by definite raising) | 2 |
| Moderate edema (area raised approximately 1 mm) | 3 |
| Severe edema (raised more than 1 mm and extending beyond area of exposure) | 4 |
| Total possible edema score | 4 |
| Total possible score for primary irritation | 8 |

A final mean score for each formulation was obtained by dividing the total score for the group by the number of animals in the group.

*Transepidermal Water Loss*

An Evaporimeter EP–1 (Servo-Med, Kinna, Sweden) with a protection cover with a screen and grid was used. Measurements were done with precautions taken according to Pinnagoda et al. [3]. The water vapor pressure gradient is measured with sensors at two different levels above the skin surface and the TEWL ($g/m^2$ h) calculated. Measurements were performed with the animal placed in an incubator for air convection control. Three measurements were carried out on each test site using printer registration. The mean TEWL value and standard deviation were calculated.

*Laser Doppler Image Scanning*

The equipment uses a low-power He-Ne laser to scan the tissue. Figure 1 shows the LDI system (Lisca Laser Doppler Perfusion Imager®).

Two step motors in the scanner head control the position of two mirrors moving the He-Ne laser beam step by step over the area of interest. At each measurement site, the

He-Ne laser beam penetrates the skin to a depth of a few hundred micrometers. The back-scattered Doppler-shifted light depending on the flux of red blood cells is detected by a photodetector. The light intensity is transformed into a signal processor for generation of an output signal (within the range 0–10 V) which is linearly proportional to tissue blood perfusion. When the scanning procedure is completed a color-coded perfusion image is displayed on a monitor screen. This image gives an informative overview of the perfusion status of the tissue under study. Appropiate data analysis and statistical calculations can be performed on the perfusion image. The equipment has been described in detail elsewhere [4].

Studies were performed with room temperature kept constantly at 23–25 °C since fluctuations in ambient temperature can have substantial influence on skin blood flow. Prior to measurements anesthesia was obtained using pentobarbital (25 mg/kg i.p.). Anesthesia was used during scanning to prevent artifacts caused by movement or shivering. The animal was placed on a heating pad with a temperature probe inserted to avoid cooling during the study. Test sites were marked with an ink pen around the area of exposure. The laser light is absorbed by the ink instead of being backscattered to the detector. The ink markings appear as nonperfusion light-gray background zones on the color-coded image, making it easier to identify and orientate the perfusion image. During scanning, the area of interest is placed right under the scanning head with the laser beam pointing at the center of the measurement area.

The scanner head was placed in parallel with the tissue surface to avoid geometrical distortion of the image. A fixed distance between the tissue and the lower part of the scanner head was used.

Sequential scanning mode was employed giving four replicative scannings within each site. The parameter settings during measurement were:

| | |
|---|---|
| Image format | $32 \times 32$ |
| Field area | $750 \text{ mm}^2$ |
| Resolution mode | high |
| Threshold | 6.1 |
| Distance to subject | 16.5 cm |
| Spatial resolution | $0.73 \text{ mm}^2$ |
| | x-axis 0.94 mm |
| | y-axis 0.78 mm |

The LDI version 2.4 software was utilized. Ambient light was switched off during scanning in order to avoid optical interference with the laser beam. In figure 2, a block diagram of the laser Doppler image scanning procedure is shown.

## Results and Discussion

In figure 3a–d examples of the perfusion imaging of skin reactions after application of placebo vehicle and the three concentrations of SLS (0.5, 1 and 2%) are shown. Each figure shows four sequential scans within the same area. The gray

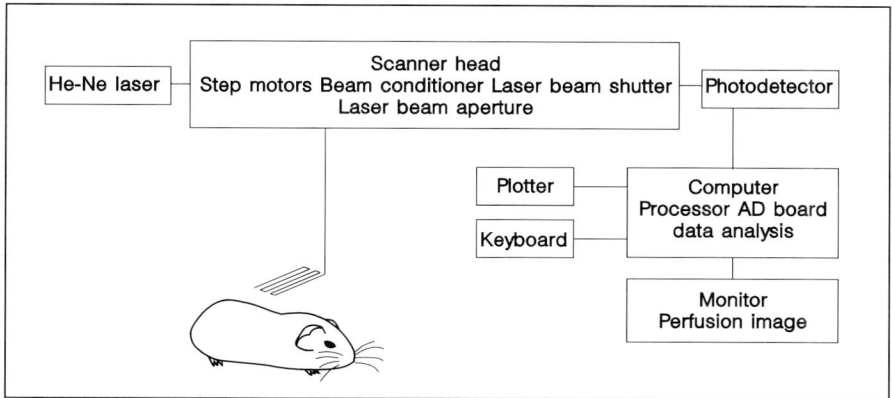

*Fig. 2.* Block diagram showing the laser Doppler image scanning procedure.

zones shown are the ink markings indicating the placement of the test sites. A dose-dependent increase in blood flow is seen on the scanning images.

From the images it is possible to study the spatial variation in perfusion within in each test area. In the color-coding scale the perfusion values from 0 to 80% are divided into five intervals with equal width. All values ranging from 80 to 100% are coded dark red. The color-codings corresponding to separate perfusion intervals are shown in the upper-left corner. For each color, the corresponding voltages are given. On each separate scan an area of interest in the perfusion image can be selected. For the framed area the mean perfusion, standard deviation of the perfusion values and number of pixels corresponding to the number of measurement sites are calculated. A single test site comprises about 255 individual measurement sites of skin blood perfusion. The bottom corner of the screen shows the mean and standard deviation of the perfusion values within the selected area and the bar-chart of the color-coding. The mean perfusion measured in this way and the mean clinical scoring of irritation are given in tables 1 and 2 for SLS and all-trans retinoic acid, respectively.

In all cases, the clinical response was primarily seen as increased erythema. A slight edema formation was observed with the highest dose of SLS and a very slight to slight edema was observed for all three doses of all-trans retinoic acid.

For SLS, a dose-dependent increase in clinical response was observed. The LDI mean values obtained were in high concordance with the clinical scorings. For all-trans retinoic acid no difference was observed between the clinical scorings obtained for the three different doses tested although there was a 25-fold difference in test concentration between the lowest and highest test concentration

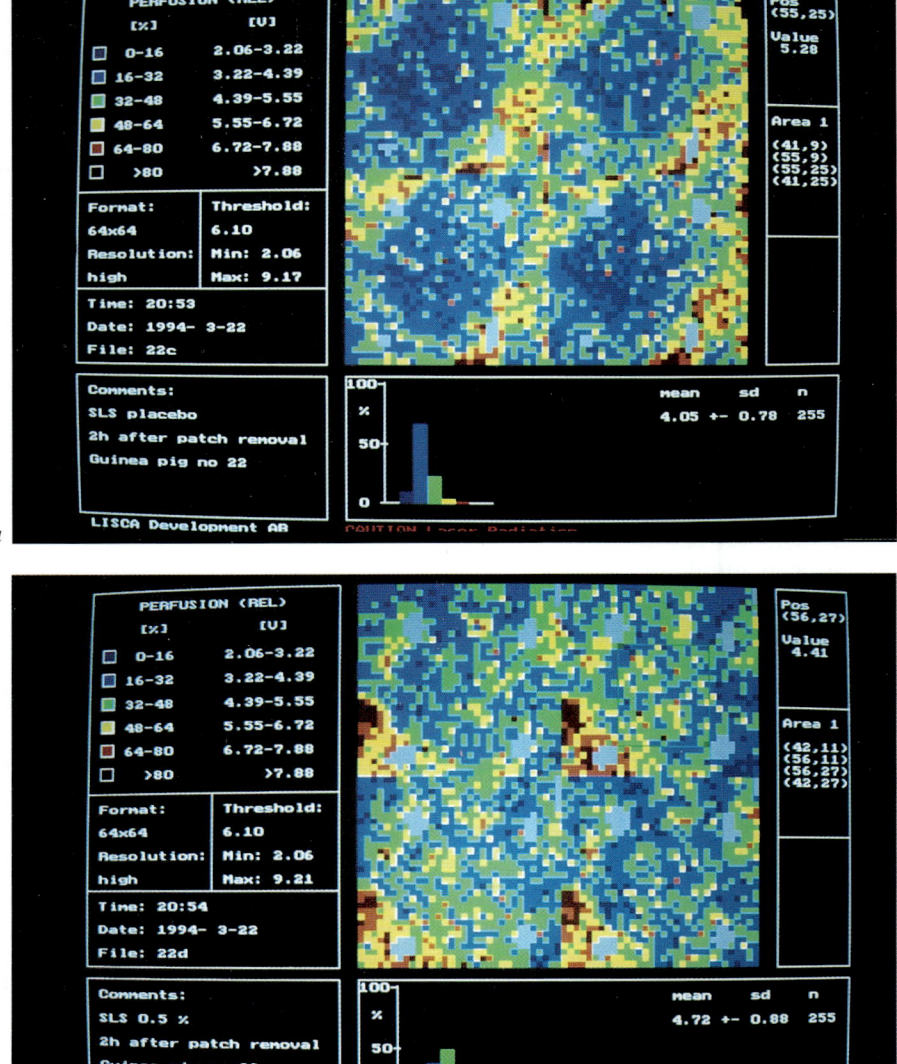

Fig. 3. Perfusion image recordings of irritative skin reactions 2 h after patch removal. For statistical evaluation, different rectangular regions of interest may be selected in the image. LDI mean values ± SD are shown in the figures. a Placebo vehicle. b SLS 0.5%. c SLS 1%. d SLS 2%.

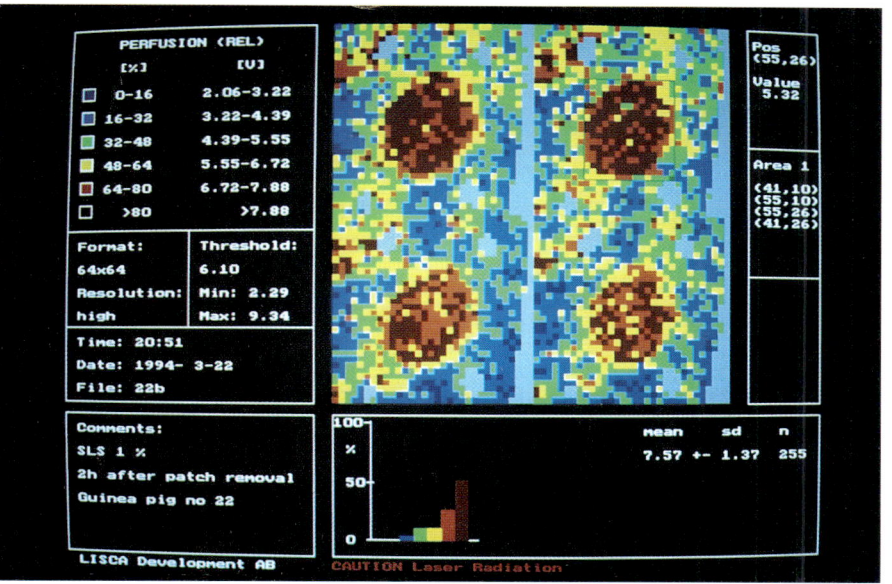

*3c*

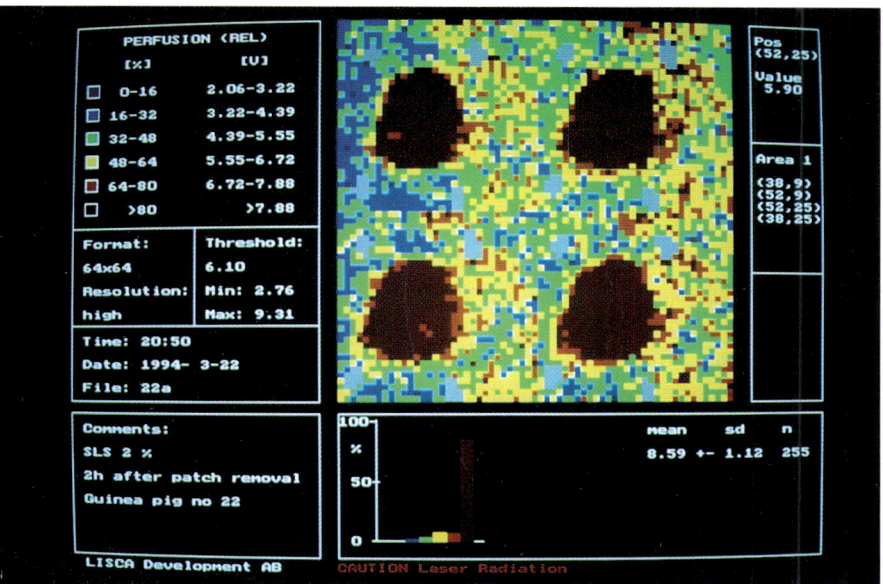

*3d*

*Table 1.* SLS patch testing: test evaluation 2 h after patch test removal (mean ± SEM, n = 4 hairless guinea pigs)

|  | Placebo | SLS 0.5% | SLS 1% | SLS 2% |
|---|---|---|---|---|
| Clinical mean score | 0.3±0.3 | 1.5±0.7 | 3.3±0.9 | 4.0±0.0 |
| LDI mean value | | | | |
| (laser Doppler image scan) | 4.89±0.67 | 5.78±0.72 | 6.71±0.73 | 8.19±0.17 |
| TEWL | 8±1 | 9±1 | 13±3 | 55±15 |

*Table 2.* All-trans retinoic acid patch testing: test evaluation 2 h after patch test removal (mean ± SEM, n = 8 hairless guinea pigs)

|  | Placebo | All-trans retinoic acid | | |
|---|---|---|---|---|
|  |  | 0.002% | 0.01% | 0.05% |
| Clinical mean score | 2.0±0.0 | 4.3±0.3 | 4.4±0.5 | 4.1±0.3 |
| LDI mean value | | | | |
| (laser Doppler image scan) | 5.92±0.27 | 7.15±0.23 | 7.00±0.37 | 7.50±0.21 |
| TEWL | 5±1 | 5±1 | 4±1 | 6±1 |

used. Still, a high concordance between clinical scoring and LDI mean value was observed. In figure 4, correlation between clinical mean scorings and LDI mean values are shown. A strong correlation between the two parameters is found giving an XY correlation of 0.915.

Measurement of TEWL is used for characterization of the skin water barrier function in irritant reactions [5–7]. SLS induced damage to the skin barrier of the hairless guinea pigs after application of 1.0 and 2.0% SLS for 24 h resulting in increased TEWL values. TEWL values were not increased compared with the mean placebo value after application of different concentrations of all-trans retinoic acid. Thus, while SLS causes barrier injuries this does not seem to be the case for all-trans retinoic acid. For all-trans retionoic acid, the cutaneous reaction is dominated by vasodilation and increased blood flow.

In conclusion, our results provide evidence that LDI employing multiple point recording and spatial mapping will become an important quantitative tool for the study of changes in microvascularization after application of irritant sub-

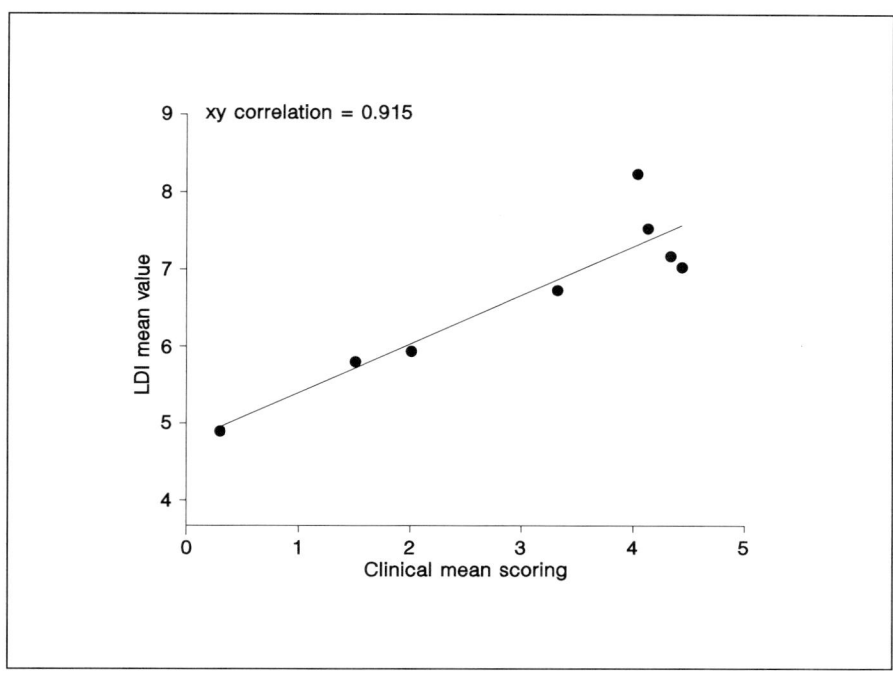

*Fig. 4.* Correlation between clinical mean scorings and LDI mean values.

stances to the skin. The LDI method overcomes the problem of spatial variation when a single-channel laser flowmeter is used and makes it possible both to quantify the intensity and the expansion of irritative skin reactions. A very high concordance is observed between this method and the subjective clinical scoring. In contrast to measurement of TEWL, this method can be used for the assessment of irritation for irritants not affecting the skin barrier.

### References

1   Wärdell, K, Jakobsson A, Nilsson GE: Laser Doppler perfusion imaging by dynamic light scatter-
    ing. IEEE Trans Biomed Eng 1993;40:309–316.
2   Wärdell K: Laser Doppler perfusion imaging. Linköping Studies in Science and technology; thesis
    308, Linköping, 1992.
3   Pinnagoda J, Tupker RA, Agner T, Serup J: Guidelines for transepidermal water loss (TEWL)
    measurement: A report from the standardization group of the European Society of Contact Derma-
    titis. Contact Derm 1990;22:164–178.

4   Wärdell K: Laser Doppler perfusion imaging. Methodology and skin applications. Linköping Studies in Science and Technology Dissertations, thesis 329, Linköping, 1994.
5   Serup J, Staberg B: Differentiation of allergic and irritant reactions by transepidermal water loss. Contact Derm 1987;16:129–132.
6   Wilhelm KP, Surber C, Maibach HI: Quantification of sodium lauryl sulphate dermatitis in man: Comparison of four techniques: Skin color reflectance, transepidermal water loss, laser Doppler flow measurement and visual scores. Arch Dermatol Res 1989;281:293–295.
7   Agner T, Serup J: Sodium lauryl sulphate for irritant patch testing: A dose-response study using bioengineering methods for determination of skin irritation. J Invest Dermatol 1990;95:543–547.
8   Draize JH, Woodard G, Calvery HO: Methods for the study of irritation and toxicity of substances applied topically to the skin and mucous membranes. J Pharmacol Exp Ther 1944;82:377–390.

Ann Fullerton, PhD, Department of Dermatological Research, Leo Pharmaceutical Products Ltd., Industriparken 55, DK–2750 Ballerup (Denmark)

Elsner P, Maibach HI (eds): Irritant Dermatitis. New Clinical and Experimental Aspects.
Curr Probl Dermatol. Basel, Karger, 1995, vol 23, pp 169–175

..............................
# Image Processing of 20 MHz B-Scan Recordings of Irritant Reactions

*Stefania Seidenari*

Department of Dermatology, University of Modena, Italy

The first studies dealing with the echographic evaluation of the responses to patch tests were carried out in 1984 and 1987 by Serup and co-workers, who employed a 15-MHz A-scanner to quantify doubtful and positive patch test reactions and to differentiate allergic and irritant responses [1, 2]. The authors demonstrated that the increase in skin thickness, due to the inflammatory edema, which appears in a positive patch test, varies with the intensification of the reaction.

Subsequently, B-scanning echography, enabling cross-sectional imaging of the skin, was developed. A positive patch test reaction appears with a thickening of the skin and a reduced echogenicity of the dermis. These aspects vary according to the strength of the inflammatory response taking place in the dermis. Looking at the computer screen, the intensity of skin reactions to irritants can be perceived at the first glance by an experienced examiner. However, in order to obtain objective information and quantification of data, computerized elaboration of images is recommended. An echographic image is a digitalized image, that means a visual representation of arrays of numbers, each pixel element or pixel representing the intensity of ultrasound which is reflected in a certain point of the image. Image processing systems achieve a rapid manipulation of portions of the image, enhancing areas of interest and removing pixels with uninteresting values. The aim of these procedures consists of improving recognition of features corresponding to tissue structures or evolutive phases of processes to be studied. and in enabling the quantification of data deriving from the image and their expression as numbers, which can be used for statistical evaluation [3, 4]. Commercially available software for elaborating 20-MHz B-scan images (Dermavision 2D, Cortex Technology) is based on the attribution of fictional values to the echoes' amplitudes. It enables the calculation of the extension of areas formed by pixels sharing similar amplitude values, after selection of amplitude bands of interest and segmentation of the image, i.e. enhancement of areas of interest. Processing of echographic images enables the assessment both of the inflammatory and of the epidermal component of irritant responses. Intervals of interest are a 0- to 30-amplitude band, marking the hyporeflecting parts of the dermis, corresponding to edema and inflammatory infiltration, and a 201–255 band, evaluating the superficial hyperreflecting part of the skin, corresponding to epidermis [5].

## Evaluation of Sodium Lauryl Sulfate-Induced Irritation

Edema and inflammatory infiltration in the dermis at sodium lauryl sulfate (SLS)-induced reactions appear echographically with a hypoechogenic area, which is subepidermal in the first phase, and spreads to the underlying dermal area as the reaction grows in intensity (fig. 1, 2). However, even in strong reactions, the hyperreflecting part of the lower dermis at volar forearm skin does not disappear completely as in allergic reactions, i.e. the inflammatory process appears more superficial. What is characteristic of SLS-induced reactions is the decrease of the superficial hyperreflecting band corresponding to the epidermis, which, in very intense reactions, disappears completely. Patch testing with SLS 0.5–5% showed that skin thickness and 0- to 30-pixel values, assessing the extension of the hyporeflecting dermal area, increase according to SLS concentration and the intensity of the inflammatory component, while a decrease of the 201- to 255-pixel values corresponds to attenuation of the epidermal reflectivity [5]. Moreover, whereas data calculated by the 0- to 30-band elaboration showed a fair correlation to clinical scoring, 201–255 values were invertly related to transepidermal water loss (TEWL).

## Evaluation of Subclinical Irritation

Instrumental assessment of irritation is particularly useful when evaluating subjects in situations where a chemical can induce a slight damage, which cannot be appreciated clinically. To evaluate subclinical irritant reactions, 63 patients affected by different types of eczematous dermatitis were challenged with a 30-min 5% SLS application on the volar aspect of the forearm [6]. In 15 cases, where no visible reactions were present at 24 h, processing of echographic images showed a decrease of the superficial hyperreflecting band, which was significant in respect to baseline values.

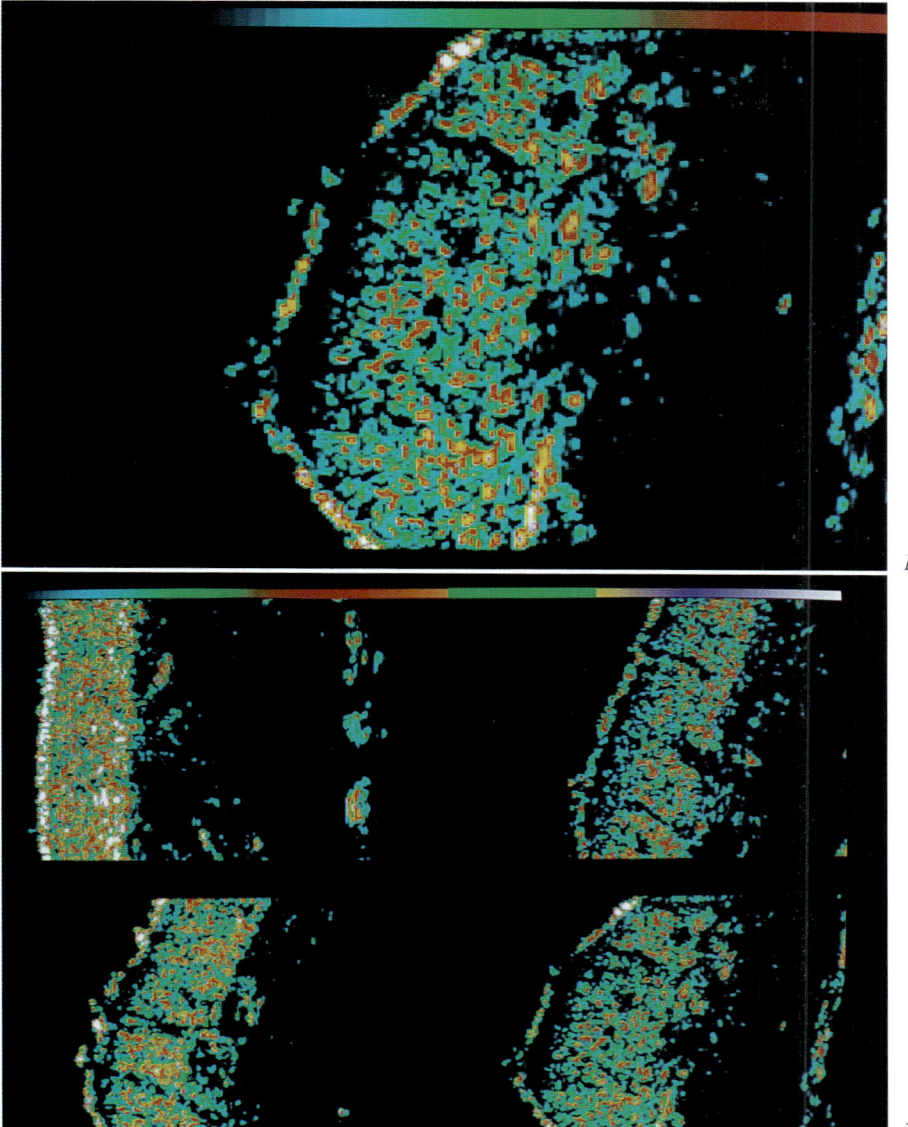

*Fig. 1.* Echographic image of a skin reaction induced by SLS: a black subepidermal area, corresponding to edema and inflammatory infiltration is clearly visible. The epidermal band is attenuated and interrupted.

*Fig. 2.* Normal volar forearm skin (upper left corner) compared to SLS-induced reactions of different intensity.

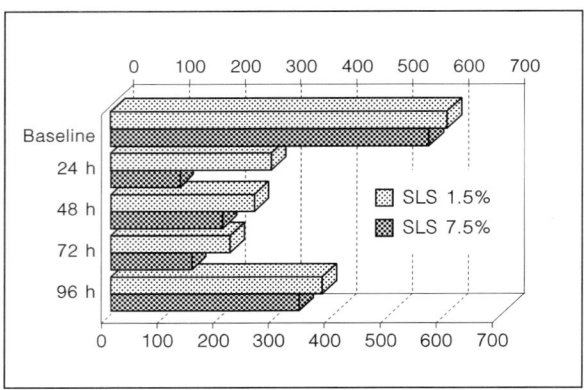

*Fig. 3.* Echographic evaluation of SLS-induced reactions in hairless mice. Attenuation of epidermal reflectivity is expressed by a decrease in the extension of areas formed by pixels reflecting within the 201–255 interval.

## Evaluation of SLS-Induced Irritation in Hairless Mice

Animal models are sometimes necessary for assessing the irritant capacity of unknown or toxic substances. The skin of hairless mice skin can be assessed by employing the same amplitude bands which are used for evaluating human skin [7]. SLS-induced irritation in mice can be appreciated by a decrease of the superficial reflectivity and an increase of the dermal echo-poor area growing with the intensity of the reaction (fig. 3). Correlations of the echographic parameters with clinical scoring and with TEWL have proved fair.

## SLS-Induced Irritation in Different Patient Groups

Ultrasound evaluation of skin damage induced by SLS can be helpful in assessing variations of skin reactivity in different patient groups. Thirty-four nickel-sensitive patients, 14 of whom were affected by atopic dermatitis, underwent a patch test with 0.05% nickel sulfate at untreated and SLS pretreated sites (5% SLS for 30 min) [8]. 30-min 5% SLS patch tests were also performed as controls. At SLS patch test sites, the intensity of the inflammatory response, as evaluated by the extension of the hypoechogenic area, was greater in atopic dermatitis patients: 0- to 30-pixel values at 24 h were significantly higher in respect to baseline in atopics alone, although at 72 h the dermal reflectivity was significantly lower in both groups (atopic dermatitis and allergic contact dermatitis patients)

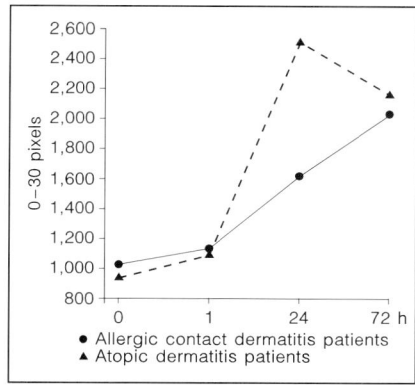

*Fig. 4.* Echographic evaluation of skin reactions induced by a 30-min 5% SLS application in atopic and nonatopic subjects. Numbers express the extension of the hypoechogenic dermal area formed by pixels reflecting within a 0–30 interval at 0, 1, 24 and 72 h.

(fig. 4). Skin barrier damage, as assessed by a 201–255 evaluation, was higher in atopics in respect to nonatopics: hyporeflectivity of the epidermis was already observable at SLS-treated skin sites at 1 h and was more evident in atopics; moreover, at 24 h a significant decrease of the 201–255 area was noticeable at SLS areas where no nickel sulfate had been applied, only in atopic subjects. These data indicate a higher reactivity to SLS in the atopic dermatitis group in respect to the allergic contact dermatitis group and a specific susceptibility of atopic skin to surfactants. Whereas statistical evaluation of the echographic data enabled a differentiation between the two subject groups, TEWL values, although characterized by a higher increase in atopics, were not able to discriminate between nickel-sensitive atopic and nonatopic subjects. Owing to a summation of immune and nonimmune mechanisms or to enhanced penetration, skin reactions to nickel sulfate were higher at SLS-pretreated skin sites in both patient groups. However, the inflammatory response, as evaluated by the extension of the 0–30 dermal area, was more intense in atopics. These data indicate that, subsequent to a slight irritant stimulus, such as a 30-min exposure to SLS, an earlier inflammatory response and a more pronounced skin damage is induced in atopics, followed by a more marked allergic reaction.

## Evaluation of Nonanoic Acid and Hydrochloric Acid-Induced Irritation

While the inflammatory component of irritant reactions has a homogeneous echographic appearance, epidermal damage caused by diverse irritants shows different sonographic patterns. By applying 40% nonanoic acid and 4% hydrochloric

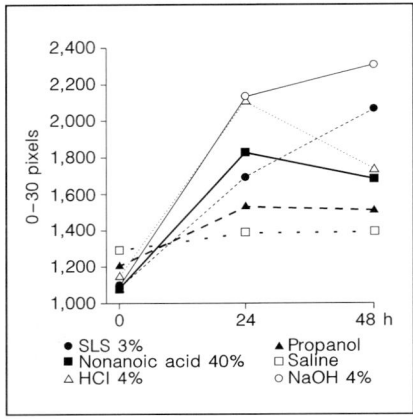

*Fig. 5.* Echographic evaluation of the inflammatory component of skin reactions induced by SLS, nonanoic acid, HCl, NaOH, propanol and saline solution. Numbers express the extension of the hypo-echogenic dermal area formed by pixels reflecting within a 0–30 interval.

*Table 1.* Echographic evaluation of irritant reactions (24-hour patch testing); comparison with other noninvasive techniques (see text, for explanations)

|  | 0–30 band | 201–255 band | Skin thickness | TEWL | Capacitance |
|---|---|---|---|---|---|
| 4% HCl | ↑↑ | ↑ | ↑ | ↑ | ↓ |
| 40% nonanoic acid | ↑ | ↗ | ↑ | ↑ | ↓↙ |
| 4% NaOH | ↑ | ↗ | ↗ | ↑↑ | ↓↓ |
| 3% SLS | ↑↑ | ↓↓ | ↑↑ | ↑↑ | ↓ |

↑ = Increase; ↑↑ = considerable increase; ↗ = slight increase; ↓ = decrease; ↓↓ = considerable decrease; ↙ = slight decrease.

acid to the skin, a thickening of the superficial hyperechogenic band was observable [9]. For these substances too, a fair correlation between clinical scores and values referring to the decrease of dermal reflectivity was present.

Figure 5 shows the intensity of the inflammatory response, as assessed by the 0- to 30-band elaboration, after a 24-hour application of diverse irritant stimuli. Responses to NaOH and to SLS are more persistent, according to the extension of the 0–30 area, with a maximum at 48 h, while reactions to nonanoic acid and HCl peak at 24 h. Decrease of the epidermal reflectivity at 24 h is typical for SLS-induced reactions, whereas other irritant substances cause either an increase or no change of the entrance echo. As already described by other authors, patch testing with irritants induces an increase in TEWL to different extents and variable degress of dehydration [10–13]. Table 1 shows the correlations between the echographic data and other instrumental parameters. Edema, as assessed by the 0- to 30-band elaboration and skin thickness measurement, is more pronounced for SLS and HCl. A decrease of the epidermal reflectivity, as assessed by the 201–255 band, is observed for SLS and an increase for HCl. Modifications of TEWL and capacitance are less pronounced for HCl and nonanoic acid, whereas NaOH 4% induces the greatest increase in TEWL accompanied by a considerable reduction in capacitance values.

In conclusion, high-frequency, high-resolution sonography associated to image analysis can contribute in characterizing skin responses to the different irritant substances, which are distinguished by a variable combination of inflammatory and epidermal aspects.

### References

1  Serup J, Staberg B, Klemp P: Quantification of cutaneous edema in patch test reactions by measurements of skin thickness with high-frequency pulsed ultrasound. Contact Derm 1984;10:88–93.
2  Serup J, Staberg B: Ultrasound for assessment of allergic and irritant patch test reactions. Contact Derm 1987;17:80–84.
3  Bamber JC, Tristam M: Diagnostic ultrasound; in Webb S (ed): The Physics of Medical Imaging. Bristol, Adam Hilger, 1988, pp 319–386.
4  Gonzalez RC, Wintz P: Digital image fundamentals; in Gonzalez RC, Wintz P (eds): Digital Image Processing. Reading, Addison-Wesley, 1987, pp 13–59.
5  Seidenari S, Di Nardo A. B scanning evaluation of irritant reactions with binary transformation and image analysis. Acta Derm Venereol (Stockh) 1992;(suppl 175):9–13.
6  Seidenari S, Belletti B: Instrumental evaluation of subclinical irritation induced by sodium lauryl sulfate. Contact Derm 1994;30:175.
7  Seidenari S, Zanella C, Pepe P: Echographic evaluation of sodium lauryl sulfate induced irritation in mice. Contact Derm 1994;30:41–42.
8  Seidenari S: Reactivity to nickel sulfate at sodium lauryl sulfate pre-treated skin sites is higher in atopics: An echographic evaluation by means of image analysis performed on 20 MHz B-scan recordings. Acta Derm Venereol (Stock) 1994;74:245–249.
9  Seidenari S: Echographic evaluation with image analysis of irritant reactions induced by nonanoic acid and hydrochloric acid. Contact Derm 1994;31:146–150.
10  Agner T, Serup J: Skin reactions to irritants assessed by non-invasive bioengineering methods. Contact Derm 1989;20:352–359.
11  Wilhelm KP, Pasche F, Surber C, Maibach HI: Sodium hydroxide-induced subclinical irritation. A test for evaluating stratum corneum barrier function. Acta Derm Venereol (Stockh) 1990;70:463–467.
12  Wilhelm KP, Maibach HI: Susceptibility to irritant dermatitis induced by sodium lauryl sulfate. J Am Acad Dermatol 1990;23:122–124.
13  Agner T: Skin susceptibility in uninvolved skin of hand eczema patients and healthy controls. Br J Dermatol 1991;125:140–146.

Stefania Seidenari, Department of Dermatology, University of Modena,
via del Pozzo 71, I–41100 Modena (Italy)

Elsner P, Maibach HI (eds): Irritant Dermatitis. New Clinical and Experimental Aspects.
Curr Probl Dermatol. Basel, Karger, 1995, vol 23, pp 176–179

# Horny Layer Thickness as Assessed Functionally Does Not Predict Sodium Lauryl Sulphate Skin Irritation

*G. Zarafonitis, P.G.M. van der Valk*

Department of Dermatology, University Hospital Nijmegen, The Netherlands

For a chemical to elicit an irritant or allergic skin reaction it has to interact with components of the skin. The horny layer is an important barrier to most chemicals, in particular by preventing chemicals penetrating the deeper layers of the skin [1,2]. So it is probable that the functional integrity of the horny layer barrier prevents interaction with components of deeper layers in a dose-dependent way. The thickness of the horny layer differs between and in individuals and may contribute to the differences in barrier function. This may explain (in part) differences in susceptibility to irritants in and between individuals [3].

If it is assumed that cellophane-tape stripping removes a layer of corneocytes of a constant thickness and that permeability constants do not vary both between or in subjects, the thickness of the horny layer can be estimated by the number of strippings needed to increase permeability according to Fick's law:

$$Js = \frac{Kp \, \Delta Cx}{\delta}$$

where $Js$ = diffusional water loss (g/m²/h); $Kp$ = permeability constant; $\Delta Cx$ = water gradient, and $\delta$ = thickness of the horny layer.

Transepidermal water loss (TEWL) is the water loss by passive diffusion through the epidermis. TEWL has been suggested as an indicator for horny layer barrier function. Stripping the skin with cellophane-tape increases transepidermal water loss according to Fick's law [4].

We studied the correlation between the number of sellotape strippings needed to remove a constant functional part of the horny layer (volar forearm) as assessed by TEWL with the response of the skin of the opposite forearm a standardized irritant stimulus. Erythema was determined semiquantitatively. A high correlation may substantiate the importance of barrier function in differences in susceptibility.

In a second experiment, we tape stripped the volar side of both arms of the same subjects to remove the same functional part as in the first experiment. The stripping eliminates differences in barrier function and may eliminate, assuming equal permeability constants, differences in horny layer thickness. Consequently, there should be fewer variations in irritant responses if barrier function is important in this respect.

## Material and Methods

We did the experiments, having obtained written informed consent, in 20 healthy volunteers, 8 men and 12 women ranging in age from 21 to 31 years of age. The study was approved by the Medical Ethics Committee. None of the subjects had a history of serious skin disease or were using topical applications or oral medicines. We tape stripped the volar side of one forearm 7 cm distally from the elbow fold with Sellotape® (Borehamwood, UK) using constant pressure. After each 5 strippings and above 30 g/m²/h after each single stripping TEWL was measured until TEWL reached 40 g/m²/h. We did a 48-hour patch test on the opposite forearm skin using sodium lauryl sulphate (SLS) 3%. (The SLS, 'electrophoresis purity', was purchased from Bio-Rad (Richmond, Calif. USA).) We pipetted 40 µl SLS on a 'silver' patch (Van der Bend, Brielle, The Netherlands), which was applied to the skin and fastened with Leukosilk (Beiersdorf, Hamburg, Germany). We measured TEWL with a Tewameter TM 210 (Courage & Khazaka, Köln, Germany) following the guidelines of the standardization group of the European Society of Contact Dermatitis [5].

Skin irritation was assessed by visual scoring on a 0–4 scale for erythema (0 = none, 1 = slight, 2 = moderate, 3 = severe, 4 = fiery). In the second experiment, in which the same volunteers participated 4 weeks later, we sellotape stripped the volar side of both forearms 7 cm distally from the elbow fold until TEWL reached 40 g/m²/h in the same manner as in the first experiment. We did a 48-hour patch test of the stripped skin on one forearm to SLS 0.03% and pipetted 40 µl SLS on the patch. The patch test procedure and the reading of skin irritation was the same as in the first experiment. In both experiments, we performed the TEWL measurements at a room temperature of 20°C and a relative humidity between 60 and 70%.

We used Pearson's correlation coefficient for calculating correlation with a one-tail probability of 5% taken as the level of significance.

## Results

The results are presented in tables 1 and 2. We found large interindividual differences between the number of strippings needed to increase TEWL to 40 g/m²/h. Correlation between the number of strippings and erythema and TEWL was small and not significant ($r = 0.12$ and $r = 0.15$ respectively).

Table 1. Correlation between erythema and TEWL of skin exposed for 24 h to 3% SLS with the number of strippings needed to increase TEWL of unexposed skin till 40 g/m²/h

| Erythema | 1.35 ± 0.67 | r = 0.12* n.s. |
| TEWL, g/m²/h | 43.97 ± 16.60 | r = 0.15 n.s. |
| Strips | 29.25 ± 12.41 | |

Mean scores ± SD.
* Pearson's coefficient of correlation. n.s. = Not significant (p > 0.05).

Table 2. Erythema and TEWL of stripped skin exposed for 24 h to 0.03% SLS and TEWL of unexposed skin 24 h after stripping till 40 g/m²/h

| Eythema | 1.55 ± 0.69 |
| TEWL1, g/m²/h | 25.91 ± 5.31 |
| TEWL2, g/m²/h | 25.04 ± 3.81 |
| Strips | 23.05 ± 6.04 |

Mean scores ± SD.
TEWL1 = Unexposed stripped skin; TEWL2 = exposed stripped skin.

In both experiments, we found large variations in erythema scores, and in the first experiment erythema correlated significantly with the TEWL of the SLS exposed skin (r = 0.86). Unexpectedly, in the second experiment we found no correlation between the erythema scores and TEWL of the SLS-exposed stripped skin (r < 0.01).

## Discussion

The number of strips needed to increase TEWL to 40 g/m²/h was not predictive of skin irritation. Exposure of stripped skin did not diminish variation in skin reaction between subjects. This may be explained by varying permeability constants being inversely related to horny layer thickness. In other words, a thin horny layer may provide a better barrier quality than a thick one. Or perhaps sello-tape stripping removes layers with varying thickness inversely related to horny layer thickness. In other words, a thin horny layer may be more difficult to strip than a thicker one. However, we hypothesize that differences in horny layer barrier function of healthy skin are not – or only slightly – responsible for differences in susceptibility to acute irritant reactions. Differences in inflammatory control mechanisms may be more important than differences in barrier function This in contrast with chronic exposure, in which dysregulation of epidermal growth and differentiation may be an important factor [6].

A good correlation between erythema scores and the increase of TEWL of *unstripped* exposed skin was shown, in contrast with the lack of correlation between erythema scores and increase of TEWL of *stripped* exposed skin. It may be concluded that the increase of TEWL and the inflammatory response (erythema) after exposure to a surfactant, although often associated, is not causally related.

### References

1   Dugard PH: Skin permeability theory in relation to measurements of percutaneous absorption in toxicology; in Marzulli FH, Maibach HI (eds): Dermatotoxicology and Pharmacology. Washington, Hemisphere, 1977, pp 525–550.
2   Malten KE: Thoughts on irritant contact dermatitis. Contact Derm 1981;7:238–247.
3   Björnberg A: Skin reactions to primary irritants in patients with hand eczema; thesis, Oscar Isac-sons Tryckeri, University of Götenborg, 1968.
4   van der Valk PGM, Maibach HI: A functional study of the skin barrier to evaporative water loss by means of repeated cellophane-tape stripping. Clin Exp Dermatol 1990;15:180–182.
5   Pinnagoda J, Tupker RA, Agner T, Serup J. Guidelines for transepidermal water loss (TEWL) measurement. Contact Derm 1990;22:164–178.
6   van der Valk PGM: Irritant contact dermatitis: A persistent dysregulation of epidermal growth? Br J Dermatol 1993;129:208–209.

Dr. P.G.M. van der Valk, Department of Dermatology, University Hospital Nijmegen,
PO Box 9101, NL–6500 HB Nijmegen (The Netherlands)

Elsner P, Maibach HI (eds): Irritant Dermatitis. New Clinical and Experimental Aspects.
Curr Probl Dermatol. Basel, Karger, 1995, vol 23, pp 180–186

..........................

# Occlusion Does Not Influence the Repair of the Permeability Barrier in Human Skin

*Julia Welzel, Klaus-Peter Wilhelm, Helmut H. Wolff*

Department of Dermatology and Venerology, Medical University of Lübeck,
Germany

Disturbances of the skin permeability barrier with increased transepidermal water loss (TEWL) lead to stimulation of DNA and lipid synthesis for regeneration of the barrier [1, 2]. Investigation in hairless mice showed that artificial restoration of the barrier by occlusion resulted in a reduction and a delay of epidermal barrier repair activities [3]. This in in contrast to the clinically documented positive effects of occlusive treatment for conditions with a disturbed permeability barrier, e.g. psoriasis, chronic hand eczema and superficial wounds [4, 5]. It was therefore our objective to evaluate the effect of occlusion on skin permeability barrier repair in human skin under controlled experimental conditions.

## Materials and Methods

*Occlusive Dressings*
Five membranes with different characteristics were used (fig. 1):
(1) OpSite: semipermeable, medium water vapor permeability, self-adhesive (Smith & Nephew Medical Ltd., Hull, UK).
(2) Tegaderm: semipermeable, medium water vapor permeability, self-adhesive (3M, St. Paul, Minn., USA).
(3) Gore-Tex: semipermeable, high water vapor permeability, nonadhesive (W.L. Gore & Associates GmbH, Putzbrunn, Germany).
(4) Polyethylene foil: nonpermeable, low water vapor permeability, nonadhesive (Melitta GmbH, Minden-Dützen, Germany).
(5) Varihesive: hydrocolloid dressing, low vapor permeability, self-adhesive (ConvaTec, Munich, Germany).

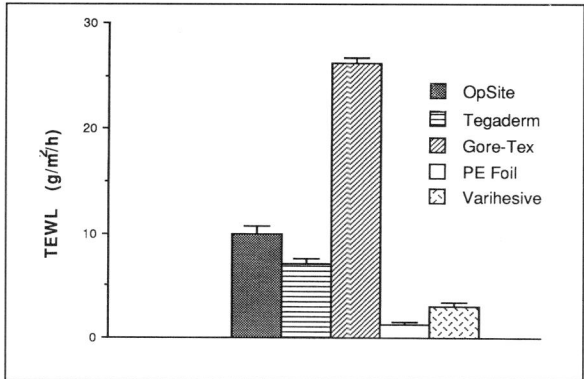

*Fig. 1.* Occlusive dressings. TEWL values of different membranes (1 h occlusion of a Petri dish filled with 10 ml water).

### Skin Irritation Model

Ten healthy female volunteers (age 22–43) were treated with sodium lauryl sulfate (SLS) to disturb stratum corneum barrier function. 200 µl of 20 m$M$ aqueous solutions were applied to six areas on the volar side of the forearm for 24 h (1st day) by means of occlusive polypropylene chambers (diameter: 25 mm, Hilltop, Cincinnati, Ohio, USA). Thereafter, treated skin was covered with the occlusive dressings for 2 × 23 h (2nd and 3rd day), or remained untreated (control). Treatments rotated on the test sites between volunteers to avoid an anatomical selection bias. Measurements of TEWL and electrical capacitance were done before irritation (day 1) and 1 h after irritation (day 2). On the 3rd and 4th days, measurements were made 1 h after removal of the membranes. The last measurements were done on the 5th day, i.e. after 24 h without any treatment.

### Tape Stripping Model

On five areas on the volar side of the forearm the stratum corneum was partially removed by repeated (30–40 ×) tape stripping, controlled by an increase of TEWL up to at least 30 g/m$^2$/h. Four areas were covered with OpSite, Gore-Tex, polyethylene foil or Varihesive for 2 × 23 h, one area was left unoccluded (control). TEWL and electrical capacitance were measured on the 1st day before and after tape stripping, on the 2nd and 3rd day 1 h after removal of the foils and on the 4th and 5th day without any treatment.

### Measurements and Statistics

We used the Tewameter (TM 210, Courage & Khazaka, Cologne, Germany) for TEWL measurements and the Corneometer (CM 820, Courage & Khazaka) for electrical capacitance measurements.

Statistical analysis was done using the Friedman rank variance analysis with consecutive Wilcoxon/Wilcox comparisons [6]. The values of the different occlusive treatments were compared with the untreated control.

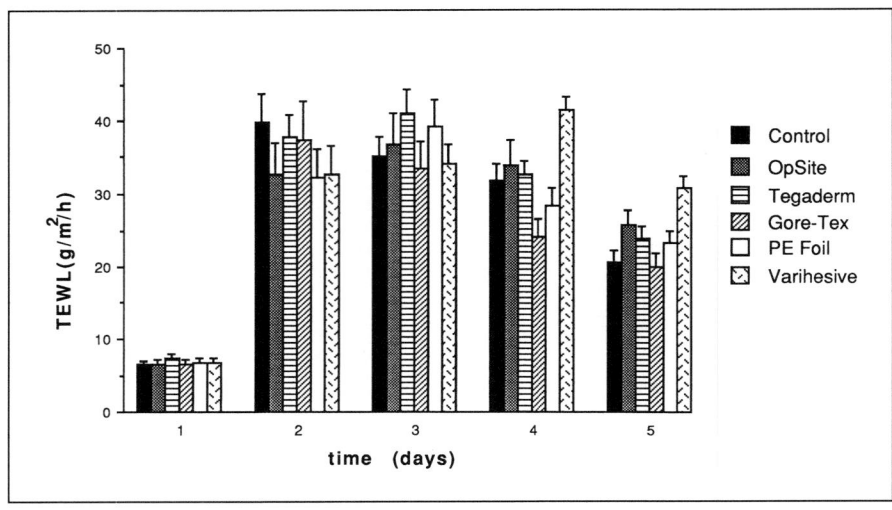

*Fig. 2.* TEWL after irritation with SLS on day 1, occlusive treatment with different membranes on days 2 and 3.

### Results

*SLS Irritation*

As expected, SLS irritation induced a marked barrier perturbation showed by a sixfold increase of TEWL after 1 day of treatment with the irritant. The barrier repair period which we observed over the next 4 days was accompanied by a stepwise decrease of TEWL which still remained elevated on the last day of investigation.

This decrease of TEWL was not influenced by the different postirritation treatments and could also be observed under occlusive conditions. There was no significant difference between the occlusive dressings and the unoccluded control site (fig. 2).

After irritation electrical capacitance increased slightly from about 60 IU (instrumental units) up to 70 IU before decreasing to 40 IU on day 5 at control sites.

This sign of desiccation of the skin surface was not as marked in the occluded areas. The electrical capacitance of these sites was higher than the control, reaching statistical significance at the 4th day for OpSite, Tegaderm and Varihesive ($p < 0.05$) (fig. 3). It stayed at higher levels even 1 h after removal of the membranes and also on the last day without any occlusive treatment.

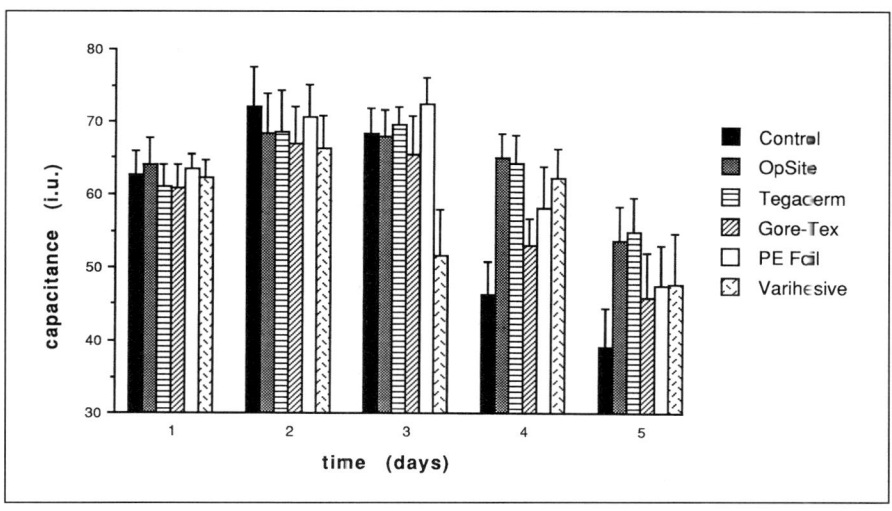

*Fig. 3.* Electrical capacitance after irritation with SLS on day 1, occlusive treatment on days 2 and 3.

*Tape Stripping*

Directly after tape stripping TEWL increased from about 7 g/m²/h up to 40 g/m²/h as a sign for partial removal of the stratum corneum. The regeneration of the barrier could be observed by a rapid decrease of TEWL on the next days down to 14 g/m²/h on day 5 at control sites. After tape stripping TEWL stayed at high levels during occlusive treatment with OpSite and Varihesive. The difference was significant on the 4th day compared to the unoccluded control (p < 0.05). TEWL decreased only after ending the occlusive treatment. Treatment with Gore-Tex and polyethylene foil did not influence the TEWL decrease despite their different water vapor permeability (fig. 4). Electrical capacitance increased after tape stripping from 60 IU up to 75 IU for 2 days, then decreased again down to 60 IU. Values for electrical capacitance were slightly higher under occlusion without a statistically significant difference to the control (fig. 5).

## Discussion

SLS irritation results in a barrier disturbance without a total loss of stratum corneum, accompanied by parakeratosis, spongiosis and exocytosis followed by acanthosis [7].

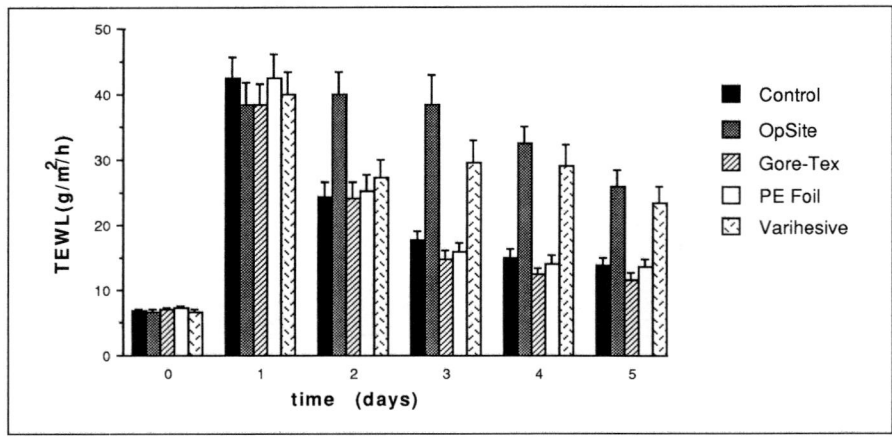

*Fig. 4.* TEWL after tape stripping on day 1, occlusive treatment on days 1 and 2.

After 1 day of SLS irritation occlusive treatment with different dressings had no effect on TEWL decrease as a sign of barrier repair. The restoration of the disturbed permeability barrier showed the same course of time irrespective of the treatment. There was no delay in repair as demonstrated by a decrease of TEWL, irrespective of the water vapor permeability of the membranes which was high for Gore-Tex, lower for OpSite and Tegaderm and very low for the hydrocolloid dressing and the polyethylene foil.

Also, self-adhesive membranes, i.e. OpSite, Tegaderm and Varihesive, showed no effect on TEWL decrease in SLS-irritated skin.

Tape stripping removes layers of stratum corneum and leads to hyperproliferation of the epidermis without severe inflammation [8]. Some of the occlusive dressings – Gore-Tex and polyethylene foil – showed no effect upon barrier recovery after tape stripping whereas other membranes like OpSite and Varihesive delayed the repair. TEWL stayed at high levels during occlusion with these membranes. It started to decrease only after ending the treatment. This effect was irrespective of the water vapor permeability of the dressings but seemed to depend on the self-adhesive capability of OpSite and Varihesive. The adhesive seemed to remove newly formed stratum corneum after tape stripping. This additional 'tape stripping-like' effect may explain the delay in TEWL decrease observed only with self-adhesive membranes. The newly formed thin stratum corneum after tape stripping seemed to be less resistant against additional disturbances than the SLS-irritated horny layer with a normal or increased thickness of stratum corneum.

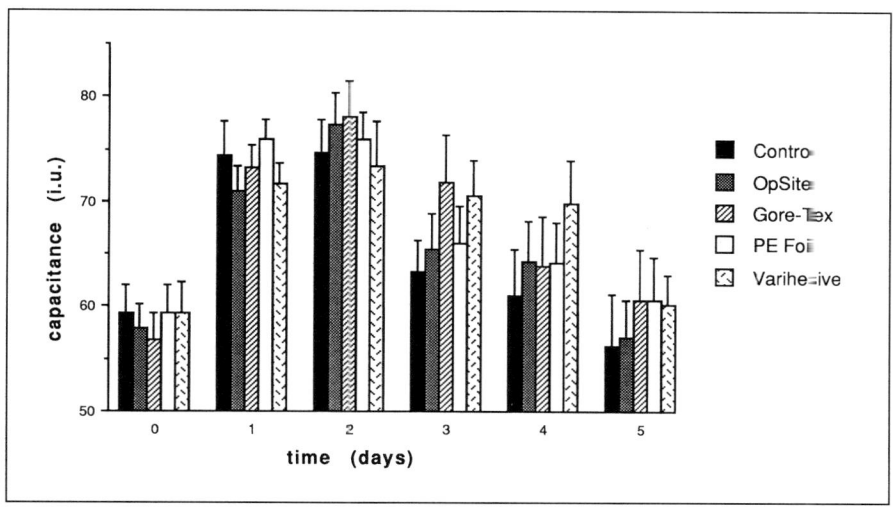

*Fig. 5.* Electrical capacitance after tape stripping, occlusive treatment on days 1 and 2.

Occlusion increased skin hydration as demonstrated by higher values for electrical capacitance. This hyperhydration could be noticed even 1 h after removal of the dressings and remained at the last day without occlusive treatment. Thus, occlusion improved the visual impression of irritated skin with a less rough and scaly surface.

In summary, our results indicate that occlusion does not delay barrier repair in human skin. This clearly contrasts previously reported data obtained in hairless mice which showed a further deterioration of the barrier function by occlusion [1]. The function of the epidermal permeability barrier is known to have an influence upon DNA and lipid synthesis [1–3] and lamellar body secretion [9]. Perturbation of the barrier leads to an increase which is interpreted as barrier repair activity. Occlusive treatment after barrier disruption as an artificial restoration of the barrier inhibits this burst of DNA and lipid synthesis.

Chronic and repeated barrier damage as in hand eczema leads to an excessive and pathologic hyperproliferation which may itself be responsible for high TEWL. Occlusion is known to have an antiproliferative and antiinflammatory effect upon hyperproliferative skin disease [4, 10] and seems to modulate, not stop, the barrier repair activities.

Investigations of the effect of occlusion should consider some side effects of the treatment. Repeated removal of self-adhesive dressings may cause an additional disruption of the barrier. The hyperhydration of the stratum corneum should be considered when determinating the TEWL, because we found high values for electrical capacitance even 1 h after removal of the membranes. The irritant should be washed off after application before starting the occlusive treatment to avoid further influence of the irritant on the barrier function. These side effects may have been responsible for some investigations which have seen a further barrier perturbation under occlusive conditions [11].

Furthermore, the stratum corneum of the hairless mouse skin was found to be more fragile and significantly less resistant against hydration than human skin [12, 13]. Thus, it was shown by Bond and Barry [12, 13] that prolonged occlusion itself was damaging the stratum corneum barrier of hairless mouse but not that of human skin. A predamaged stratum corneum either by surfactant exposure or by tape stripping will be even more susceptable to this hydration damage.

Our results suggest that the hairless mouse may be a limited model for the study of barrier function of human skin.

## References

1 Proksch E, Feingold KR, Mao-Qiang M, Elias PM: Barrier function regulates epidermal DNA synthesis. J Clin Invest 1991;87:1668–1673.
2 Proksch E: Regulation der epidermalen Permeabilitätsbarriere durch Lipide und durch Hyperproliferation. Hautarzt 1992;43:331–338.
3 Proksch E, Feingold KR, Elias PM: Epidermal HMG CoA Reductase activity in essential fatty acid deficiency: Barrier requirements rather than eicosanoid generation regulate cholesterol synthesis. J Invest Dermatol 1992;99:216–220.
4 Volden G: Successful treatment of chronic skin diseases with clobetasol propionate and a hydrocolloid occlusive dressing. Acta Derm Venerol (Stockh) 1992;72:69–71.
5 Mennen U, Wiese A: Fingertip injuries management with semiocclusive dressing. J Hand Surg 1993;18:416–422.
6 Sachs L: Angewandte Statistik. Berlin, Springer, 1991, pp 673–675.
7 Willis CM, Stephens CJM, Wilkinson JD: Preliminary findings on the pattern of epidermal damage induced by irritants in man; in Frosch PJ, Dooms-Goossens A, Lachapelle JM, Rycroft RJG, Scheper RJ (eds): Current Topics in Contact Dermatitis. Berlin, Springer, 1989, pp 42–45.
8 Hashimoto Y, Tsutsui M, Iizuka H: Flow cytometric analysis of pig epidermal keratinocytes: Effects of tape stripping. J Dermatol Sci 1992;4:193–201.
9 Menon GK, Feingold KR, Elias PM: Lamellar body secretory response to barrier disruption. J Invest Dermatol 1992;98:279–289.
10 Friedmann JJ: Management of psoriasis vulgaris with a hydrocolloid occlusive dressing. Arch Dermatol 1987;123:1046–1052.
11 Van der Valk PGM, Maibach HI: Post-application occlusion substantially increases the irritant response to the skin to repeated short-term sodium lauryl sulfate (SLS) exposure. Contact Derm 1989;21:335–338.
12 Bond JR, Barry BW: Limitations of hairless mouse skin as a model for in vitro permeation studies through human skin: Hydration damage. J Invest Dermatol 1988;90:486–489.
13 Bond JR, Barry BW: Hairless mouse skin is limited as a model for assessing the effect of penetration enhancers in human skin. J Invest Dermatol 1988;90:810–813.

Dr. Julia Welzel, Department of Dermatology and Venerology, Medical University of Lübeck, Ratzeburger Allee 160, D–23538 Lübeck (Germany)

Elsner P, Maibach HI (eds): Irritant Dermatitis. New Clinical and Experimental Aspects.
Curr Probl Dermatol. Basel, Karger, 1995, vol 23, pp 187–197

..........................
# Efficacy of Barrier Creams

*A.M. Grunewald*[a], *M. Gloor*[a], *W. Gehring*[a], *P. Kleesz*[b, 1]

[a] Dermatological Clinic of Karlsruhe Municipal Hospital (Director: Prof. *M. Gloor*), and
[b] Arbeitsmedizinischer Dienst der Berufsgenossenschaft Nahrungsmittel und
Gaststätten (Directors: Prof. *S. Radandt*, Dr. *R. Grieshaber*), Mannheim, Germany

Chronic irritant dermatitis due to repetitive contact with surfactants cutting oils, acids and alkalis is one of the most common occupational diseases. The role of repetitive irritation due to surfactants onto different skin function parameters (corneometry, laser Doppler flow, pH-metry, transepidermal water loss (TEWL), colorimetry, sebumetry) has been evaluated recently by our group with in vivo experiments in a human model [1]. As a protective device gloves are not always suitable or effective [2]. Barrier creams offer an alternative, and recently published data showed a detectable efficacy of commercially available barrier creams in vivo [3]. Referring to our human model of irritant contact dermatitis by repetitive washing with a surfactant [1], we were investigating the irritation-inhibiting effect of 3 different commercially available barrier creams.

## Materials and Methods

### Experimental Subjects
The investigations were carried out on 15 healthy volunteer subjects for each barrier cream tested. The criterion for exclusion was the presence of skin diseases. The age of the subjects was between 19 and 35 years. The gender distribution was 67% female and 33% male. All subjects avoided cosmetic treatment of the areas of skin under investigation with soaps or creams for the 2 days preceding the investigation and for the duration of the experiment.

### Wash Solution
Sodium lauryl sulfate (SLS; Texapon K 12®, Caesar & Loretz GmbH, Hilden, Germany) was used as an irritant in this study. The selected concentration of SLS was 0.01 mol/l. Referring to former studies, this concentration is low enough to prevent an acute irritative

[1] Particular thanks are due to Frau Barbara Wasik for carrying out the investigations.

response but still has sufficient washing activity [1]. The pH of the selected surfactant solutions was adjusted to 6.5.

### Wash Procedure, Time Points of Measurement and Application of Barrier Creams

To avoid interactions between the different test methods (e.g. pH and TEWL or corneometry and sebumetry), we decided to have 4 independent test areas on each forearm. These 4 areas were marked by means of a stencil, so that measurements could be carried out at identical sites before and after the wash procedure. The order of tests was corneometry, sebumetry, pH-metry, TEWL, laser Doppler and chromametry. For the standardized washing procedure on days 1 and 8, a 250-gram foam rubber roller soaked in the wash solution was used for washing. This was rolled forwards and backwards over the forearm 50 times, with no pressure being applied. The roller was always resoaked with wash solution after it had been rolled 15 times. Subsequently, the skin was rinsed for 2 min in still water at body temperature. Measurements were performed on day 1 to obtain baseline values and 30-, 60-, 120- and 180-min values after carrying out the standardized washing procedure. The subjects were then asked to wash their arms for 3 min 5 times daily for 7 days in a nonstandardized fashion. Between washings, the appropriate arm was treated (5 times daily) with one barrier cream as determined on the first day. After day 7, the subjects were not allowed to wash anymore for exactly 12 h and measurements were performed on day 8 to obtain baseline values and 30-, 60-, 120- and 180-min values after carrying out a final standardized washing procedure.

### Barrier Creams

The following barrier creams were tested: Tactosan® (Stockhausen, Krefeld, Germany), water in oil emulsion containing Eucoriol (sodium bischlorophenyl sulfamine), petrolatum, mineral oil, ozokerite, glycerol oleate, lanolin alcohol, magnesium sulfate, fragrance and Bronopol.

Saniwip® (Basotherm GmbH, Biberach an der Riss, Germany), a water in oil emulsion, containing ozokerite, castor oil, glycerol oleate, vaseline, isopropylmyristate, lactic acid and lanolin wax.

Marly Skin® (Quinta GmbH, Freiburg, Germany), appearing as a creamy foam, contains stearic acid, propylene glycol, glycerol, dimethylpolysiloxan and deodorant. Its protection is supposed to be due to the formation of an invisible stearin net layer.

### Measurement Procedures

Experiments were performed during the same time of the year (autumn 1993). All measurements took place in a climatized room of our clinic. Room temperature was 22 ± 1°C and relative humidity 45 ± 5%. The subjects had to acclimatize to these conditions for 30 min before the experiment. The following methods of investigation were used:

*Corneometry.* Corneometry (Corneometer CM 820, Courage & Khazaka, Cologne, Germany) measures the capacitative resistance of the skin, which correlates with the moisture content of the deeper layers of the stratum corneum. The values obtained are relative. Eight measurements were carried out at closely spaced sites, and their mean was calculated [4, 5].

*Measurement of Skin-Surface Lipids with Sebumetry.* The principle of sebumetry (Sebumeter SM 810, Courage & Khazaka), is based on changes in the light transmission of a transparent film that arises from the adhesion of lipids on pressing the film against the skin. Results are given as μg lipids per cm² skin surface. Measurements were carried out at 3 closely spaced sites, and the mean of the individual values was taken [6].

*Determination of Transepidermal Water Loss with an Evaporimeter.* TEWL is a measure of the barrier function of the stratum corneum [7]. The measurements are made by 2 moisture sensors in an open-topped chamber (Tevameter TM 210, Courage & Khazaka). Temperature dependencies of the measured values are automatically allowed for. The measured signals are only recorded when a constant value has been detected for 30 s (variation allowed was $<0.5$ g/m$^2$ h).

*Measurement of the pH of the Skin Surface.* A pH electrode (Skin pH meter PH 900, Courage & Khazaka) was used. Measurements were carried out in a droplet of distilled water [8]. The mean of 3 measurements was registered.

*Laser Doppler Flowmetry with a Commercial Instrument.* The circulation in the vessels of the subpapillary capillary plexus was measured (Laser Doppler Flowmeter Pf2, Perimed, Sweden). Relative values were obtained. The value was only recorded after a constant signal had been observed for at least 1 min [9].

*Measurement of Skin Color with a Chromameter.* This method (Chromameter CR 200, Minolta, Ahrensburg, Germany) allows the determination of the brightness, color tone and color saturation of the skin [1, 3]. Our measurements were based on the L*a*b* system. The color coordinate a* measures the red coloration of the skin, and was used for assessment. Each value is the average of 3 individual measurements. Relative values were also obtained in this investigation.

*Statistical Assessment.* Comparison of values was carried out between the barrier cream treated and untreated site as well as between day 1 and day 8. The Wilcoxon test for paired differences was used in all cases. Since values show a normal distribution, they are expressed as mean $\pm$ SD. With a high number of statistical tests being performed, false-positive significances on the 0.05% level can be expected. Therefore, single p values of only $p < 0.05$ in the time course of measurements (i.e. 30, 60, 120 and 180 min measurements) were not interpreted as being significant and are presented graphically in brackets (fig. 1–3). Nevertheless, multiple p values of $<0.05$ in one time course of one test series confirm the correctness of the statistical assessment. Therefore, they were interpreted as a significant result.

(For fig. 1, 2, 3 see pages 190–192.)

*Fig. 1.* Efficacy of Tactosan treatment against deterioration of skin function parameters due to repetitive washings with 0.01 mol/l SLS. Bioengineering method values before (0 min) and after (30, 60, 120 and 180 min) standardized washing procedure on days 1 and 8. * $p < 0.01$, + $p < 0.05$, (+) $p < 0.05$, but not interpreted as significant (for explanation, see 'Statistical assessment'). For SD and exact values, see table 1.

*Fig. 2.* Efficacy of Saniwip treatment against deterioration of skin function parameters due to repetitive washings with 0.01 mol/l SLS. Bioengineering method values before (0 min) and after (30, 60, 120 and 180 min) standardized washing procedure on days 1 and 8. * $p < 0.01$, + $p < 0.05$, (+) $p < 0.05$, but not interpreted as significant (for explanation, see 'Statistical assessment'). For SD and exact values, see table 1.

*Fig. 3.* Efficacy of Marly Skin treatment against deterioration of skin function parameters due to repetitive washings with 0.01 mol/l SLS. Bioengineering method values before (0 min) and after (30, 60, 120 and 180 min) standardized washing procedure on days 1 and 8. * $p < 0.01$, + $p < 0.05$, (+) $p < 0.05$, but not interpreted as significant (for explanation, see 'Statistical assessment'). For SD and exact values, see table 1.

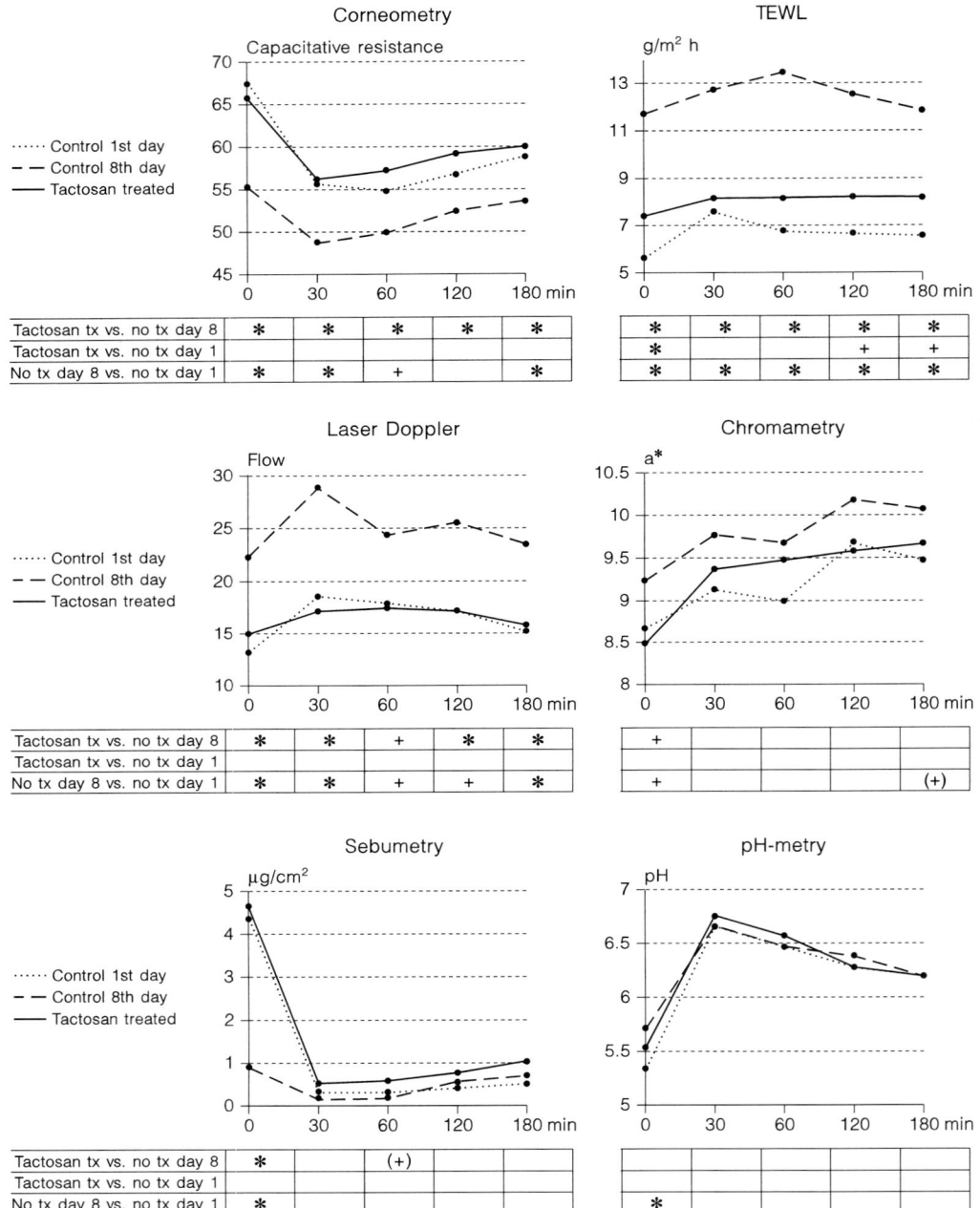

Fig. 1. (For legend see p. 189.)

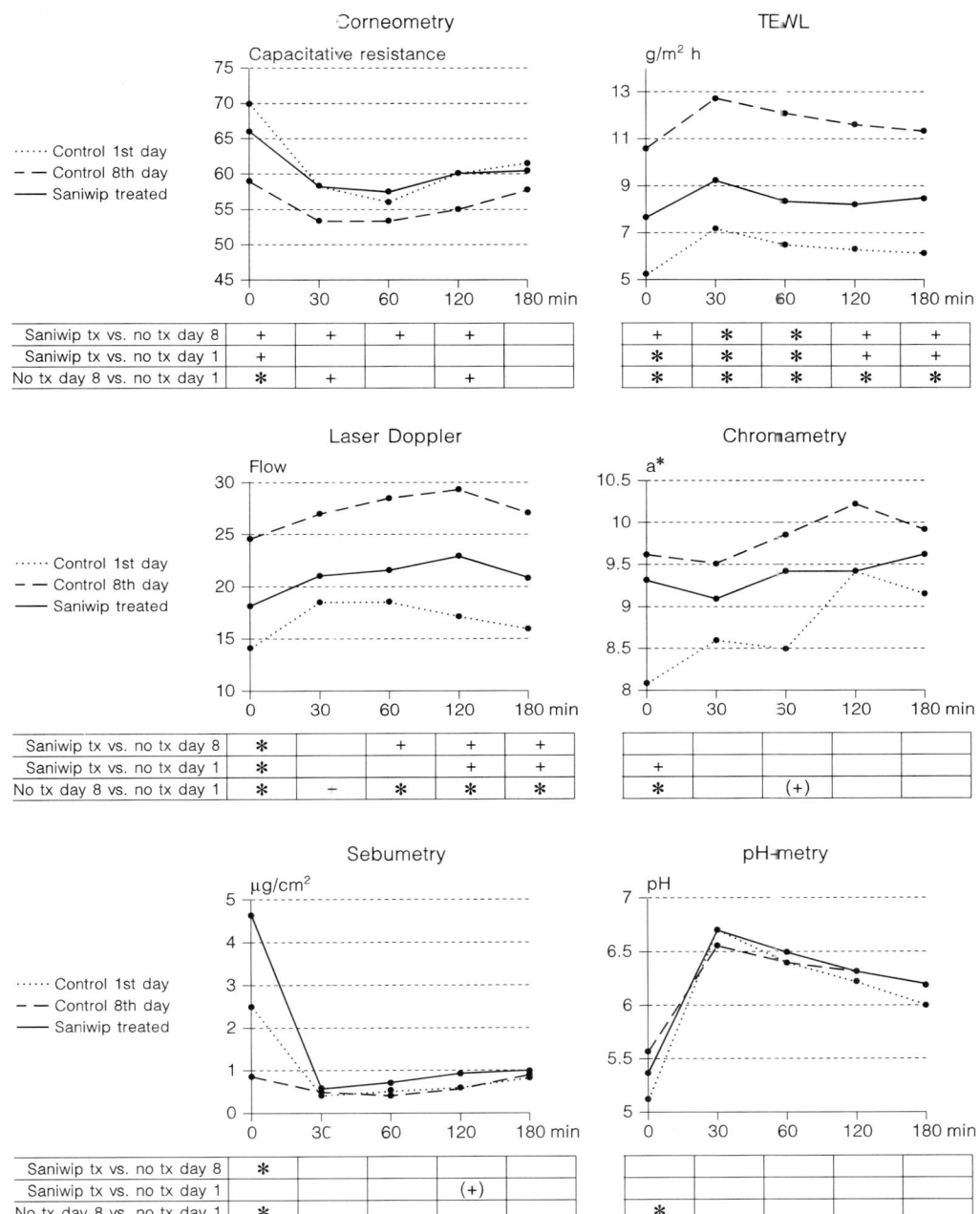

## Corneometry

### Capacitative resistance

···· Control 1st day
– – Control 8th day
—— Saniwip treated

| | | | | |
|---|---|---|---|---|
| Saniwip tx vs. no tx day 8 | + | + | + | + |
| Saniwip tx vs. no tx day 1 | + | | | |
| No tx day 8 vs. no tx day 1 | * | + | | + |

## TEWL

g/m² h

| | | | | |
|---|---|---|---|---|
| + | * | * | + | + |
| * | * | * | + | + |
| * | * | * | * | * |

## Laser Doppler

Flow

···· Control 1st day
– – Control 8th day
—— Saniwip treated

| | | | | |
|---|---|---|---|---|
| Saniwip tx vs. no tx day 8 | * | | + | + | + |
| Saniwip tx vs. no tx day 1 | * | | | + | + |
| No tx day 8 vs. no tx day 1 | * | – | * | * | * |

## Chromametry

a*

| | | | | |
|---|---|---|---|---|
| | | | | |
| + | | | | |
| * | | (+) | | |

## Sebumetry

μg/cm²

···· Control 1st day
– – Control 8th day
—— Saniwip treated

| | | | | |
|---|---|---|---|---|
| Saniwip tx vs. no tx day 8 | * | | | | |
| Saniwip tx vs. no tx day 1 | | | | (+) | |
| No tx day 8 vs. no tx day 1 | * | | | | |

## pH-metry

pH

| | | | | |
|---|---|---|---|---|
| | | | | |
| | | | | |
| * | | | | |

*Fig. 2.* (For legend see p. 189.)

---

Efficacy of Barrier Creams

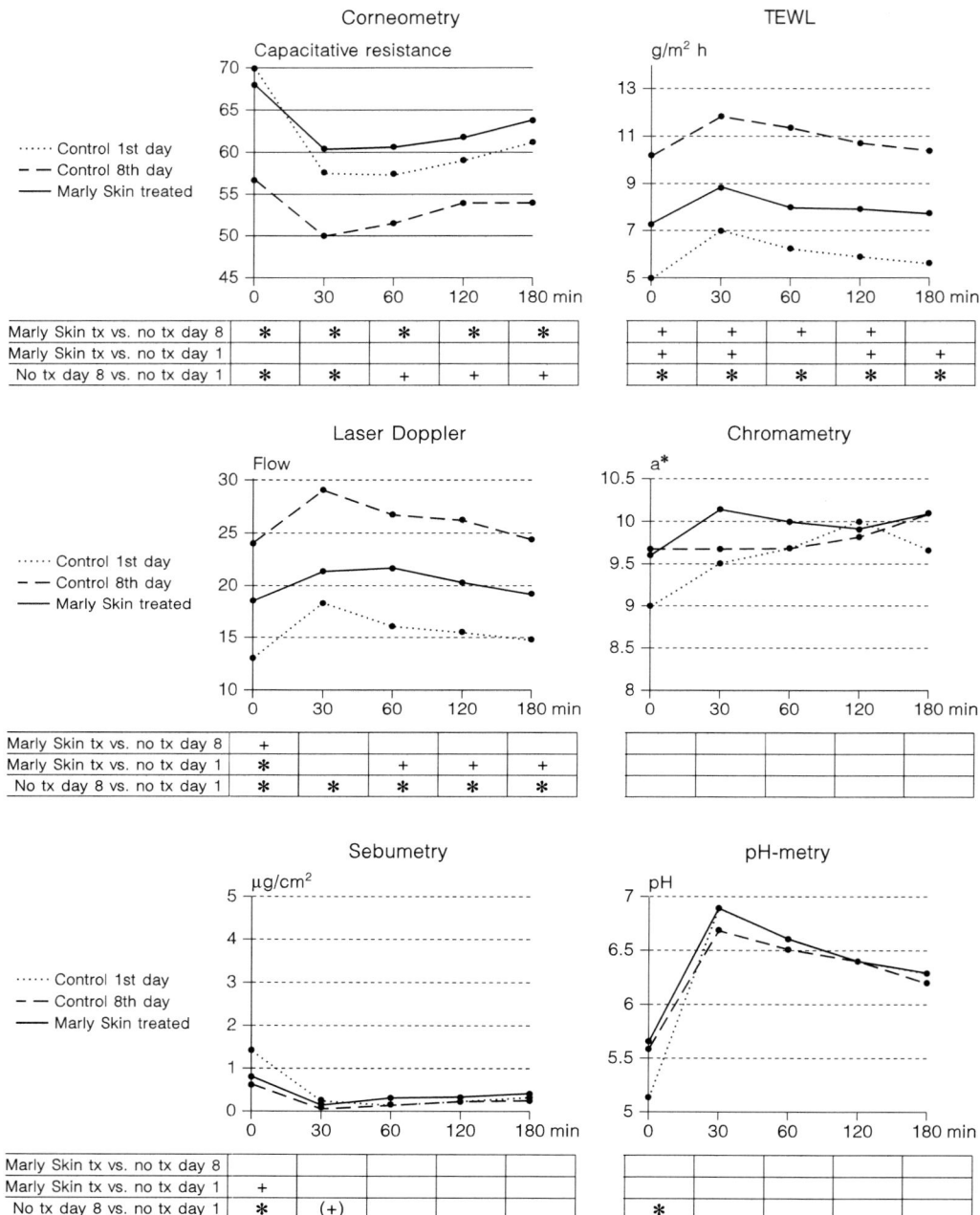

**Corneometry**

Capacitative resistance

····· Control 1st day
– – Control 8th day
—— Marly Skin treated

| | 0 | 30 | 60 | 120 | 180 min |
|---|---|---|---|---|---|
| Marly Skin tx vs. no tx day 8 | * | * | * | * | * |
| Marly Skin tx vs. no tx day 1 | | | | | |
| No tx day 8 vs. no tx day 1 | * | * | + | + | + |

**TEWL**

g/m² h

| | 0 | 30 | 60 | 120 | 180 min |
|---|---|---|---|---|---|
| | + | + | + | + | |
| | + | + | | + | + |
| | * | * | * | * | * |

**Laser Doppler**

Flow

····· Control 1st day
– – Control 8th day
—— Marly Skin treated

| | 0 | 30 | 60 | 120 | 180 min |
|---|---|---|---|---|---|
| Marly Skin tx vs. no tx day 8 | + | | | | |
| Marly Skin tx vs. no tx day 1 | * | | + | + | + |
| No tx day 8 vs. no tx day 1 | * | * | * | * | * |

**Chromametry**

a*

| | 0 | 30 | 60 | 120 | 180 min |
|---|---|---|---|---|---|
| | | | | | |
| | | | | | |
| | | | | | |

**Sebumetry**

µg/cm²

····· Control 1st day
– – Control 8th day
—— Marly Skin treated

| | 0 | 30 | 60 | 120 | 180 min |
|---|---|---|---|---|---|
| Marly Skin tx vs. no tx day 8 | | | | | |
| Marly Skin tx vs. no tx day 1 | + | | | | |
| No tx day 8 vs. no tx day 1 | * | (+) | | | |

**pH-metry**

pH

| | 0 | 30 | 60 | 120 | 180 min |
|---|---|---|---|---|---|
| | | | | | |
| | | | | | |
| * | | | | | |

*Fig. 3.* (For legend see p. 189.)

## Results

The results are given in detail in table 1 and figures 1–3. All 3 barrier creams seem to prevent skin dehydration to a high degree. The barrier creams are able to reduce but not completely prevent barrier damage as evaluated by TEWL measurements. Skin inflammation as evaluated by laser Doppler flow seems to be completely prevented by Tactosan and is reduced only by Saniwip and Marly Skin. Concerning sebumetry, chromametry and pH-metry, the results were less distinct. Skin surface lipids were significantly reduced after repetitive washings on the 8th day. This effect was completely prevented by Tactosan and partially prevented by Saniwip. Marly Skin did not show any inhibition of skin delipidisation after repetitive washings. Concerning pH and colorimetry, no significant protective effect of the three barrier creams was detectable.

*Table 1.* Efficacy of Tactosan, Saniwip and Marly Skin treatment against deterioration of skin function parameters due to repetitive washings with 0.01 mol/l SLS: for significance, see fig. 2–4

|  | 8th day barrier cream treated | Control 1st day mean (SD) | Control 8th day mean (SD) |  | 8th day barrier cream treated | Control 1st day mean (SD) | Control 8th day mean (SD) |
|---|---|---|---|---|---|---|---|
| *Corneometry* |  |  |  | *Chromametry* |  |  |  |
| Tactosan |  |  |  | Tactosan |  |  |  |
| 0 min | 65.9 (8.2) | 68.1 (7.2) | 55.6 (5.9) | 0 min | 8.5 (1.6) | 8.7 (1.2) | 9.3 (1.4) |
| 30 min | 56.2 (8.7) | 55.8 (5.8) | 48.6 (6.1) | 30 min | 9.4 (1.6) | 9.1 (1.2) | 9.8 (1.6) |
| 60 min | 57.1 (7.4) | 54.7 (5.8) | 49.7 (6.1) | 60 min | 9.5 (1.3) | 9.0 (1.3) | 9.7 (1.6) |
| 120 min | 58.9 (7.0) | 56.6 (5.2) | 52.3 (6.4) | 120 min | 9.6 (1.4) | 9.7 (1.2) | 10.2 (1.7) |
| 180 min | 60.0 (7.6) | 58.5 (6.3) | 53.4 (4.9) | 180 min | 9.7 (1.1) | 9.5 (1.3) | 10.1 (1.4) |
| Saniwip |  |  |  | Saniwip |  |  |  |
| 0 min | 66.1 (6.6) | 70.2 (6.5) | 59.3 (7.4) | 0 min | 9.3 (1.9) | 8.1 (1.4) | 9.6 (1.9) |
| 30 min | 58.6 (4.2) | 58.5 (6.0) | 53.6 (7.2) | 30 min | 9.1 (1.9) | 8.6 (1.9) | 9.5 (2.2) |
| 60 min | 57.3 (4.1) | 56.1 (6.5) | 53.4 (6.6) | 60 min | 9.4 (1.9) | 8.5 (1.8) | 9.8 (1.8) |
| 120 min | 59.9 (4.8) | 60.1 (5.2) | 55.1 (6.5) | 120 min | 9.4 (1.5) | 9.4 (1.7) | 10.2 (1.6) |
| 180 min | 60.5 (4.8) | 61.1 (7.3) | 57.8 (7.2) | 180 min | 9.6 (1.8) | 9.2 (1.4) | 9.9 (1.6) |
| Marly Skin |  |  |  | Marly Skin |  |  |  |
| 0 min | 67.8 (9.0) | 69.8 (6.4) | 56.6 (9.9) | 0 min | 9.6 (2.5) | 9.0 (1.5) | 9.7 (2.1) |
| 30 min | 60.3 (7.1) | 57.3 (6.6) | 50.0 (8.2) | 30 min | 10.1 (2.3) | 9.5 (1.6) | 9.7 (2.7) |
| 60 min | 60.5 (6.5) | 57.2 (6.5) | 51.1 (8.3) | 60 min | 10.0 (1.8) | 9.7 (1.5) | 9.7 (2.4) |
| 120 min | 61.5 (6.1) | 58.7 (6.6) | 53.1 (8.7) | 120 min | 9.9 (1.6) | 10.0 (1.6) | 9.8 (2.1) |
| 180 min | 63.5 (6.5) | 61.1 (6.8) | 53.9 (9.6) | 180 min | 10.1 (1.4) | 9.7 (1.6) | 10.1 (1.9) |

(Table 1 continued next page.)

*Table 1* (continued)

| | 8th day barrier cream treated | Control 1st day mean (SD) | Control 8th day mean (SD) | | 8th day barrier cream treated | Control 1st day mean (SD) | Control 8th day mean (SD) |
|---|---|---|---|---|---|---|---|
| *TEWL* | | | | *Sebumetry* | | | |
| Tactosan | | | | Tactosan | | | |
| 0 min | 7.5 (2.4) | 5.6 (2.3) | 11.8 (3.7) | 0 min | 4.6 (4.2) | 4.4 (6.0) | 0.9 (0.7) |
| 30 min | 8.3 (2.3) | 7.6 (2.4) | 12.8 (4.1) | 30 min | 0.5 (0.7) | 0.3 (0.4) | 0.2 (0.4) |
| 60 min | 8.3 (2.7) | 7.0 (2.3) | 13.5 (4.7) | 60 min | 0.6 (0.5) | 0.3 (0.4) | 0.2 (0.4) |
| 120 min | 8.4 (2.7) | 6.9 (2.0) | 12.6 (4.3) | 120 min | 0.8 (0.6) | 0.4 (0.4) | 0.5 (0.4) |
| 180 min | 8.4 (2.6) | 6.8 (2.2) | 12.0 (4.2) | 180 min | 1.0 (0.8) | 0.5 (0.4) | 0.7 (0.4) |
| Saniwip | | | | Saniwip | | | |
| 0 min | 7.6 (2.0) | 5.2 (1.2) | 10.5 (3.4) | 0 min | 4.6 (5.0) | 2.5 (2.1) | 0.8 (0.6) |
| 30 min | 9.3 (2.3) | 7.2 (1.4) | 12.7 (3.8) | 30 min | 0.5 (0.5) | 0.3 (0.4) | 0.4 (0.5) |
| 60 min | 8.3 (2.0) | 6.5 (1.2) | 12.1 (3.6) | 60 min | 0.7 (0.5) | 0.5 (0.5) | 0.4 (0.4) |
| 120 min | 8.2 (2.3) | 6.3 (1.5) | 11.6 (4.4) | 120 min | 0.9 (0.6) | 0.6 (0.4) | 0.6 (0.6) |
| 180 min | 8.4 (2.6) | 6.1 (1.2) | 11.3 (4.2) | 180 min | 1.0 (0.9) | 0.8 (0.5) | 0.9 (0.7) |
| Marly Skin | | | | Marly Skin | | | |
| 0 min | 7.4 (2.8) | 4.9 (1.4) | 10.2 (4.7) | 0 min | 0.7 (1.0) | 1.4 (1.0) | 0.5 (0.8) |
| 30 min | 8.9 (2.6) | 7.1 (2.1) | 11.9 (4.9) | 30 min | 0.2 (0.3) | 0.3 (0.3) | 0.1 (0.3) |
| 60 min | 8.1 (2.8) | 6.4 (1.9) | 11.4 (4.9) | 60 min | 0.4 (0.6) | 0.2 (0.2) | 0.2 (0.3) |
| 120 min | 8.0 (2.5) | 6.0 (1.5) | 10.8 (4.9) | 120 min | 0.4 (0.6) | 0.2 (0.3) | 0.3 (0.6) |
| 180 min | 7.9 (2.8) | 6.8 (1.7) | 10.5 (4.8) | 180 min | 0.5 (1.0) | 0.3 (0.3) | 0.4 (0.4) |
| *Laser Doppler* | | | | *pH-metry* | | | |
| Tactosan | | | | Tactosan | | | |
| 0 min | 15.1 (3.4) | 13.5 (3.7) | 22.5 (8.0) | 0 min | 5.5 (0.6) | 5.3 (0.6) | 5.7 (0.6) |
| 30 min | 17.1 (4.5) | 18.5 (4.1) | 28.7 (14.8) | 30 min | 6.8 (0.4) | 6.7 (0.3) | 6.7 (0.5) |
| 60 min | 17.0 (4.4) | 17.5 (3.5) | 24.4 (9.8) | 60 min | 6.6 (0.5) | 6.5 (0.3) | 6.5 (0.5) |
| 120 min | 17.1 (4.6) | 17.0 (3.3) | 25.3 (12.1) | 120 min | 6.3 (0.5) | 6.3 (0.4) | 6.4 (0.5) |
| 180 min | 15.7 (4.0) | 15.4 (3.9) | 23.3 (9.9) | 180 min | 6.2 (0.6) | 6.2 (0.3) | 6.2 (0.5) |
| Saniwip | | | | Saniwip | | | |
| 0 min | 18.1 (5.1) | 14.4 (2.5) | 24.6 (8.4) | 0 min | 5.4 (0.7) | 5.1 (0.7) | 5.6 (0.6) |
| 30 min | 21.1 (6.5) | 18.8 (3.1) | 27.0 (11.3) | 30 min | 6.7 (0.4) | 6.7 (0.4) | 6.6 (0.4) |
| 60 min | 21.7 (7.4) | 18.5 (2.6) | 28.4 (11.3) | 60 min | 6.5 (0.5) | 6.4 (0.4) | 6.4 (0.5) |
| 120 min | 22.9 (9.7) | 17.3 (2.6) | 29.1 (14.3) | 120 min | 6.3 (0.5) | 6.2 (0.4) | 6.3 (0.5) |
| 180 min | 20.6 (7.5) | 16.0 (2.4) | 27.2 (11.9) | 180 min | 6.2 (0.5) | 6.0 (0.5) | 6.2 (0.6) |
| Marly Skin | | | | Marly Skin | | | |
| 0 min | 18.7 (5.3) | 13.0 (2.9) | 23.8 (8.2) | 0 min | 5.6 (0.8) | 5.1 (0.5) | 5.7 (0.4) |
| 30 min | 21.3 (6.7) | 18.3 (4.4) | 29.0 (14.9) | 30 min | 6.9 (0.3) | 6.9 (0.4) | 6.7 (0.3) |
| 60 min | 21.5 (7.2) | 16.1 (3.6) | 26.8 (10.8) | 60 min | 6.6 (0.3) | 6.6 (0.3) | 6.5 (0.3) |
| 120 min | 20.2 (6.8) | 15.5 (3.1) | 26.1 (12.7) | 120 min | 6.4 (0.3) | 6.4 (0.4) | 6.4 (0.3) |
| 180 min | 19.2 (6.4) | 14.8 (3.0) | 24.5 (9.3) | 180 min | 6.3 (0.3) | 6.3 (0.4) | 6.2 (0.4) |

*Table 2.* Comparison of the efficacy of Tactosan, Saniwip and Marly Skin

| | Tactosan | Saniwip | Marly Skin |
|---|---|---|---|
| **TEWL** | | * | |
| 0 min | −4.5 (1.8) | * −2.9 (1.8) | −2.8 (2.6) |
| 30 min | −4.5 (2.9) | + −3.4 (2.5) | −2.8 (3.1) |
| 60 min | −5.1 (4.2) | + −3.8 (1.9) | −3.3 (2.6) |
| 120 min | −4.1 (3.7) | −3.5 (2.5) | −2.9 (2.8) |
| 180 min | −3.6 (3.2) + | −2.9 (2.2) * | −2.7 (2.9) |
| **Corneometry** | | + | + |
| 0 min | 10.4 (7.5) | 6.8 (6.9) + | 11.1 (3.7) |
| 30 min | 7.6 (8.5) * | 4.9 (7.2) * | 10.3 (4.9) |
| 60 min | 7.4 (7.3) + | 3.9 (4.8) * | 9.4 (4.8) |
| 120 min | 6.7 (8.3) | 4.7 (5.0) | 8.5 (5.0) |
| 180 min | 6.6 (8.3) | 2.7 (6.2) | 9.6 (4.7) |

Difference between barrier-cream-treated site and untreated site on day 8 after 7 days of repetitive washings before (0 min) and after (30, 60, 120 and 180 min) standardized washing procedure. Statistically significant differences between two barrier creams are stated with $^+ p < 0.05$ and $* p < 0.01$. Laser Doppler, colorimetry, sebumetry and pH values did not show any statistical significant differences between the three barrier creams tested (data not shown in the table).

Comparing the efficiency of the three barrier creams tested, only the corneometry and TEWL values show significant differences (table 2). Tactosan and Marly Skin produce a stronger inhibition of skin dehydration after 8 days of repetitive washing than Saniwip There is no statistically significant difference between Marly Skin-treated and Tactosan-treated sites concerning corneometry values. Inhibition of barrier damage due to repetitive washings is more efficient with Tactosan than with Saniwip or Marly Skin treatment.

## Discussion

The efficacy of various barrier creams has been evaluated with very different methods in the past [10–13]. With barrier cream treatment and irritant stress having been only once, those investigations were not comparable to physiological conditions. Normal working place conditions lead to cumulative skin damage due to repetitive irritation [1, 14]. To our knowledge, there is only one published investigation that evaluated barrier cream efficacy against repetitive irritation by surfactants [3]. Nevertheless, SLS was used at a concentration of 5–10% which is known to be toxic already at a single application [15] and which seems to be higher than found physiologically. Additionally SLS irritation was done by a 30-min patch occlusion which enhances irritation in an unphysiological manner. In our investigation, we were able to evaluate the efficacy of all three barrier creams against skin irritation due to repetitive washing with a surfactant. Nevertheless, the study also showed that the three barrier creams are not able to offer complete skin protection. Protection of barrier function seems to be most effective with the use of Tactosan, protection of skin dehydration seems to be most effective with the use of Tactosan and Marly Skin. The demonstration of a partial protection of skin function by barrier creams confirms the results of Frosch et al. [3] and concerning different irritants, of Gloor et al. [2] and Schwanitz [12]. Nevertheless, the example of Marly Skin indicates that with Frosch's and Schwanitz's experimental arrangements irritation by occlusion is too intense to be able to show any more protective efficacy of Marly Skin. Comparing the predicative sense of the different bioengineering methods being used, our investigations additionally let us assume that corneometry, TEWL and laser Doppler flow measurements are especially sensitive to detect barrier cream efficacy.

For future investigations about barrier creams, the importance of acceptance has to be taken into consideration. Evaluations under real working-place conditions are necessary to definitely determine the efficacy of barrier creams.

## References

1   Grunewald AM, Gloor M, Gehring W, Kleesz P: Damage to the skin by repetitive washing. Contact Derm, in press.
2   Gloor M, Köhler St, Gehring W: Hautschutzmassnahmen, ein Stiefkind prophylaktisch dermatologischer Tätigkeit. Z Hautkr 1991;66:201–207.
3   Frosch PJ, Schulze-Dirks A, Hoffmann M, Axthelm I, Kurte A: Efficacy of skin barrier creams. Contact Derm 1993;28:94–100; 29:74–77; 29:113–118.
4   Mosler U: Hautfeuchtigkeitsmessung – kein Problem mit dem Corneometer CM 420. Parf Kosm 1983;64:375–379.
5   Triebskorn A, Gloor M, Greiner F: Comparative investigations on the water content of the stratum corneum using different methods of measurement. Dermatologica 1983;167:64–69.

6 Tronnier W, Kohn-Bussius H: Zur Brauchbarkeit optischer Methoden für die Bestimmung des Hautoberflächenfettes. Kosmetologie 1972;4:230–234.

7 Pinnagoda J, Tupker RA, Agner T, Serup J: Guidelines for transepidermal water loss measurement. Contact Derm 1990;22:164–178.

8 Post A, Gloor M, Gehring W: Über den Einfluss der Hautwaschung auf den Haut-pH-Wert. Derm Monatsschr 1992;178:216–222.

9 Mahmoud G, Lachapelle JM: Evaluation of the protective value of an antisolvent gel by laser Doppler flowmetry and histology. Contact Derm 1985;13:14–19.

10 Gehring W, Dördelmann C, Gloor M: Effektivitätsnachweis von Hautschutzpräparaten. Allergologie 1994;17:97–101.

11 Mahmoud G, Lachapelle JM, van Neste D: Histological assessment of skin damage by irritants: Its possible use in the evaluation of a 'barrier cream'. Contact Derm 1984;11:179–185.

12 Schwanitz HJ: Hautphysiologische Untersuchungsmethoden. TW Dermatologie 1993;23 19–24.

13 Tronnier H: Methodische Ansätze zur Prüfung von Hautschutzmitteln. Dermatosen 1993;41:100–107.

14 Malten KE: Thoughts on irritant contact dermatitis. Contact Derm 1981;7:238–247.

15 Willis Carolyn M, Stephens Catherine JM, Wilkinson JD: Experimentally-induced irritant contact dermatitis: Determination of optimum irritant concentrations. Contact Derm 1988;18:20–24.

Dr. med. A.M. Grunewald, Städt. Hautklinik, Moltkestrasse 14, D-76133 Karlsruhe (Germany)

Elsner P, Maibach HI (eds): Irritant Dermatitis. New Clinical and Experimental Aspects.
Curr Probl Dermatol. Basel, Karger, 1995, vol 23, pp 198–206

..........................

# Effect of N-Acetylcysteine, an Inhibitor of Tumor Necrosis Factor, on Irritant Contact Dermatitis in the Human

*Florence Pasche-Koo, Ana Arechalde, Jean-François Arrighi,*
*Conrad Hauser*

Department of Dermatology, University Hospital, Geneva, Switzerland

Cytokines are involved in the cutaneous response to the application of irritants. In the mouse, tumor necrosis factor-α (TNF-α) has been shown to be induced in epidermal basal keratinocytes and in the dermal infiltrate after induction of contact hypersensitivity reactions and irritant reactions [1]. Seemingly, TNF plays a key role in these two types of reactions, and administration of antibodies to TNF or soluble TNF receptors blocked irritant reaction and contact hypersensitivity in the mouse [1].

In addition, UVB has been shown to induce TNF-α release from keratinocytes in vitro [2]. Since solar exposure can cause a significant inflammatory response in the skin, TNF might be involved in the mediation of this reaction.

N-acetylcysteine (NAC) is an antioxidant reagent which inhibits the action of the nuclear transcription factor kB, which promotes the transcription of the gene for TNF [3].

Pretreatment of the skin with topical NAC has been shown to antagonize the development of irritant and contact hypersensitivity reactions induced by the application of trinitrochlorobenzene in the mouse [4].

In this study, we decided to evaluate in a human model the effect of a 10% NAC-containing cream on irritant contact dermatitis. In addition, we studied the effect of pretreatment with an NAC cream on UVB-induced erythema.

## Material and Methods

On the back of 6 healthy nonatopic volunteers, serial dilutions of sodium dodecyl sulfate (SDS), 1, 2, 5, 7 and 10% in water, and serial dilutions of dimethylsulfoxide (DMSO), 30, 50, 70, 80 and 90% in water, were applied twice on both sides of the back for 4 h. Large

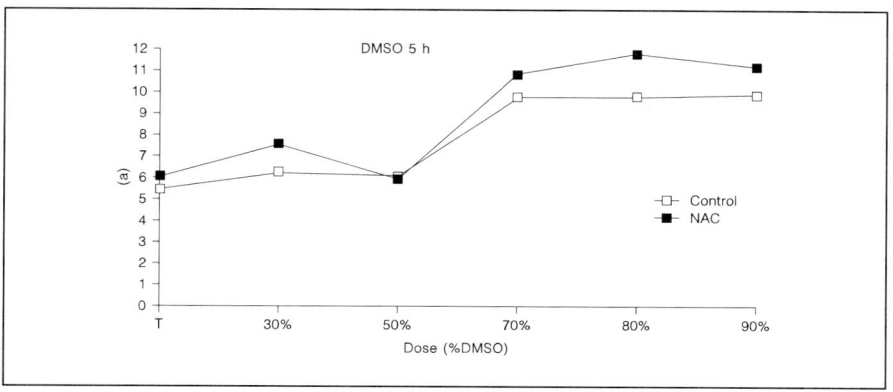

*Fig. 1.* Effect of NAC on DMSO-induced erythema.

Finn Chambers on Scanpor® were used. Pretreatment with a 10% N-acetylclysteine cream (Inpharzam, Cadempino, Switzerland) was done on the left side of the back, and the excipient cream was applied on the right side. Readings were done using a Minolta CR 300 chromameter, 3 measurements per site. Visual scoring was also done at every site of test using the following scale: 0 = no reaction; 0.5 = very weak erythema; 1 = weak erythema; 2 = strong erythema; 3 = very strong erythema; 4 = erythema and edema (papule); 5 = erythema, edema and vesicles; 6 = erythema, edema and vesicles exceeding the limits of the test; 7 = erythema, edema, vesicles and cutaneous necrosis.

On the back of 5 nonatopic volunteers, UVB irradiation was done twice on both sides of the back. Eight surfaces of 2 cm² were irradiated with a Waldmann UV 800 lamp using a different dose in each square: 5, 10, 15, 20, 25, 30, 35 and 40 mJ/cm². Before irradiation, pretreatment with a 10% NAC cream was done on the left side of the back, and on the right side, the excipient cream was applied. Statistical analysis was done on the chromametric measures using analysis of variance for repeated measures with the Huynh-Feldt adjustment for p value [5].

### Results

Although the irritants DMSO and SDS induced dose-dependent erythema, as assessed by chromametry (fig. 1) and by visual scoring (data not shown), there were no statistically significant effects when the NAC-treated sites were compared to symmetrically located vehicle-treated sites (fig. 2–13). Similarly, UVB light-induced erythema was not affected by the NAC-containing cream (fig. 14–18).

The only significant result was the increase of the cutaneous erythema when the dose of the irritant or UVB was increased (p = 0.005 for SDS; p < 0.001 for DMSO; p = 0.0094 for UVB).

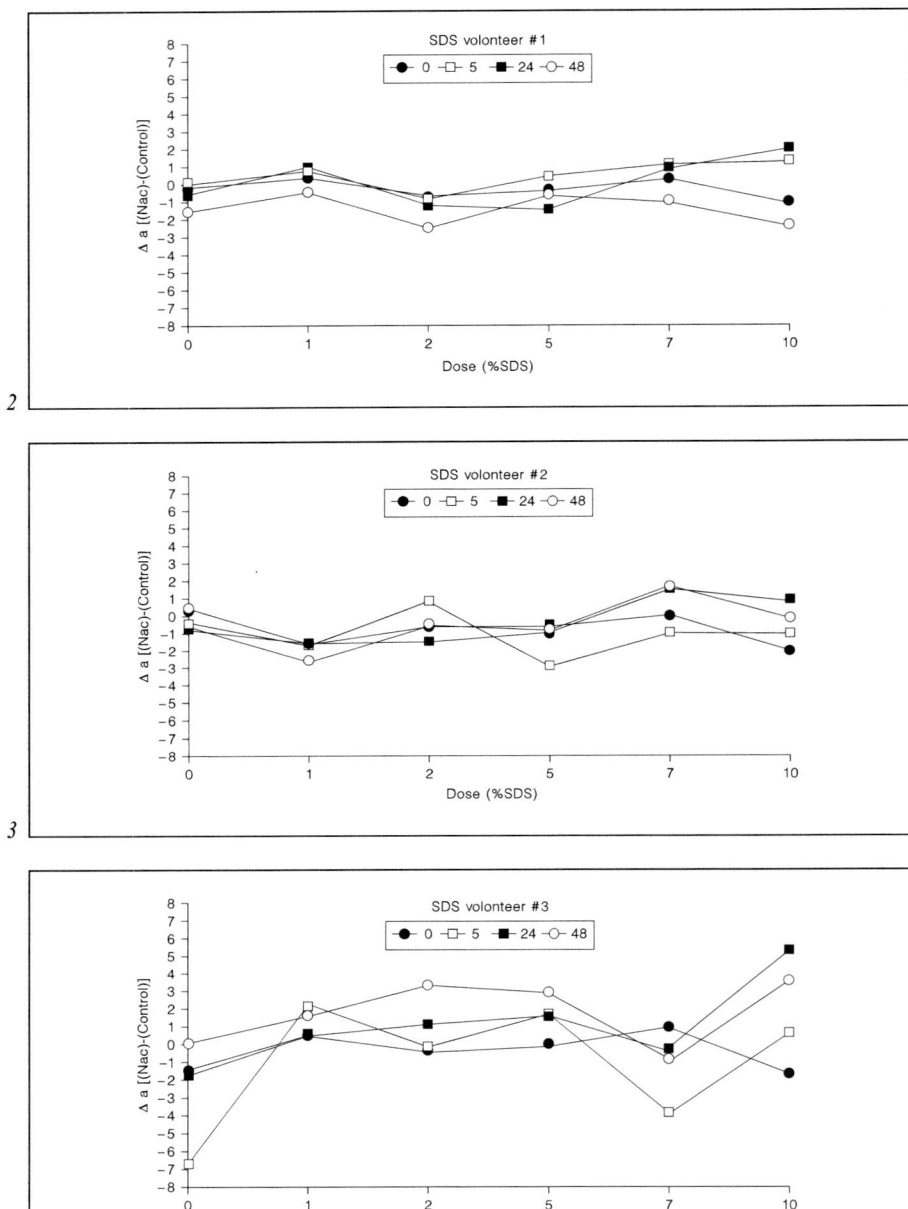

*Fig. 2–13.* Difference of erythema induced by SDS or DMSO with and without NAC treatment.

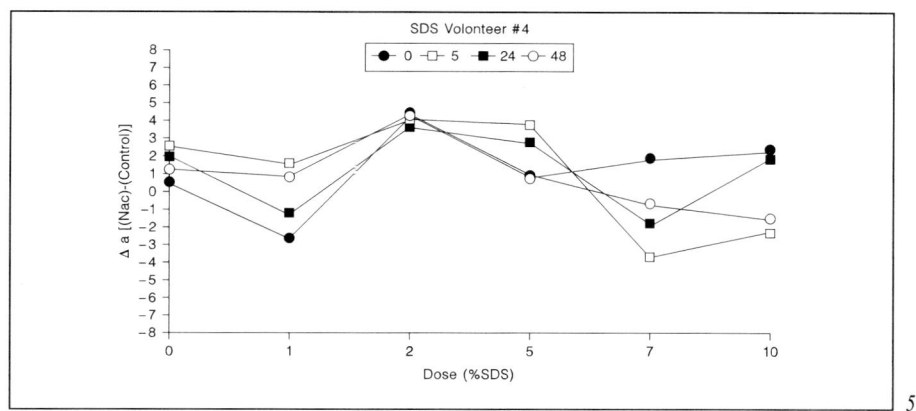

5

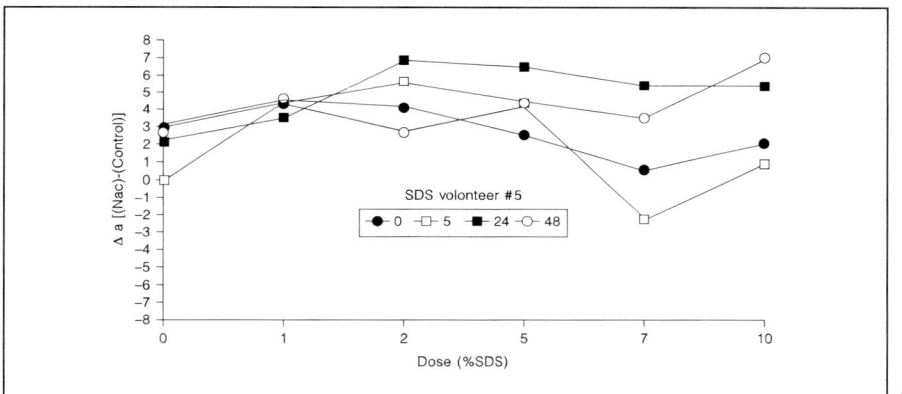

6

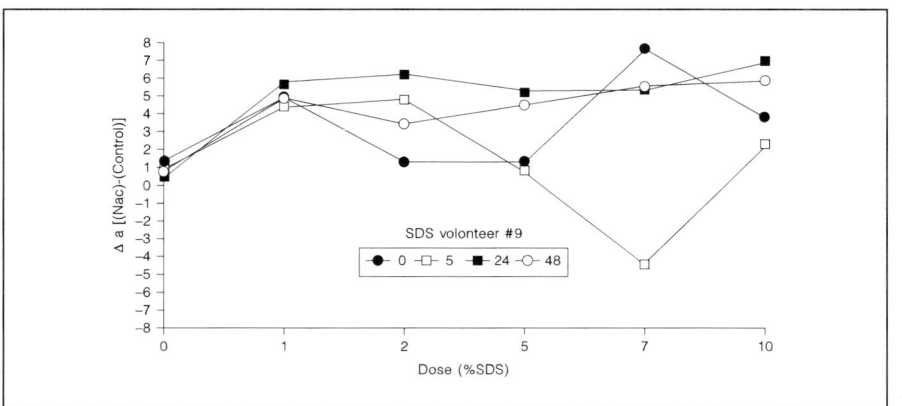

7

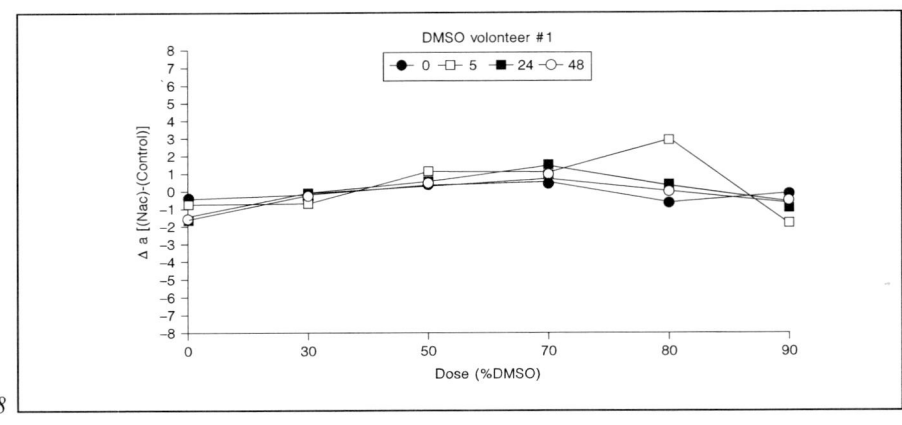

8

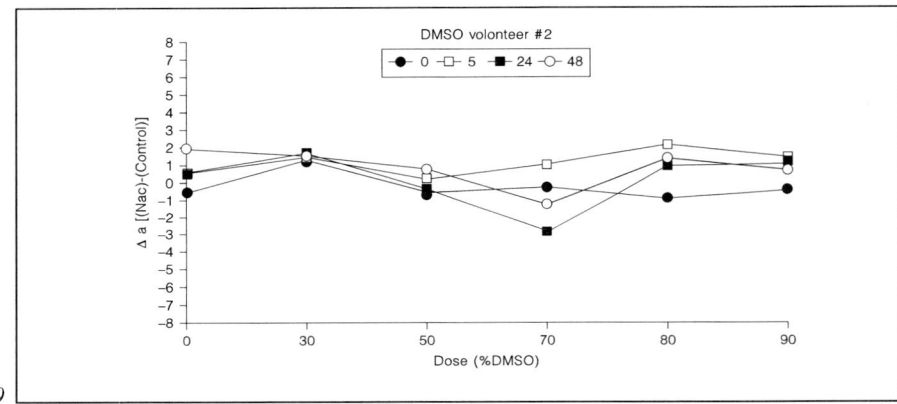

9

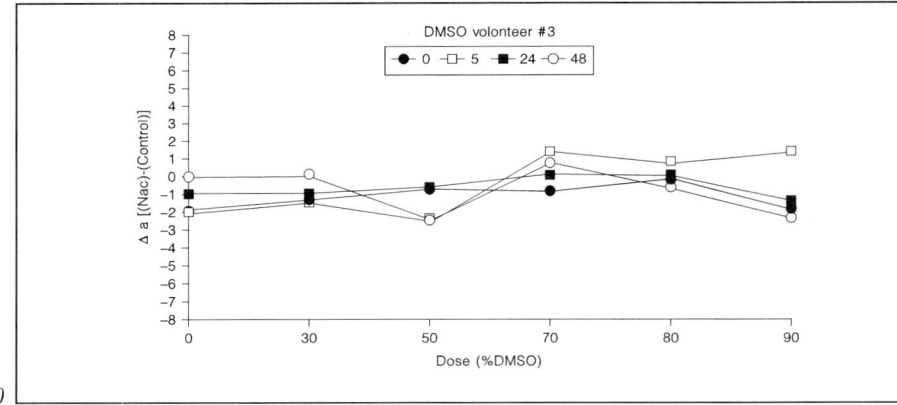

10

(For legend see p. 200.)

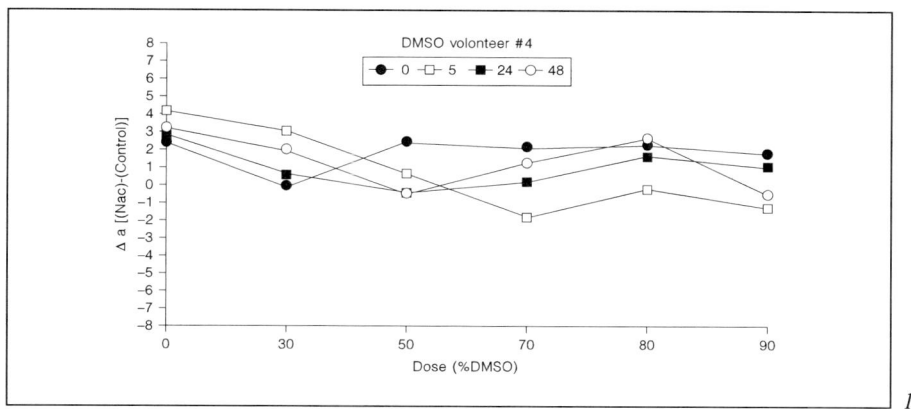

11

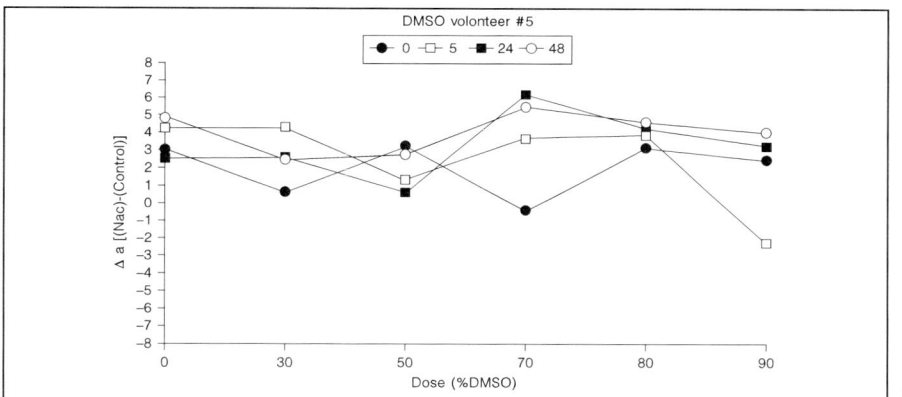

12

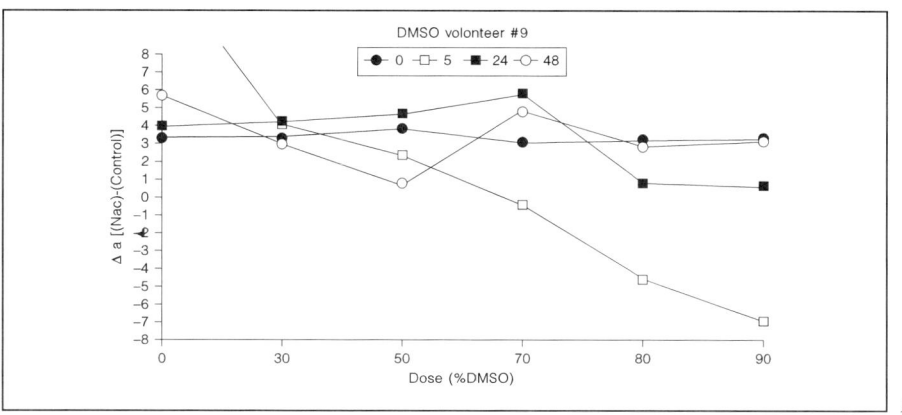

13

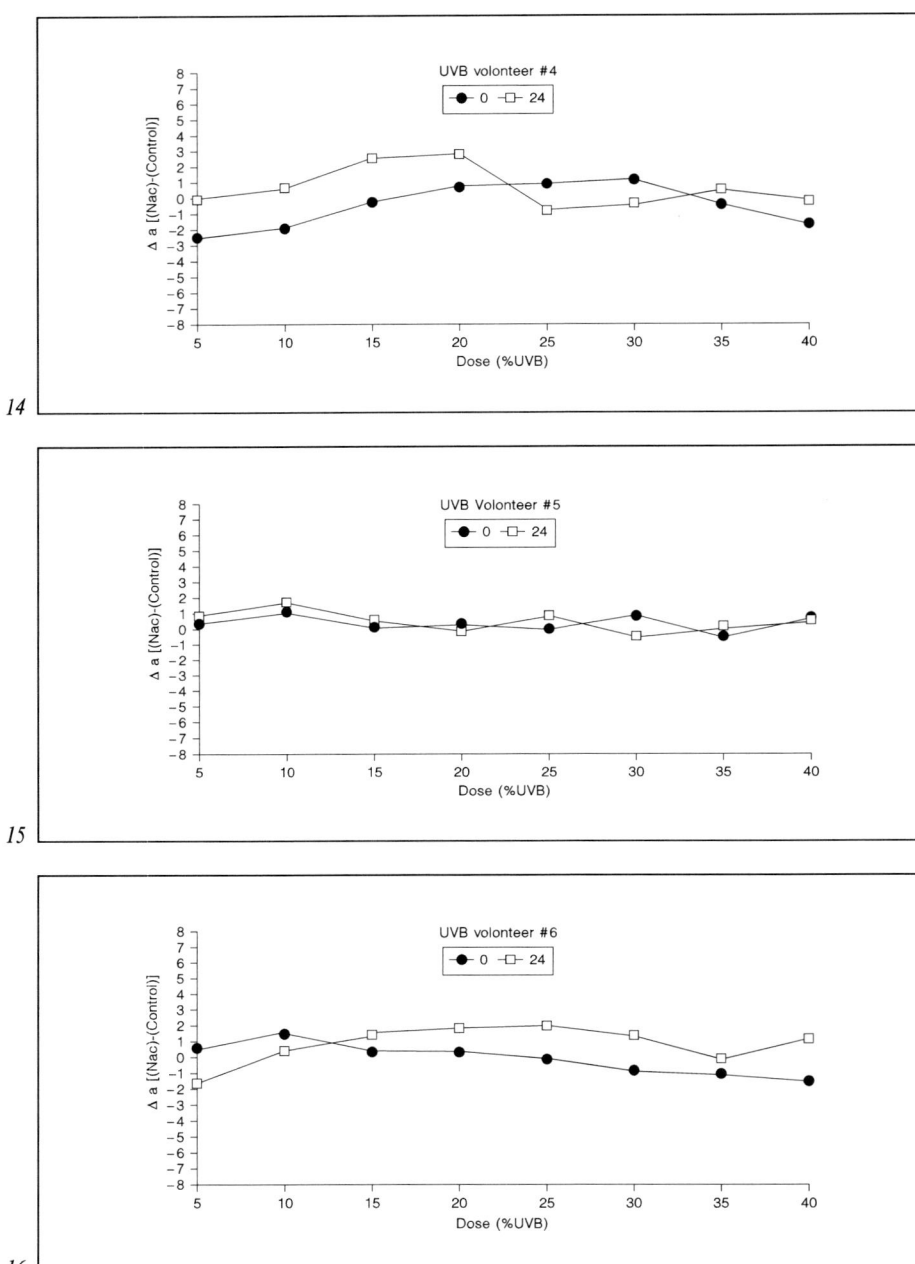

*Fig. 14–18.* Difference of erythema induced by UVB with and without NAC treatment.

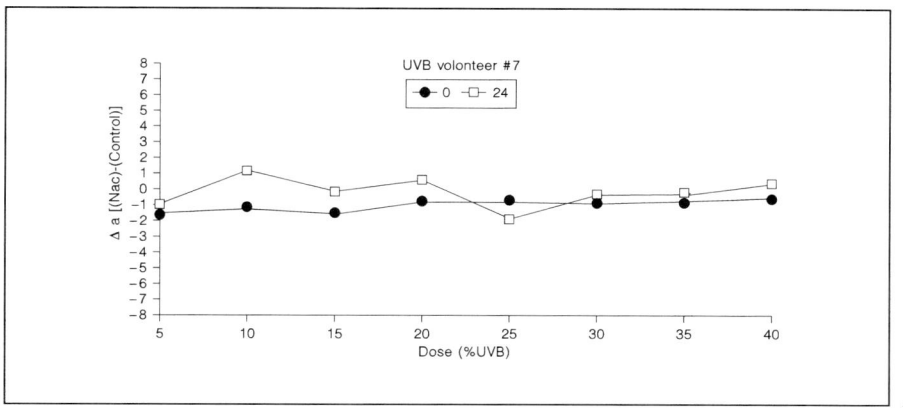

17

18

## Discussion

In contrast to what was shown in the mouse, the application of a cream containing 10% NAC failed to inhibit irritant dermatitis induced by either SDS or DMSO. Moreover, topical NAC had no significant effect on the UVB induced erythema.

It further needs to be determined if, in the human, TNF-α is induced in the epidermis after application of irritants such as SDS or DMSO, or after irradiation by UVB. And if it is induced, it is possible that the NAC molecule cannot penetrate human epidermis, in contrast to what happens with mouse epidermis.

# References

1   Piguet PF, Grau GE, Hauser C, Vassalli P: Tumor necrosis factor is a critical mediator in hapten-induced irritant and contact hypersensitivity reactions. J Exp Med 1991;173:673–679.
2   Köck A, Schwarz T, Kirnbauer R, Urbanski A, Perry P, Ansel JC, Luger TA: Human keratinocytes are a source for tumor necrosis factor x: Evidence for synthesis and release upon stimulation with endotoxin or ultraviolet light. J Exp Med 1990;172:1609–1614.
3   Lenardo MJ, Baltimore D: NF-kB: A pleiotropic mediator of inducible and tissue-specific gene control. Cell 1989;58:227–229.
4   Senaldi G, Pointaire P, Piguet PF, Grau G: Protective effect of N-acetyl-cysteine in hapten induced irritant and contact hypersensitivity reactions. J Invest Dermatol, in press.
5   BMDP Statistical Software: University of California Berkeley.

Florence Pasche-Koo, Department of Dermatology, University Hospital,
CH–1211 Geneva 14 (Switzerland)

Elsner P, Maibach HI (eds): Irritant Dermatitis. New Clinical and Experimental Aspects.
Curr Probl Dermatol. Basel, Karger, 1995, vol 23, pp 207–223

# Topical Corticosteroids:
# Experience with Mometasone Furoate

*Edwin A. Peets*

Schering-Plough Corporation, Kenilworth, N.J., USA

Eczematous dermatitis constitutes a significant dermatological problem having attendant societal and occupational consequences. A recent Quality of Life survey of 92 adults, aged 16–67, with severe eczema conducted by the United Kingdom National Eczema Society found family life for 80% of people with this devastating disease was significantly affected [1]. Over half of the people surveyed felt that their eczema had detrimentally affected their personal relationships. Those surveyed had an average of 11 outpatient and 3 inpatient hospital visits per year due to their skin disease. Forty-two percent lost income over the previous year, with an average loss of 10 weeks of work.

Irritant and allergic contact dermatitis are among the most prevalent economically and socially important forms of the disease, affecting significant numbers of adults and children. It has been estimated that allergic contact dermatitis alone accounts for 7% of all occupational illness in the United States, and that treatment, disability payments, and lost productivity amounts to approximately $250 million a year [2]. European epidemiological studies have estimated the prevalence of contact dermatitis to be between 1 and 10% of the general population, with irritant dermatitis occurring in approximately 1–4% of the population [3]. Atopic dermatitis is one of the most common dermatoses of infancy and childhood, and recent data suggests that its incidence is increasing [4]. One report estimated that atopic dermatitis accounts for approximately 20% of pediatric dermatology visits [5].

While these diseases are responsive to topical corticosteroid therapy their treatment presents a challenge for the physician, particularly in view of their often chronic nature, the necessity for long-term intermittent treatment, the risk of local and systemic corticosteroid side effects, and the development of sensitization to topical corticosteroids. This latter complication of therapy has been documented in recent years in a significant number of patients [6–13].

Mometasone furoate (Elocon©, primary trademark) 0.1% is a structurally unique medium potency topical corticosteroid developed by Schering-Plough Corporation for the once daily treatment of these corticosteroid-responsive dermatoses. Cream, ointment, and lotion formulations have been developed and marketed in the United States since 1987 and in Europe since 1990. The mometasone furoate molecule has a steroid nucleus similar to beclomethasone, with chlorine at the 9 and 21 carbon positions of the molecule and a 17(2') furoate ester, unique among marketed corticosteroids.

The successful dissociation of clinical potency and unwanted side effects of this unique molecule has resulted in the development of a corticosteroid having an improved efficacy and safety profile [14–16]. Available in its various formulations, for once a day application, this corticosteroid is highly efficacious in the treatment of psoriasis, contact dermatitis, atopic dermatitis, and a variety of other dermatoses, while maintaining a favorable safety profile, similar to that of hydrocortisone 1% [17–26]. The efficacy and safety documented in preclinical and clinical experience with mometasone over the past decade are reviewed in this chapter.

**Preclinical Studies**

In vitro and in vivo studies in mice and rats have assisted in understanding the mechanisms underlying the clinical potency and safety of mometasone. These studies measured mometasone's effects on proinflammatory cytokines, corticosteroid receptor binding affinity, and effects on Langerhans cells. Results of these studies are presented here. These findings should be viewed with caution however and cannot be directly extrapolated to predict results in humans using the drug clinically.

*Cytokines*
Recent evidence implicates cytokines, produced by activated leukocytes, as chemical mediators responsible for initiating and maintaining inflammatory disease. Interleukin 6 (IL-6) has been particularly implicated in the development of allergic dermatitis [5]. Inhibition of cytokine synthesis and release is one mechanism of the anti-inflammatory action of corticosteroids. In a recent in vitro study Barton et al. [27] used the inhibition of production of three proinflammatory cytokines, interleukin 1 (IL-1), tumor necrosis factor (TNF-α), and IL-6, in murine cells to rank the potency of seven commonly used corticosteroids, including mometasone. Mometasone was found to be the most potent inhibitor of production of all three cytokines. The $IC_{50}$s (concentration of compound that inhibits 50% of the control response) of mometasone were 0.05 n$M$ (IL-1), 0.15 n$M$ (IL-6), and 0.25 n$M$ (TNF-α). The rank order of potencies for the inhibition of IL-6 production was found to be mometasone furoate > beclomethasone dipropionate >

betamethasone valerate > betamethasone dipropionate > dexamethasone > hydrocortisone > betamethasone alcohol. This order parallels the rank order of clinical potency of these compounds [28].

### Receptor Binding

Topically applied corticosteroids produce their local effects by binding with specific glucocorticosteroid receptors located in the dermis and epidermis. The binding affinity of mometasone and its metabolites to glucocorticoid receptors in rat epidermis and dermis have been measured in comparison to that of alclometasone dipropionate, betamethasone dipropionate, and betamethasone valerate [29]. The binding affinities of mometasone furoate and most of its metabolites were greater than those of alclometasone dipropionate (low potency) and betamethasone dipropionate, and equivalent to that of betamethasone valerate; the latter corticosteroids are of at least moderate potency. For all compounds tested, the binding affinities were greater in the epidermis than in the dermis. This finding may imply a low potential for mometasone to produce detrimental effects on dermal collagen.

### Langerhans Cells

Epidermal and dermal Langerhans cells, dendritic cells derived from the bone marrow, are known to be antigen-presenting cells in allergic dermatitis [28, 30]. Topically applied corticosteroids reduce the numbers of these Langerhans cells, both locally and at distant untreated sites in the skin of mice and guinea pigs [5, 30]. In a mouse Ia+ Langerhans cell density assay, mometasone was compared to fluocinolone following local application to the ear. Both corticosteroids caused a reduction in numbers of Langerhans cells at the sites of application, but, unlike fluocinolone, mometasone applied once daily for 5 days at a concentration of 0.001% was devoid of systemic effects, as determined by lack of reduction in Langerhans cell numbers at distant untreated sites [30]. Recovery time of Langerhans cell numbers at the application sites was also shorter for mometasone than for fluocinolone. Histological evidence of dermal atrophy was less pronounced following mometasone treatment, suggesting that its absorption into dermal and subcutaneous tissue was less than that of fluocinolone; a depot effect of mometasone, i.e. occupancy of a reservoir in some skin compartment, was further suggested.

## Clinical Pharmacology Studies

The pharmaceutical formulation of mometasone furoate has been discussed previously [16]. A strongly lipophilic molecule, mometasone furoate is dissolved in hexylene glycol, which increases penetration of the stratum corneum. Evidence of the depot/reservoir in human skin was seen in a recent study comparing mometasone cream to its vehicle and betamethasone dipropionate and betamethasone valerate creams in UV-B-induced inflammation [31]. Single applications of these materials were made without occlusion to normal skin immediately following irradiation. Laser Doppler blood flowmetry showed mometasone to be significantly better than the comparator steroids in reducing inflammation after 5, 12, and 24 h ($p < 0.05$). Only mometasone maintained a statistically significant difference over vehicle in reduction of the UV-B-induced inflammation after 24 h, suggesting reservoir occupancy which correlates with its efficacy in a variety of dermatoses after once daily application.

This reservoir effect is also consistent with the observed low percutaneous absorption of radiolabeled mometasone into the systemic circulation (0.7% of the applied dose for ointment and 0.4% for cream in normal volunteers) [11, 32]. This phenomenon along with rapid biotransformation in the liver may account for the low systemic activity of mometasone, as has been demonstrated by its minimal effect on hypothalamic-pituitary-adrenal (HPA) axis function [17, 22–24].

## Clinical Efficacy in Patients with Contact, Atopic, or Allergic Dermatitis

Mometasone has been studied widely in large groups of patients with contact, atopic, or allergic dermatitis. Studies comparing mometasone cream 0.1% to hydrocortisone butyrate cream in adults with atopic dermatitis were conducted in Denmark [22] and Sweden [20]. Adults with a variety of dermatoses, including atopic and allergic dermatitis, were subjects of further study in Argentina [21] and New Zealand [26], comparing mometasone cream 0.1% to betamethasone valerate cream (table 1). In addition, mometasone cream has been studied in atopic dermatitis and a variety of other dermatoses in pediatric populations in the United States and Mexico [19, 24, 33], in comparison to hydrocortisone 1% and 2.5%, alclometasone dipropionate, and clobetasone butyrate (table 2).

In each of these clinical trials mometasone cream 0.1% was applied once daily while the comparator corticosteroid was applied twice daily. For this reason the trials were single-blind, with only the investigator blinded to the identity of the treatment. All studies were randomized, multicenter, parallel group comparisons. Patients were evaluated at weekly visits, and in some studies also after 3–4 days of treatment. The severities of three or more of the following disease signs/symptoms were evaluated: erythema, crusting, scaling, excoriation, induration, pruritus, and pain. Efficacy assessments included sign/symptom severity scores of a target lesion and a global evaluation of overall change in disease status of all treated lesions. Individual and total sign/symptom scores were evaluated, as well as the percent change from baseline in sign/symptom scores. Some studies also included a patient evaluation of treatment effects.

*Table 1.* Mometasone cream 0 1% in adults

| Study | Disease; patient ages | Comparator | n | Treatment duration | Efficacy results |
|---|---|---|---|---|---|
| Viglioglia et al. [21] Argentina 1990 | dermatoses, varied[1]; 12 years up | betamethasone valerate | 35/34 | 3 weeks | no significant differences; improvement in sign/symptom scores after 3 weeks: mometasone 94%; betamethasone 97% |
| Wishart [26] New Zealand 1993 | dermatoses, varied[2]; adults | betamethasone valerate | 28/30 | 4 weeks | no significant differences; improvement in sign/symptom scores after 4 weeks: mometasone 93%; betamethasone 90% |
| Hoybye et al. [22] Denmark 1991 | atopic dermatitis; 18–70 years | hydrocortisone butyrate | 49/45 | 3 weeks + 3 weeks of 3 ×/week | no significant difference after first 3 weeks of daily treatment, percent of patients cleared or markedly improved after 3 weeks: mometasone 88%; hydrocortisone 78%; mometasone significantly better (p < 0.01) at end of second 3 weeks (3 × week): mometasone 85%; hydrocortisone 71% |
| Gip et al. [20] Sweden 1990 | atopic dermatitis and seborrheic dermatitis[3], 12 years up | hydrocortisone butyrate | 107/109 | 3 weeks | mometasone significantly better (p < 0.05), improvement in total disease sign/symptom scores at treatment endpoint: mometasone 86%; hydrocortisone 77% |

n = Number of efficacy evaluable patients: mometasone/comparator.
[1] Allergic contact dermatitis, atopic dermatitis, seborrheic dermatitis, eczema, neurodermatitis.
[2] Atopic dermatitis, eczema, seborrheic dermatitis, psoriasis, allergic dermatitis.
[3] Majority of patients had atopic dermatitis.

## Adult Studies, Varied Dermatoses

In a 3-week study in 77 patients with a variety of dermatoses, mometasone cream applied once daily was compared to betamethasone valerate cream applied twice daily [21]. The most frequent diseases diagnosed in the 69 efficacy evaluable patients were allergic contact dermatitis, present in 37 patients (54%), and atopic dermatitis, which occurred in 18 patients (26%). Other diagnoses were eczema

*Table 2.* Mometasone cream 0.1% in children

| Study | Disease; patient ages | Comparator | n | Treatment duration | Efficacy results |
|---|---|---|---|---|---|
| Vernon et al. [24] USA 1991 | atopic dermatitis; 6 months to 12 years | hydrocortisone 1% | 24/24 | 6 weeks | mometasone significantly better (p = 0.01); improvement in total sign/symptom scores at treatment endpoint: mometasone 95%; hydrocortisone 75% |
| Lane [unpubl.] USA 1991 | atopic dermatitis; 4 months to 13 years | alclometasone dipropionate 0.05%; hydrocortisone 2.5% | 69/68/69 | 4 weeks | mometasone significantly better (p < 0.05); improvement in total sign/symptom scores at treatment endpoint: mometasone 94%; alclometasone 77%; hydrocortisone 80% |
| Dominguez et al. [19] Mexico 1990 | dermatoses, varied[1]; 6–12 years | clobetasone butyrate 0.05% | 29/32 | 3 weeks | no significant differences; improvement in total sign/symptom scores at treatment endpoint; mometasone 92%; clobetasone 87% |

n = Number of efficacy evaluable patients: mometasone/comparator.

[1] Atopic dermatitis, allergic contact dermatitis, seborrheic dermatitis, actinic prurigo.

(7 patients), seborrheic dermatitis (5 patients), and neurodermatitis (2 patients). Both treatment groups showed rapid, progressive improvement throughout the study. At each of the post-baseline study visits (days 3, 7, 14, and 21) no statistically significant differences were found between the two treatment groups with regard to mean percent improvement in total target lesion sign/symptom scores, physician's global evaluation of overall change in disease status, and the patients' evaluation of treatment effects. At the end of the study mometasone-treated patients achieved an average of 94% improvement in signs and symptoms and betamethasone-valerate-treated patients showed an average of 97% improvement.

A second study also demonstrated that once daily mometasone cream was as effective as twice daily betamethasone valerate cream in 58 patients with a variety of dermatoses [26]. In this 4-week multicenter study the most common diagnoses were atopic dermatitis (31 patients) and eczema (15 patients). By the end of the 4 weeks of treatment the once-daily mometasone treatment group demonstrated a 93% improvement in target lesion sign/symptom scores, compared to a 90% improvement in the twice-daily betamethasone valerate group.

## Adult Studies, Atopic Dermatitis

In a three-center study comparing mometasone cream to hydrocortisone butyrate cream in adults with atopic dermatitis, mometasone was applied once daily for 3 weeks and hydrocortisone butyrate was applied twice daily for 3 weeks, followed by 3 weeks of treatment for 3 consecutive days per week (maintenance treatment), once and twice daily, respectively [22]. In the global evaluation of change in disease status after the initial 3 weeks of treatment, 43 of 49 patients (88%) using mometasone were judged cleared or markedly improved, compared with 35 of 45 patients (78%) using hydrocortisone butyrate (p = 0.28). At the end of the 3-week maintenance period, improvement in the mometasone group was significantly better, with 41 of 48 patients (85%) in the mometasone treatment group cleared or markedly improved, compared to 27 of 38 patients (71%) in the hydrocortisone butyrate group (p = 0.0025).

The superior efficacy of once-daily mometasone cream over twice daily hydrocortisone butyrate cream was further demonstrated in a 3-week study in 216 patients, 12 years of age and older, including 205 with atopic dermatitis [20]. Differences between the two treatment groups in mean percent improvement in total disease sign/symptom severity scores of target lesions, as well as global evaluation of overall change in disease status, were statistically significant (p < 0.05), favoring mometasone, as early as 3 days after treatment initiation, and remained significant throughout the study.

## Pediatric Studies in Atopic Dermatitis and Allergic Dermatitis

The advantage of once-daily treatment with mometasone in children was demonstrated in three studies, including two in patients with moderate-to-severe atopic dermatitis, and one in a variety of dermatoses. In the first study, 48 children between the ages of 6 months and 12 years with atopic dermatitis were treated with mometasone cream 0.1% once daily or hydrocortisone cream 1.0% twice daily for up to 6 weeks [24]. Fifteen patients in each group were cleared completely of their disease prior to the scheduled completion of the study and were discontinued early (median duration of treatment 3 weeks). At treatment endpoint, the percent improvement in total sign/symptom score averaged 95% for mometasone cream-treated patients and 75% for patients applying hydrocortisone cream (p = 0.01). A subgroup of patients with more than 25% of body surface area involved with disease showed an even wider difference in improvement in disease (92 vs. 62%, respectively, p = 0.01), favoring mometasone.

In the second pediatric study in atopic dermatitis patients, a much larger trial, the efficacy and safety of once-daily mometasone furoate cream 0.1%, twice-

*Table 3.* Mometasone cream or ointment 0.1%: special safety studies

| Study | Formulation; disease or normal | Design; objective | Comparator; dosage regimen | n | Treatment duration | Safety results |
|---|---|---|---|---|---|---|
| Katz [34] USA 1988 | cream; normal skin | double-blind, bilateral paired comparison; determination of atrophogenic potential | hydrocortisone cream 1% b.i.d.; mometasone q.d. | 30/30 | 6 weeks | no overt signs of atrophy; covert signs (minimal), skin thinning (slight), reversible – both treatments; slightly greater – mometasone |
| Brasch [35] Germany 1991 | cream; normal skin | open label; determination of atrophogenic potential | none; q.d. | 6 | 12 months | no signs of local atrophy clinically or histologically; no systemic effects |
| Lebwohl et al. [25] USA 1991 | ointment; psoriasis and atopic dermatitis | open label; steroid sensitive areas; determination of atrophogenic potential | none; q.d. | 38 | 2 weeks | no signs of local atrophy |
| Katz et al. [18] USA 1989 | ointment; psoriasis | bilateral paired comparison; determination of atrophogenic potential | hydrocortisone ointment 1%; both q.d. | 51/51 | 6 weeks | mometasone: 1 patient mild thinning and 1 moderate telangiectasia; hydrocortisone: 1 patient mild thinning |
| Bressnick et al. [17] USA 1988 | ointment; psoriasis | double-blind, parallel group; determination of effects on HPA axis | hydrocortisone ointment 1%; 15 g q.d. | 24/24 | 3 weeks | no measurable effect on HPA axis function – no differences from hydrocortisone; mometasone: 1 patient moderate burning, 2 patients mild telangiectasia; hydrocortisone: 1 patient mild itching |
| Higashi and Katagiri [32] Japan 1990 | ointment; normal skin | open label; determination of effects on HPA axis | none; 10 g q.d.; occluded | 5 | 5 days | serum cortisol levels did not decrease |

n = Number of safety evaluable subjects: mometasone/comparator.

daily hydrocortisone cream 2.5%, and twice-daily alclometasone dipropionate cream 0.05% were compared in 206 children, aged 6 months to 12 years (data on file, Schering-Plough Corporation). Once-daily mometasone demonstrated statistically significant superiority over twice-daily hydrocortisone and alclometasone in improvement in disease sign/symptom scores at all weekly study visits and at treatment endpoint (p < 0.05 for both pairwise comparisons).

In the third pediatric study, once-daily treatment with mometasone cream was compared to twice-daily applications of clobetasone butyrate cream 0.05% [19]. Sixty-one children between the ages of 6 and 12 years were treated for 3 weeks. Seventy-five percent of the patients had atopic dermatitis, 18% had allergic contact dermatitis, 5% had seborrheic dermatitis and 2% actinic prurigo. By day 4, sign/symptom scores indicated an average improvement of 45% in the patients in the group treated with mometasone and 36% in those treated with clobetasone. At treatment endpoint the corresponding numbers were 92 and 87%, respectively. These differences, although not statistically significant, favored mometasone. Global evaluations of overall change in disease status from baseline were similar in the two groups, although improvement occurred more rapidly in mometasone-treated patients.

A case history of an irritant dermatitis successfully treated with mometasone cream has been reported recently [33]. The eruption appeared on the hands of a child 1 day after contact with the contents inside a toy 'Sqwish Ball'. Twice-daily treatment with mometasone cream cleared the condition in 5 days. The presumed irritant was diisononyl phthalate, and the manufacturer reported that 5 previous cases of dermatitis following rupture of the ball had been reported.

## Clinical Safety of Mometasone

Historically, the severity of side effects of topical corticosteroid therapy has been proportional to the clinical potency of the corticosteroid. During the development program leading to mometasone, a dissociation of potency and these side effects, both local and systemic, was sought. In order to determine if this dissociation was realized, mometasone's safety was studied extensively both in normal volunteers, where large quantities were applied for extended durations or under occlusion to maximize absorption, and in patients with diseased skin, including atopic dermatitis and psoriasis, where drug absorption was increased due to loss of the normal skin barrier. One study examined the effects of mometasone when used in corticosteroid-sensitive areas of diseased skin. Table 3 lists mometasone studies in which determination of safety (and investigation of separation of potency and side effect potential) was the primary objective.

**Local Effects**

Four studies, two in normal volunteers and two in patients, were conducted with the specific objective of evaluating the potential of mometasone ointment to cause skin atrophy [18, 25, 34, 35]. One special study involved application of mometasone cream once daily and hydrocortisone cream twice daily to the foreheads of 30 normal volunteers in a 6-week, double-blind, bilateral paired comparison [34]. Results of this study have been reviewed previously [16], and indicated neither hydrocortisone nor mometasone induced overt signs of skin atrophy. Covert signs of atrophy, i.e. telangiectasia, loss of normal skin markings, and skin thinning, were observed at both treatment sites (3 × magnification with or without oil/glass interface). Slight, though measurable skin thinning was also observed at both sites. The mometasone-treated sites demonstrated a slightly greater degree of thinning (p < 0.04) after 6 weeks of application; both groups demonstrated a return to nearly normal thickness by 2 weeks posttreatment.

An even more challenging study was one which studied the long-term atrophogenic potential of mometasone cream in 6 normal volunteers, who applied the cream to a 3-cm area on the upper arm once a day for 12 months [35]. No signs of atrophy following this extended application were observed either overtly or histologically, nor was there evidence of systemic (clinical laboratory) effects. The author warned, however, that with diseased skin greater absorption could occur, leading to the possibility of systemic side effects. While one would not expect that mometasone would be used for 12 consecutive months in patients with irritant (contact, or allergic) dermatitis because of the generally rapid response of this condition to treatment, it is nevertheless important to know that mometasone may be applied for somewhat extended periods with a low risk of local or systemic side effects, as demonstrated under these exaggerated conditions.

A special study designed to observe atrophogenic potential of mometasone in corticosteroid-sensitive or thin-skinned areas was conducted in 24 psoriasis patients and 14 patients with atopic dermatitis [25]. In this open-label study, 15 patients applied mometasone ointment to the face and intertriginous areas once daily for 2 weeks. Examinations for signs of skin atrophy, striae, and telangiectasia were made on study days 1, 2, 3, 4, 8, and 15. None of these signs were noted in any of the treated sites. The author cautions, however, that even though mometasone was without effect, these results should not be used to justify prolonged use of any topical steroid on the face or intertriginous areas, as atrophy and striae may develop.

A second study designed with the objective of assessing mometasone's atrophogenic potential in patients was a bilateral paired comparison of mometasone ointment with hydrocortisone ointment 1.0%, 6 weeks of treatment, in 51 patients with psoriasis [18]. Both treatments were applied once per day. After 6 weeks 1 patient had evidence of mild skin thinning at both treatment sites and another moderate telangiectasia at the mometasone treatment site.

In addition to these special studies, local effects were monitored in the clinical efficacy studies in atopic or contact dermatitis listed in tables 1 and 2. No signs of atrophy were reported in any of the patients in these studies.

In a 3-week study designed to measure HPA axis effects, described below, mometasone ointment 0.1% was compared to hydrocortisone ointment 1.0% [17]. Patients in both treatment groups applied 15 g of ointment daily. Two of 24 patients using mometasone developed mild telangiectasia. No signs of atrophy were noted among the hydrocortisone patients.

Other local adverse experiences, commonly associated with corticosteroid formulations, were reported in limited numbers in the clinical trials presented in tables 1–3. Events occurring in patients using mometasone included local stinging, burning, pruritus, furunculosis, tenderness, increased sweating, and erythema. Most were mild in severity and occurred with similar frequency in the comparator groups.

### HPA Axis Effects

Suppression of the HPA axis is a concern for any patient using corticosteroids frequently or for extended periods of time or on damaged skin. Because patients with irritant and allergic contact dermatitis often become chronic users of topical corticosteroids, the results of studies examining the systemic effects of long-term use in any disease condition are relevant. Diseased skin, due to its compromised barrier function, maximizes systemic absorption. The systemic effects of mometasone were studied both in patients with dermatitis, and in normal volunteers, where in one instance occlusive dressings were applied to maximize the potential for systemic absorption (table 3).

In one such study the effects of mometasone ointment 0.1% on HPA axis function were compared to those of hydrocortisone ointment 1.0% in patients with resistant psoriasis [17]. Forty-eight patients, 24 in each treatment group, applied 15 g of mometasone or hydrocortisone (without occlusion) once daily for 21 days in this randomized, double-blind, parallel group study. Morning plasma cortisol levels were measured three times during the 5 days prior to treatment initiation, and on 2 consecutive days at the end of each week of treatment. Baseline and posttreatment 24-hour urinary free cortisol and 17-hydroxycorticosteroid (17-OHC) levels were also monitored. Overall changes in plasma cortisol levels for both treatment groups were minor and at the end of 3 weeks of treatment, mean levels were slightly increased above those pretreatment; no differences were observed between the two groups (fig. 1). Mean levels of urinary free

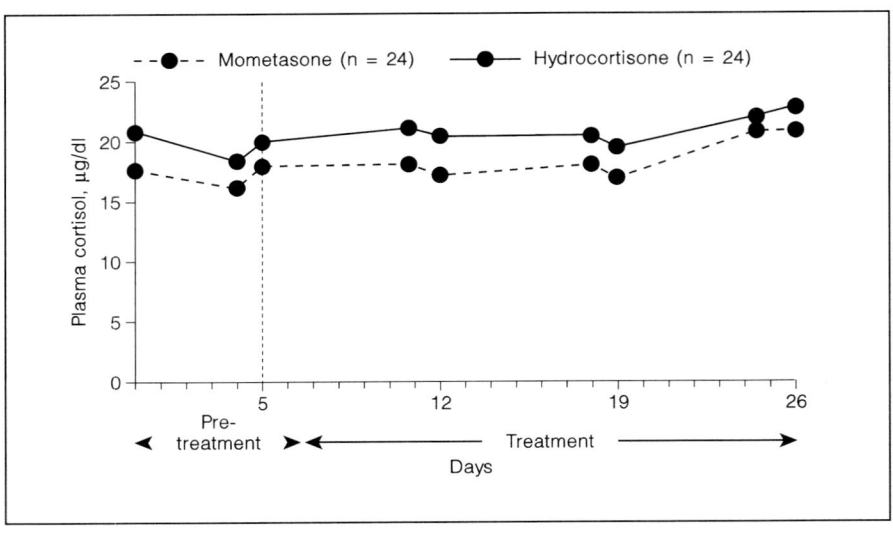

*Fig. 1.* Mean plasma cortisol levels following treatment with mometasone ointment 0.1% or hydrocortisone ointment 1%. Normal range: 5–20 µg/dl.

cortisol and 17-OHC were normal throughout the study in both treatment groups as well.

In the pediatric study in atopic dermatitis discussed above, Vernon et al. [24] found no differences in mean plasma cortisol levels ($p < 0.63$ at day 8, $p < 0.29$ at treatment endpoint) or change in mean cortisol levels between the mometasone cream and the hydrocortisone cream treatment groups (fig. 2). During the study one patient – in the hydrocortisone treatment group – demonstrated a subnormal plasma cortisol value, however (5.0 µg/dl on study day 8).

HPA axis effects were also examined in the previously described study comparing mometasone cream to hydrocortisone butyrate cream in adults with atopic dermatitis [22]. Nineteen of 96 patients had plasma cortisol levels measured throughout the study. No significant differences in cortisol levels were found between the two treatment groups at the end of 3 and 6 weeks of treatment. The range of cortisol values obtained included one or more values below the normal range for the testing laboratory at 6 weeks for the mometasone group and at baseline, 3 weeks, and 6 weeks for the hydrocortisone butyrate group.

In another study designed to measure the effects of mometasone as well as its absorption, levels in plasma, and excretion, under conditions which maximized absorption, 5 healthy male volunteers applied 10 g of mometasone ointment 0.1%

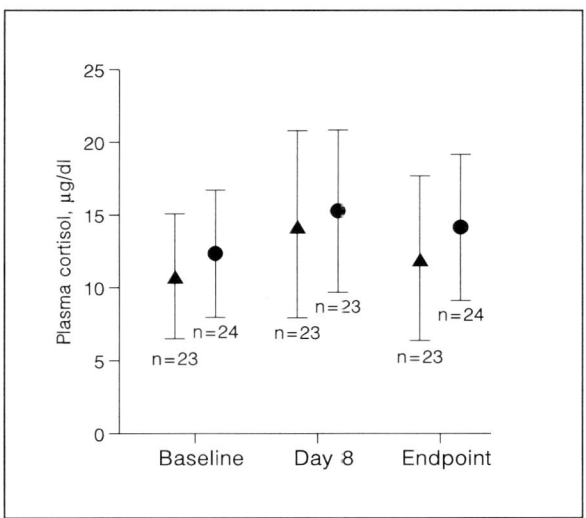

*Fig. 2.* Mean plasma cortisol levels for mometasone cream 0.1% applied once daily (▲) and hydrocortisone cream 1% applied twice daily (●). Vertical bars represent ± one standard deviation.

daily to the upper trunk for 5 days [32]. Application sites were occluded for 20 h/day. Plasma and urinary mometasone levels and serum cortisol levels were measured daily for 13 days. A plasma concentration of 100 pg/ml of mometasone was detected 15 h after application; these levels decreased rapidly after removal of the ointment. Metabolites of mometasone were not detected. Cumulative urinary excretion of mometasone was 0.00076% of the total dose administered. Even under these exaggerated conditions, serum cortisol levels did not decrease throughout the study.

## Contact Sensitization

Much recent attention has been given to the development of contact sensitization to topical corticosteroids. Once thought to be a rare phenomenon, it is now recognized as occurring in a significant number of patients with irritant and contact dermatitis, eczemas of the hand and lower leg, and atopic dermatitis [6, 8, 9, 11, 12, 36]. It should be considered as a differential diagnosis in any cases of dermatoses usually responsive to steroid treatment, which fail to respond as predicted. Many attempts have been made to characterize the nature and occurrence of the phenomenon [7, 9, 36], to characterize the sensitizing moiety of the cortico-steroid molecule, and breakdown products, and metabolites most likely to be allergenic [6, 7, 36–38], and to find markers for testing corticosteroids for their potential to induce these reactions [6–11]. Classical patch testing is generally used, but there is some evidence that intradermal testing is more diagnostic [10, 12, 36]. The vehicle, concentration, penetration, anti-inflammatory effects of the steroid, and timepoints for reading reactions affect the sensitivity and specificity of the patch testing procedure [7, 8, 11, 12, 36]. Importantly, cross-sensitization occurs, and has been documented with the reaction even to unmarketed compounds in naive individuals [7, 9–11, 38].

Contact sensitivities have now been reported to over 50 corticosteroids [38]. Dooms-Goossens [13] observed recently, after 7 years of studying this phenomenon in thousands of patients, that contact allergic reactions to steroids seem to occur mainly to the more recently developed, nonfluorinated corticosteroids. Molecules which most frequently cause allergic reactions are tixocortol pivalate, budesonide, hydrocortisone butyrate, alclometasone dipropionate, and predni-carbate [13]. Other compounds which frequently give positive patch tests are hydrocortisone, prednisolone, hydrocortisone acetate, cortisone acetate, and methyl prednisolone [8, 9]. Less frequently reported are betamethasone valerate, clobetasol propionate, clobetasol butyrate, beclomethasone dipropionate, fluran-drenolone, dexamethasone, amcinonide, fluocinonide, fluocortinbutyl ester, fluo-cortolone, triamcinolone, desonide, and desoximethasone [8, 38].

Primary sensitivity to mometasone has not been reported, and unlike the case with the other corticosteroids, only rare cases of cross-sensitivity have been reported [9, 10]. In one study, Dooms-Goossens and Morren [9] reported results of 2,073 patients patch tested over a 34-month period at their clinic in Belgium. Sixty-one of these patients (2.9%) responded to one or more corticosteroid mark-ers in a standard series, including budesonide, hydrocortisone alcohol, tixocortol pivalate, hydrocortisone butyrate, and others. Of these 61 patients, 48 were subse-quently tested with an expanded series, including corticosteroids presumably to which they had not been exposed – drugs not available on the market. Corticoste-roids from the expanded series causing positive reactions in one or more of these 48 patients included alclometasone dipropionate, clobetasone butyrate, dexa-methasone sodium phosphate, fluocinonide, hydrocortisone acetate, hydrocorti-sone butyrate, prednisolone, prednicarbate, and triamcinolone acetonide. Those most frequently found to provoke positive reactions were alclometasone dipro-pionate, prednisolone and its caproate ester, hydrocortisone acetate, cortisone acetate, methyl prednisolone, and prednicarbate (not on the Belgian market). Only one patient demonstrated a positive reaction (cross-reaction) to mometa-sone (also not marketed in Belgium at the time).

The only other report of a cross-reaction to mometasone to date involved 1 of 15 female patients exhibiting corticosteroid contact allergy who were patch tested with up to six different corticosteroids, including mometasone, betamethasone valerate, alclometasone dipropionate, hydrocortisone butyrate, tixocortol pivalate, and hydrocortisone alcohol [10]. All patients reacted to at least one, and some to as many as five, of the six agents. The one patient who reacted to mometasone had a history of chronic topical, oral, and intra-articular corticosteroid use. She developed a positive patch test to mometasone prior to ever having used the drug. A follow-up use test on normal skin was negative, but false negatives are not uncommon in corticosteroid use tests due to the barrier to penetration provided by healthy skin [10, 12]. The author concluded that cross-sensitization to mometasone is rare.

### Summary and Conclusions

Seven years of postmarketing clinical experience with mometasone furoate, a unique medium potency topical corticosteroid developed for the once daily treatment of dermatoses, have demonstrated its high degree of efficacy coupled with minimal safety liability. Its benefits have been shown in numerous clinical studies since its introduction in 1987. In studies in patients with atopic dermatitis, allergic contact dermatitis, and a variety of other dermatoses, mometasone has shown efficacy which is superior to hydrocortisone and hydrocortisone butyrate, and not different from betamethasone valerate and betamethasone dipropionate, in spite of mometasone being applied only once per day, one-half the frequency of the comparative agents. The inhibitory effects of mometasone on proinflammatory cytokines has been shown to be superior to other corticosteroids tested, including betamethasone dipropionate. There is evidence of a skin reservoir effect, which correlates with its QD effectiveness.

The clinical data suggest a dissociation of side effects from efficacy. Despite its high degree of efficacy, the safety profile of mometasone is similar to that of hydrocortisone. Low percutaneous absorption and rapid hepatic biotransformation may contribute in part to its low systemic activity, as has been noted by its minimal effects on HPA axis function.

Local side effects of mometasone are minimal. It has been used safely in children and on the face and intertriginous areas for limited periods of time. Intermittent and long-term use for up to 6 or 12 weeks in patients with diseased skin and up to 12 months in normal volunteers have shown minimal potential for causing cutaneous atrophy; any effects that were observed were rapidly reversed after discontinuation of therapy. These data suggest that patients may use mometasone repeatedly with minimal risk of local side effects. However, as with any corticosteroid, long-term and repeated use and any application to the face or intertriginous areas should be monitored closely for early signs of atrophy, and discontinued prior to irreversible striae formation.

Importantly, contact sensitization, which has become widely recognized as a problem with the chronic use of topical steroids, complicating their use, has not been reported with mometasone, and cross-sensitization appears to be rare.

Thus, the effectiveness and safety of the once daily regimen has made mometasone treatment a valuable tool for the physician in the treatment of irritant and allergic dermatitis.

## References

1   Funell C: Quality of Live Survey. UK National Eczema Society, January 1994.
2   Belsito DV: Eczematous dermatitis; in Fitzpatrick TB (ed): Dermatology in General Medicine, ed 3. New York, McGraw-Hill, 1993, pp 1531–1542.
3   Smit HA, Coenraads PJ: Epidemiology of contact dermatitis. Epidemiology of clinical allergy. Monogr Allergy. Basel, Karger, 1993, vol 31, p 32.
4   Rystedt I: Hand eczema and long-term prognosis in atopic dermatitis; in Menne T, Maibach HI (eds): Hand Eczema. Boca Raton, CRC Press, 1994, pp 105–114.
5   Sampson HA: Atopic dermatitis. Ann Allergy 1992;69:469–479.
6   Coopman S, Degreef H, Dooms-Goossens A: Identification of cross-reaction patterns in allergic contact dermatitis from topical corticosteroids. Br J Dermatol 1989;121:27–34.
7   Rivara GP, Tomb RR, Foussereau J: Allergic Contact Dermatitis from topical corticosteroids. Contact Derm 1989;21:83–91.
8   Burden AD, Beck MH: Contact hypersensitivity to topical corticosteroids. Br J Dermatol 1992;127: 497–500.
9   Dooms-Goossens A, Morren M: Results of routine patch testing with corticosteroid series in 2,073 patients. Contact Derm 1992;26:182–191.
10  Rasanen L, Tuomi ML: Cross-sensitization to mometasone furoate in patients with corticosteroid contact allergy. Contact Derm 1992;27:323–325.
11  Degreef H, Dooms-Goossens A: The new corticosteroids: Are they effective and safe? Dermatol Clin 1993;11:155–160.
12  Lauerma AI, Reitamo S: Contact allergy to corticosteroids. J Am Acad Dermatol 1993;28:618–622.
13  Dooms-Goossens A: Sensitization to corticosteroids, an increasing problem. 1994 Postgrad Course Allergological Aspects of Dermatology, ICACI XV EAACI, June 28, 1994, pp 79–88.
14  Korting HC, Kerscher MJ, Schafer-Korting M: Topical glucocorticoids with improved benefit/risk ratio: Do they exist? J Am Acad Dermatol 1992;27:87–92.
15  Liden S: Optimal efficacy of topical corticoids in psoriasis. Semin Dermatol 1992;11:275–277.
16  Samson C, Peets E, Winter-Sperry R, Wolkoff H: Mometasone furoate – Elocon© – a medium potency topical corticosteroid with favorable efficacy/safety profile: in Maibach HI, Surber C (eds): Topical Corticosteroids. Basel, Karger, 1992, pp 462–479.
17  Bressinck R, Williams J, Peets E: Comparison of the effect of mometasone furoate ointment 0.1%, and hydrocortisone ointment 1%, on adrenocortical function in psoriasis patients. Today's Ther Trends 1988;5:25–35.
18  Katz HI, Prawer S, Watson M, Scull T, Peets E: Mometasone furoate ointment 0.1% vs. hydrocortisone ointment 1.0% in psoriasis. Int J Dermatol 1989;29:342–344.
19  Dominguez L, Hojyo T, Vega E, Jones M, Peets E: Comparison of the safety and efficacy of mometasone furoate cream 0.1% and clobetasone butyrate cream 0.05% in the treatment of children with a variety of dermatoses. Curr Ther Res 1990;48:128–139.

Peets

20 Gip L, Lindberg L, Nordin P, Jones ML, Peets E: Clinical study of mometasone furoate cream 0.1% compared to hydrocortisone butyrate cream 0.1% in treatment of atopic and seborrheic dermatitis. Today's Ther Trends 1990;8:21–34.
21 Viglioglia P, Jones M, Peets E: Once-daily 0.1% mometasone furoate cream versus twice-daily 0.1% betamethasone valerate cream in the treatment of a variety of dermatoses. J Int Med Res 1990;18: 460–467.
22 Hoybye S, Moller SB, Bang FDC, Ottevanger V, Veien NK: Continuous and intermittent treatment of atopic dermatitis in adults with mometasone furoate versus hydrocortisone 17-butyrate. Curr Ther Res 1991;50:67–72.
23 Kelly JW, Cains GD, Rallings M, Gilmore SJ: Safety and efficacy of mometasone furoate cream in the treatment of steroid responsive dermatoses. Australas J Dermatol 1991;32:85–91.
24 Vernon HJ, Lane AT, Weston W: Comparison of mometasone furoate 0.1% cream in the treatment of childhood atopic dermatitis. J Am Acad Dermatol 1991;24:603–607.
25 Lebwohl M, Peets E, Chen V: Limited application of mometasone furoate on the face and intertriginous areas: analysis of safety and efficacy. Int J Dermatol 1993;32:830–831.
26 Wishart JM: Mometasone versus betamethasone creams: a trial in dermatoses. NZ Med J 1993; 106:203–205.
27 Barton BE, Jakway JP, Smith SR, Siegel MI: Cytokine inhibition by a novel steroid, mometasone furoate. Immunopharmacol Immunotoxicol 1991;13:251–261.
28 Stoughton RB, Cornell RC: Corticosteroids; in Fitzpatrick TB (ed): Dermatology in General Medicine, ed 3. New York, McGraw-Hill, 1993, pp 2846–2850.
29 Isogai M, Shimizu H, Esumi Y, Terasawa T, Okada T, Sugeno K: Binding affinities of mometasone furoate and related compounds including its metabolites for the glucocorticoid receptor of rat skin tissue. J Steroid Biochem Mol Biol 1993;44:141–145.
30 Belsito DV, Baer RL, Schultz JM, Thorbecke GJ: Relative lack of systemic effects of mometasone furoate on Langerhans cells of mice after topical administration as compared with other glucocorticosteroids. J Invest Dermatol 1988;91:219–223.
31 Bjerring P: Comparison of bioactivity of mometasone furoate 0.1% fatty cream, betamethasone dipropionate 0.05% cream and betamethasone valerate 0.1% cream in humans. Skin Pharmacol 1993;6:187–192.
32 Higashi N, Katagiri K: Percutaneous absorptionof 0.1% mometasone furoate ointment fate, excretion and adrenocortical suppression. Skin Res 1990;32:395–402.
33 Brodell RT, Torrence BP: Sqwish ball dermatitis. J Am Acad Dermatol 1992;26:641–642.
34 Katz HI: In vivo model to assay the atrophogenicity of topical steroids (poster). 47th Ann Meet Am Acad Dermatol, Washington, 1988.
35 Brasch J: The atrophogenic potential of mometasone furoate in a clinical long-term study. Z Hautkr 1991;66:785–787.
36 Lauerma AI, Maibach HI: The role of corticosteroid allergy in hand eczema; in Menne T, Maibach HI (eds): Hand Eczema. Boca Raton, CRC Press, 1994, pp 324–327.
37 Boujnah-Khouadja A, Brandle I, Reuter G, Foussereau J: Allergy to 2 new corticoid molecules. Contact Derm 1984;11:83–87.
38 Hisa T, Katoh J, Yoshioka K, et al: Contact allergies to topical corticosteroids. Contact Derm 1993; 28:174–179.

Edwin A. Peets, PhD, Sr. Director, Presidential Fellow, Dermatology, Clinical Research
Schering-Plough Research Institute, 2015 Galloping Hill Road, Kenilworth, NJ 07033 (USA)

Elsner P, Maibach HI (eds): Irritant Dermatitis. New Clinical and Experimental Aspects.
Curr Probl Dermatol. Basel, Karger, 1995, vol 23, pp 224–229

..........................
# Quantitative Structure-Activity Relationship and Cytotoxicity

*Nancy K. Mize* [a], *Juanita A. Johnson* [a], *Corwin Hansch* [b], *Michel Cormier* [a]

[a] ALZA Corporation, Palo Alto, Calif.,
[b] Pomona College, Chemistry Department, Claremont, Calif., USA

Quantitative structure-activity relationship (QSAR) is a method for objectively comparing the interaction of chemical compounds with biological systems [1]. It is used to describe the relative importance of physicochemical characteristics of compounds for a specified biological activity. With QSAR analysis, a group of compounds can be mathematically ranked according to their relative potency. This mathematical description has openend new avenues of discovery.

QSAR is used for prediction of activity of drugs and chemicals on biological systems [2, 3]. QSAR analysis can also be used to direct the synthesis of new therapeutic entities based on the predicted beneficial physicochemical properties of the drug [1, 4].

Here we examine a group of 92 drug compounds for cytotoxicity in isolated human fibroblasts. We have correlated physicochemical properties with cytotoxicity.

## Methods

The viability of human dermal fibroblasts in the presence of the drug compounds was assayed as previously described using MTT (3-[4,5-dimethylthiazol-2-yl]-2,5-diphenyltetrazolium bromide) reduction to formazan by mitochondrial enzymes as an indicator of cytotoxicity [5]. Briefly, confluent fibroblast cells were incubated for 16 h in the presence of increasing concentrations of drug. At the end of the incubation period, medium was replaced for 4 h by medium containing the tetrazolium salt, MTT. Metabolism of MTT to a blue formazan was quantitated by spectrophotometry at 540 nm. Inhibitory concentration 50% ($IC_{50}$) values represent the molar concentration of the drug compound required to kill 50% of fibroblast cells.

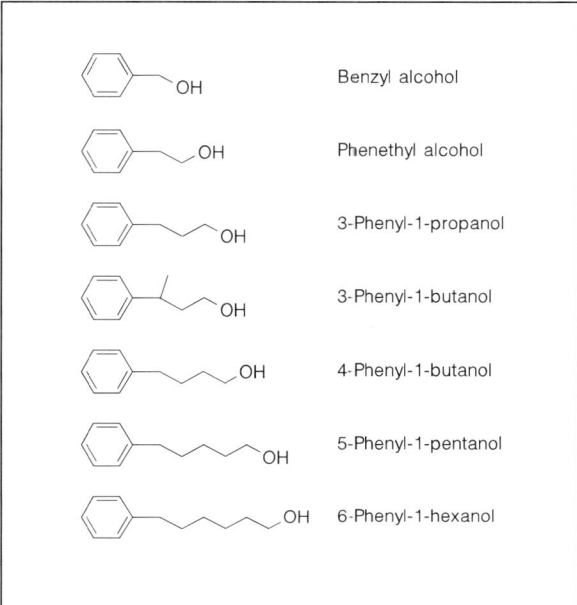

*Fig. 1.* Structures of 7 phenyl alcohols. The alcohols have the same hydroxy and phenyl groups; the increase in hydrophobicity is a result of increasing the length of the aliphatic chain.

log P (P = octanol/water partition coefficent) values were obtained from the Pomona College Chemistry department database and include both estimated and experimentally obtained measurements for the neutral species. log P values at pH 7.4 (log $P_{7.4}$) for figure 4b were all experimentally derived and obtained from the Pomona database. The QSAR equations were derived from the Pomona College QSAR program.

**Results**

*Cytotoxicity of Phenylalcohols*

Seven phenylalcohols were tested for human dermal fibroblast cytotoxicity as described in 'Methods'. These alcohols constitute a closely related series of compounds as the structures in figure 1 demonstrate. As expected, increasing length of side chain is associated with increased hydrophobicity as measured by log P. Hydrophobicity ranged from log P = 0.98 for benzyl alcohol, to 3.62 for 6-phenyl-1-hexanol, which represents more than 2 orders of magnitude range for hydrophobicity.

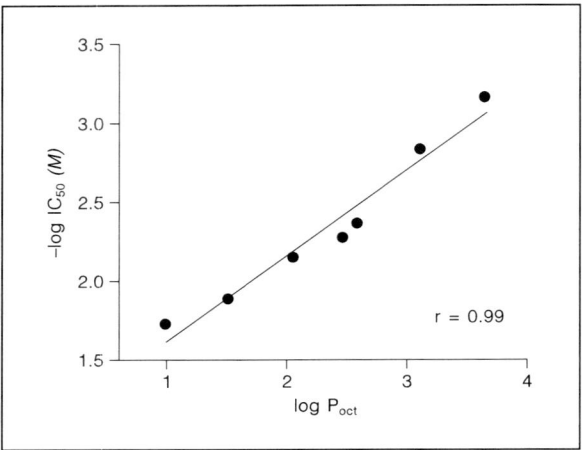

*Fig. 2.* Correlation of octanol/water partitioning ($P_{oct}$) and cytotoxicity ($IC_{50}$, molar concentration) for the 7 alcohols shown in figure 1. Increasing hydrophobicity is correlated with increasing cytotoxicity.

Cytotoxicity of these 7 alcohols as measured by MTT is shown in figure 2 as a function of hydrophobicity. The correlation between cytotoxicity and hydrophobicity for these 7 compounds is shown by the following equation:

$$\log 1/IC_{50} = 0.64\ (\pm\ 0.17)\ \log P + 0.97\ (\pm\ 0.38) \tag{1}$$
$$r = 0.99,\ SD = 0.18,\ n = 7.$$

This equation is comparable with other equations derived with alcohols in general [6].

### Cytotoxicity of Heterogeneous Compounds

Ninety-two heterogeneous compounds were tested for toxicity using the MTT assay on fibroblasts as above. The correlation between the $IC_{50}$ and log P values for these compounds is shown in figure 3 (r = 0.85; SD = 0.46). Initially, 121 compounds were tested for cytotoxicity; 29 compounds were eliminated from the QSAR study. The eliminated compounds include surfactants, aldehydes, compounds that are toxic at hyperosmotic concentrations, and agents acting on the ionic balance of the cell.

The compounds used in the computations were from the following therapeutic classes: 1 analgesic, 2 antianginal, 5 anticholinergic, 8 antidepressant, 1 antidiabetic, 3 antiemetic, 8 antihistaminic, 6 antihypertensive, 1 antimalarial, 1 antioxidant, 3 antiparkinsonian, 2 antipruritic, 7 antipsychotic, 1 antispasmodic, 1 anxyolitic, 1 bronchodilator, 1 decongestant, 1 insecticide, 3 local anesthetic, 1

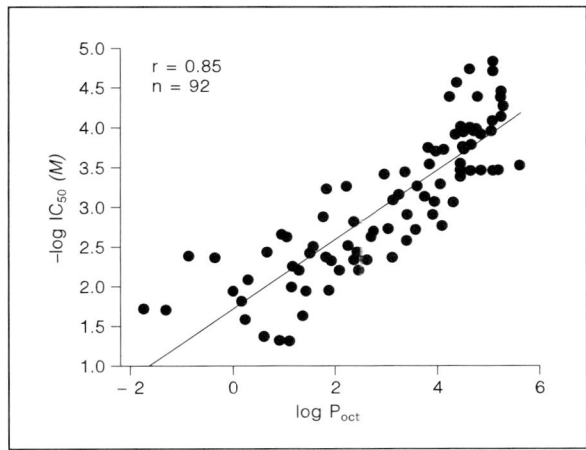

*Fig. 3.* Correlation of octanol/water partitioning ($P_{oct}$) and cytotoxicity ($IC_{50}$) for 92 drug compounds. The log $P_{oct}$ values range from $-1.75$ to $+5.54$.

muscle relaxant, 3 narcotic analgesic, 1 narcotic antagonist, 7 NSAID (nonsteroidal antiinflammatory drugs), 3 antimicrobial, 3 tranquilizer, 1 ultraviolet screen. Seventeen miscellaneous compounds, including drug metabolites, pharmacological aids, as well as various alcohols were also used for the computation.

Statistical evaluation of the $IC_{50}$ and log P data has led to the derivation of the following equation:

$$\log 1/IC_{50} = 0.43\,(\pm\,0.06)\log P + 1.71\,(\pm\,0.19) \tag{2}$$
$$r = 0.85,\ SD = 0.46,\ n = 92.$$

### Cytotoxicity of Charged vs. Uncharged Species

From the Pomona database, we found 48 compounds from the group of 92 with published log P values at ph 7.4. Comparison of the correlation obtained between cytotoxicity and log P of the neutral species or log P at pH 7.4, revealed a decreased correlation at pH 7.4 (fig. 4a, b).

### Discussion

Previous studies on closely chemically related compounds clearly demonstrate a strong correlation between toxicity and log P [6, 7]. In isolated human fibroblasts, we verified this phenomenon. When we use a very small set of closely

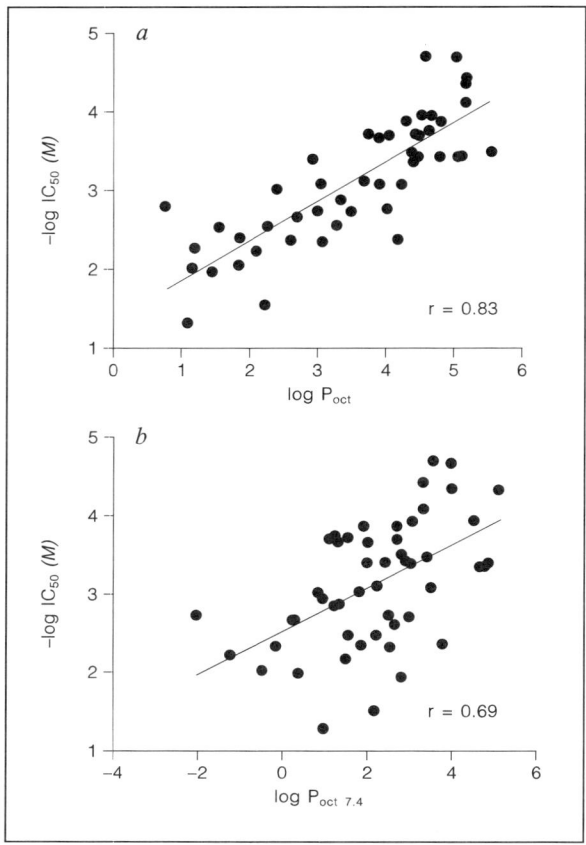

*Fig. 4. a* Correlation of octanol/water partitioning ($P_{oct}$) and cytotoxicity ($IC_{50}$) for 48 compounds that have published values for log P at pH 7.4. *b* Correlation of log $P_{7.4}$ and cytotoxicity ($IC_{50}$) for the same 48 compounds.

related compounds such as the 7 phenyl alcohols in figure 1, we get almost perfect correlation between cytotoxicity and log P (r = 0.99).

A good correlation is still observed with the heterogeneous 92 compounds (r = 0.85). Nevertheless, a good correlation was obtained only if quite a few compounds are removed from the original list of 121. Aldehydes were removed from the QSAR study because they are particularly reactive molecules, therefore, their stability in the presence of fibroblasts is questionable. We eliminated agents that act on the ionic balance of cells because their cytotoxic action may be due to their specific pharmacological activity which may be unrelated to their hydrophobicity, e.g. ionophores. It is somewhat difficult to obtain accurate log P values for surfactant molecules, so these were also eliminated from the list. Highly polar salts and other compounds likely to produce osmotic shock were not included because these $IC_{50}$ values did not measure the cytotoxicity of the drug, but rather the cytotoxicity due to their osmotic effect. The 92 compounds remaining comprise more than 25 therapeutic classes (as listed in 'Results').

As the mechanisms of cytotoxicity may not totally rest on hydrophobicity, the decreased correlation with 92 compounds ($r = 85$) compared to 7 alcohols ($r = 0.99$) is not unexpected. Indeed, the alcohols used in our example have the same hydroxy and phenyl groups; the change in hydrophobicity is obtained only by increasing the length of the aliphatic chain. As suggested previously [6], it is expected that substitution of the alcohol group with another functional group would yield a similar, but parallel curve to that in figure 2. The 92 compounds (fig. 3), which offer a good representation of many chemical structures available in therapeutics, may represent families of curves dependent on the functional groups present on the molecules. Therefore, it is not surprising that the correlation obtained with the 92 compounds was lower than that obtained with the 7 alcohols. Overall, the correlation is still very significant, suggesting that these compounds may have a common cytotoxic mechanism.

Of these 92 compounds, log P at pH 7.4, which could be obtained for 48 compounds, proved to be less predictive of cytotoxicity than log P of the unbuffered species (fig. 4a, b). At pH 7.4, the ionized form of the drug is the predominant species. This may indicate that interaction or binding of the neutral species to hydrophobic components of cells, such as cellular membranes, is the trigger for cytotoxicity. Alternatively, the data may indicate that penetration into the cell is the rate-limiting step in the cytotoxic action, and that log P of the neutral species is a good measure of that penetration.

### References

1  Hansch C: The QSAR paradigm in the design of less toxic molecules. Drug Metabol Rev 1985;15: 1279–1294.
2  Babich H, Borenfreund E: Structure-activity relationship (SAR) models established in vitro with the neutral red cytotoxicity assay. Toxicol In Vitro 1987;1:3–9.
3  Hansch C: Quantitative structure-activity relationships and the unnamed science. Acc Chem Res 1993;26:147–154.
4  Dunn W, III: QSAR approaches to predicting toxicity. Toxicol Lett 1988;43:277–283.
5  Swisher D, Cormier M, Johnson J, Ledger P: A cytotoxicity assay using normal, human keratinocytes: characterization and applications. Models Dermatol 1989;4:131–137.
6  Hansch C, Kim D, Leo AJ, Novellino E, Silipo C, Vittoria A: Toward a quantitative comparative toxicology of organic compounds. CRC Crit Rev Toxicol 1989;19:185–226.
7  Kubinyi H, Kehrhahn O-H: Quantitative structure-activity relationships. VI. Non-linear dependence of biological activity on hydrophobic character: Calculation procedures for the bilinear model. Arzneim-Forsch 1978;28:598–601.

Nancy K. Mize, ALZA Corporation, Palo Alto, CA 94303 (USA)

Elsner P, Maibach HI (eds): Irritant Dermatitis. New Clinical and Experimental Aspects.
Curr Probl Dermatol. Basel, Karger, 1995, vol 23, pp 230–242

..........................
# Measurement of Proinflammatory Mediator Production by Cultured Keratinocytes

*R. Roguet, C. Cohen, A. Rougier, J. Leclaire[1]*

L'Oréal Basic Research Center, Aulnay-sous-Bois, France

Skin irritancy has been defined as 'a nonimmunologic local inflammatory reaction characterized by erythema, edema, or corrosion, following single or repeated application of a substance to the identical cutaneous site' [1]. After percutaneous absorption, the interaction of xenobiotics with the different layers of the living epidermis is the first step in the onset of primary skin irritation. It has been suggested that keratinocytes may be the main target of potential irritants [2, 3]. Both constitutively and after activation, keratinocytes produce in vivo [4–7] and in vitro [3, 8–10] a large number of inflammatory mediators such as arachidonic acid derivatives, cytokines and growth factors. Various agents can activate keratinocytes. For example, UV irradiation leads to the expression and release, both in vitro and in vivo, of cytokines such as IL-1$\alpha$ [3, 11, 12], IL-6 [13, 14] and THF-$\alpha$ [15]. Similarly, topical application in vivo and incubation of keratinocytes in vitro with irritants induces, among other phenomena, synthesis of IL-1$\alpha$ and IL-1$\beta$ [3, 5, 16–18], IL-6 [19, 20] and TNF$\alpha$ [21].

In this chapter, we summarize previously published work [3, 20] on the synthesis and/or secretion of IL-1$\alpha$ and IL-6 by various keratinocyte lines with that of normal human keratinocytes in conventional culture. The action of chemical irritants and UVB is also assessed in terms of these parameters. In addition, using a reconstituted epidermis model, we report the release of IL-1$\alpha$ by keratinocytes after treatment with various surfactants and compare the results to historical skin irritancy data obtained in animals.

[1] We thank M.H. Grandidier and E. Popovic for their excellent technical assistance and Madame Boissier for preparing the manuscript.

## Materials and Methods

*Surfactants*
The test surfactants were purchased from Sigma, Lever, ICI, Marchon, Hoechst, Witco and Seppic, or synthesized and purified by L'Oréal.

*In vivo Data*
Historical cutaneous Draize test data were used for correlation studies.

*Cell Cultures*
*Normal Human Keratinocytes in Primary Culture.* Human keratinocytes were isolated as described elsewhere [22] from breast skin obtained during plastic surgery on women aged 40–60 years. Isolated cells were cultured in modified Eagle's medium supplemented with 10% fetal calf serum, penicillin (50 U/ml), streptomycin (50 µg/ml), amphotericin (2.5 µg/ ml), sodium pyruvate (1 m$M$) and glutamine (2 m$M$). Primary cultures of human keratinocytes were initiated at $2 \times 10^5$ viable cells/cm$^2$. The cultures were grown in a humidified incubator at 37°C with 5% CO$_2$ supplementation.

*Culture of Established Keratinocyte Cell Lines.* Keratinocytes from the spontaneously immortalized HaCaT cell line (a gift from Dr. Fusenig, Heidelberg, Germany) and keratinocytes from the NCTC 2544 cell line (Flow Laboratories, McLean, Va., USA) were seeded at $10^5$ cells/cm$^2$ in 35-mm culture dishes or 96-well microtiter tissue-culture plates containing DMEM. Transformed SVK 14 keratinocytes (a gift from Dr. Lyne) were seeded at the same density in the same plastic dishes in MEM medium. Both media were supplemented with 10% fetal calf serum, penicillin (50 U/ml), streptomycin (50 µg/ml), amphotericin (2.5 µg/ ml), Na-pyruvate (1 m$M$) and glutamine (2 m$M$).

*Three-Dimensional Cultures.* Reconstructed Episkin epithelia were supplied by Imedex (Chaponost, France). Briefly, they were obtained by seeding human keratinocytes (3 $\times 10^5$ cells/cm$^2$) on a collagen matrix (1.12 cm$^2$), consisting of collagen types I and III coated with a thin layer of collagen IV, which were fixed at the bottom of plastic chambers by toric rings. The culture medium was DMEM/Ham F12 containing hormones, growth factors and fetal calf serum. After 3 days of proliferation, the cultures were placed flush with the surface of the culture medium (emerged cultures) to induce differentiation of the epithelium. The Episkin cultures were deposited on a nutrient gel during transport to maintain viability. Quality-control tests (pH, temperature, etc.) were carried out in accordance with the manufacturer's recommendations. At reception, the plastic chambers containing the epithelia were transferred aseptically to sterile 12-well culture dishes containing 2 ml/well of maintenance medium (DMEM, supplied with the kit).

*Surfactant Treatment*
*Monolayer Cultures.* When confluency was reached (2–3 days), the culture medium was replaced by fresh medium containing various dilutions of the test surfactants for 18 h at 37°C. Then the monolayers were rinsed three times with 200 µl of PBS before measuring MTT-converting activity.

*Three-Dimensional Cultures.* After maintaining the cultures at 37°C in a humidified incubator with 5% CO$_2$ overnight, the maintenance medium was replaced with 2 ml of test medium at 37°C (provided with the kit). Aliquots (150 µl) of dilutions of the test surfactants in sterile distilled water were applied to the surface of the cultures with an automatic pipette, and incubated for 18 h at 37 °C with 5% $CO_2$.

## Irradiation

The UVB source was a fluorescent sunlamp (Philips TL 20/12) emitting in the 270- to 380-nm range with a maximum at 310 nm. Irradiance was measured by using an Osram Centra UV meter. Before irradiation the medium of confluent cultures in 35-mm plastic dishes was eliminated, cells were washed twice with 1 ml of PBS, and a film of PBS was left during irradiation. Fresh medium was then added and the cells were cultured for 24 h.

## Cytotoxicity Assay

MTT conversion was measured as described by Mossmann [23] in the case of monolayer cultures. With the Episkin cultures, the epidermis fixed on the plastic chambers was placed in 12-well culture dishes containing 2 ml/well of MTT solution (0.3 mg MTT/ml test medium) and incubated at 37 °C for 3 h. A punch biopsy (diameter = 8 mm) was made in the center of the epidermis and the formazan precipitates were extracted with 300 µl of acidified isopropanol (0.04 $N$ HCl) at room temperature overnight. Aliquots (200 µl) of each extract were transferred to 96-well culture plates and optical density was measured at 570 nm.

The $IC_{50}$ values (concentration of surfactant or UVB dose inhibition MTT conversion by 50%) were graphically determined from dose-effect curves.

## Sample Preparation

Culture supernatants were collected, centrifuged to eliminate detached cells, and stored at −80 °C until use. Two milliliters of fresh medium was added to the cell layer, which was immediately frozen in liquid nitrogen. Frozen cells were mechanically harvested and sonicated with a Sonifier B-20 (Bioblock Scientific Co., Illkirch, France). Cell lysates were centrifuged at 5,000 $g$ for 30 min, and the supernatant was stored at −80 °C.

## IL-1α and IL-6 Determination

IL-1α and IL-6 were assayed in the extracellular (culture supernatant) and intracellular (cell lysate) compartments using RIA kits (cat. No. 528 and 537; Radiochemical Center, Amersham, UK). IL-1α was assayed in the medium underlying Episkin cultures by the same method.

Interactions between surfactants and the two mediators were investigated by measuring mediator standards after 24 h of incubation with various surfactant concentrations.

## Protein Determination

Protein content was determined in each control culture by using the Biorad kit (Biorad, cat. No. 5000006) after solubilization of the cell layer in 2 ml of 0.3 $N$ KOH. Standard curves were established with calf serum albumin.

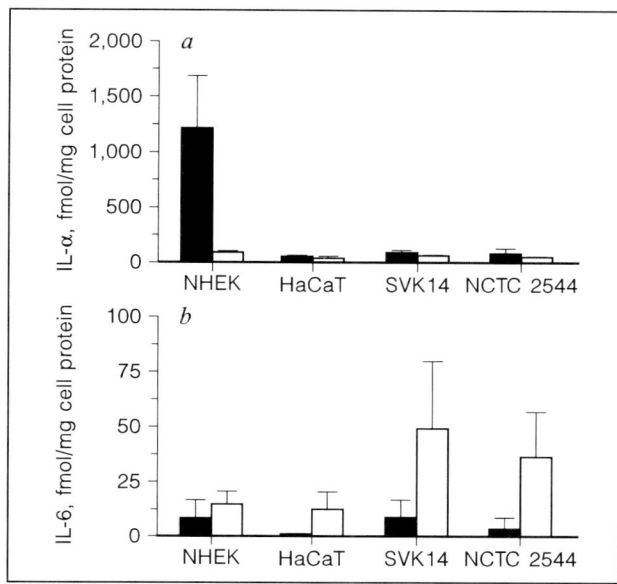

Fig. 1. Basal level of IL-1α and IL-6 in medium and cell lysate of normal and immortalized human keratinocytes. Levels of IL-1α (*a*) and IL-6 (*b*) present in the medium (□) or in the cell lysate (■) were determined as described in 'Materials and Methods' (mean of three independent determinations ± SE).

## Results

No biochemical modification of the IL-1α and IL-6 standards were observed with the range of surfactant concentration tested (data not shown).

### Baseline IL-1α and IL-6 Levels in Normal Human Keratinocytes and Immortalized Cell Lines

Figure 1 shows the baseline levels of these two mediators, expressed per milligram of protein, in normal (NHEK) and immortalized keratinocytes (HaCaT, SVK 14 and NCTC 2544 lines). Normal keratinocytes produced more IL-1α than the established cell lines. The cell-associated fraction of IL-1α was always larger than the extracellular fraction. This was especially marked in the case of normal keratinocytes (cell-associated/extracellular IL-1α ratio = 16).

In the case of IL-6, the distribution was different, as this cytokine was present in both compartments. With the exception of the SVK 14 cell line, baseline IL-6 values were comparable in the different types of keratinocytes.

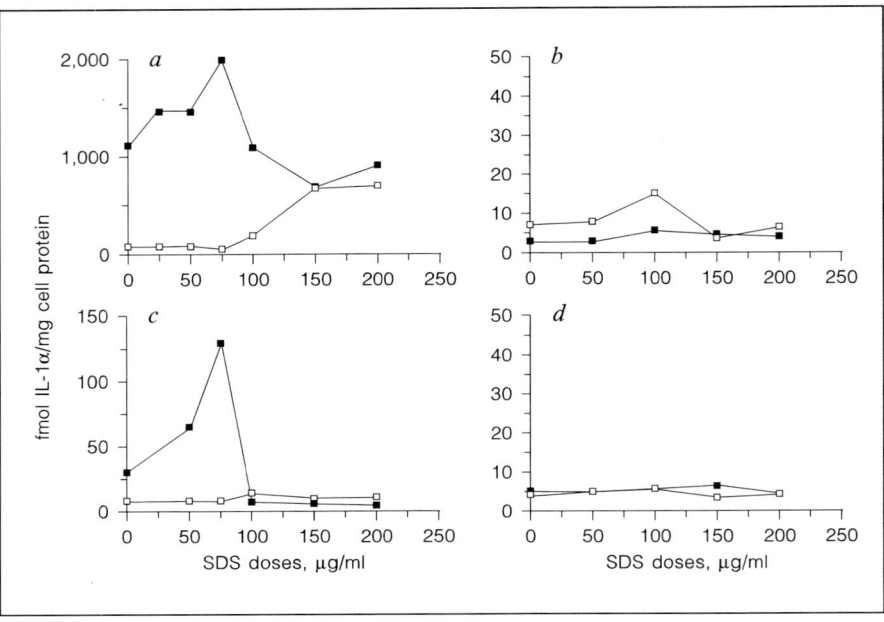

*Fig. 2.* Effect of SDS on the production of IL-1α by normal and immortalized human keratinocytes. Keratinocytes were treated with SDS and IL-1α present in the medium (open symbols) or cell lysate (full symbols) was determined as decribed in 'Materials and Methods'. *a* Normal human keratinocytes. *b* HaCaT keratinocytes. *c* SVK 14 keratinocytes. *d* NCTC 2544 keratinocytes.

*Effect of Surfactants on IL-1α and IL-6 Production by Normal and Immortalized Keratinocytes*

The effect of sodium dodecyl sulfate (SDS) on the synthesis and release of IL-1α by the different cell lines is shown in figure 2. At sublethal surfactant concentrations, intracellular IL-1α content was increased in the case of normal keratinocytes and the SVK 14 cell line. Higher surfactant concentrations induced IL-1α release into the extracellular medium. The HaCaT and NCTC 2544 cell lines appeared to be less sensitive to IL-1α induction by SDS.

IL-6 release increased strongly in the presence of SDS in the case of SVK 14 cells and, to a lesser extent, HaCaT cells. SDS did not appear to induce IL-6 expression by normal keratinocytes or NCTC 2544 cells (fig. 3).

Table 1 summarizes the action of the different surfactants tested at their $IC_{50}$ (determined by MTT conversion; table 2) on the release of IL-1α and IL-6 by the keratinocyte lines. In general, the four surfactants induced stronger IL-1α expres-

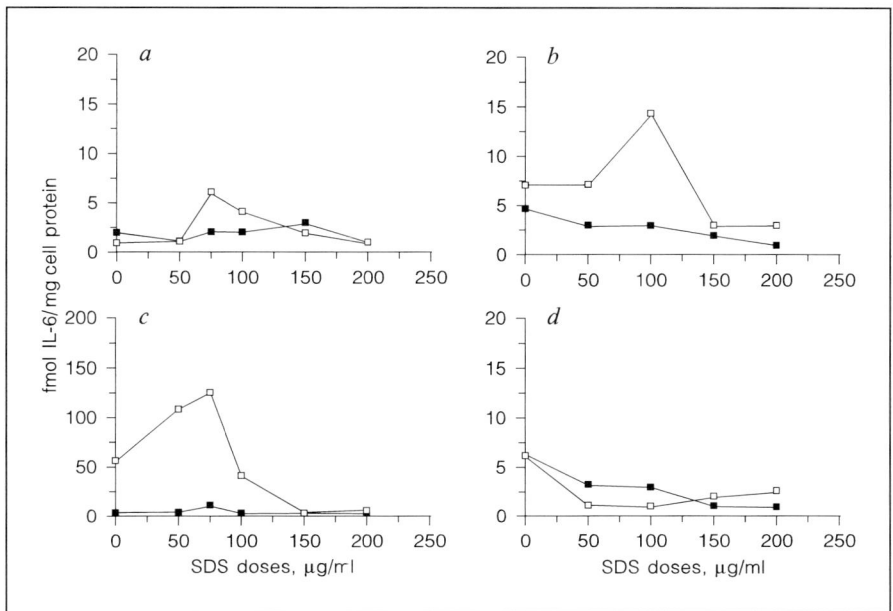

Fig. 3. Effect of SDS on the production of IL-6 by normal and immortalized human keratinocytes. Keratinocytes were treated with SDS and IL-6 present in the medium (open symbols) or cell lysate (full symbols) was determined as described in 'Materials and Methods'. *a* Normal human keratinocytes. *b* HaCaT keratinocytes. *c* SVK 14 keratinocytes. *d* NCTC 2544 keratinocytes.

sion by normal keratinocytes than by the established cell lines. With the exception of miranol, they induced an expression of IL-1α by SVK 14 cells. In the case of IL-6, synthesis was slightly increased by SDS in normal keratinocytes. SDS, miranol and, above all, Tween 20 increased IL-6 production by SVK 14 cells. Only SDS induced IL-6 expression by HaCaT cells.

### *Effect of UVB Irradiation on IL-1α and IL-6 Expression by Normal and Immortalized Keratinocytes*

UVB-induced IL-1α expression by normal keratinocytes at 0.050 J/cm², a dose below the cytotoxic dose (IC$_{50}$: 0.2 J/cm²) (table 2). Doses close to the IC$_{50}$ induced dose-dependent IL-1α release into the extracellular medium. Synthesis of IL-1α was induced little if at all by UVB in the established cell lines (fig. 4; table 1). IL-6 was induced by UV (fig. 5), especially in the established cell lines but only at doses close to those which were cytotoxic.

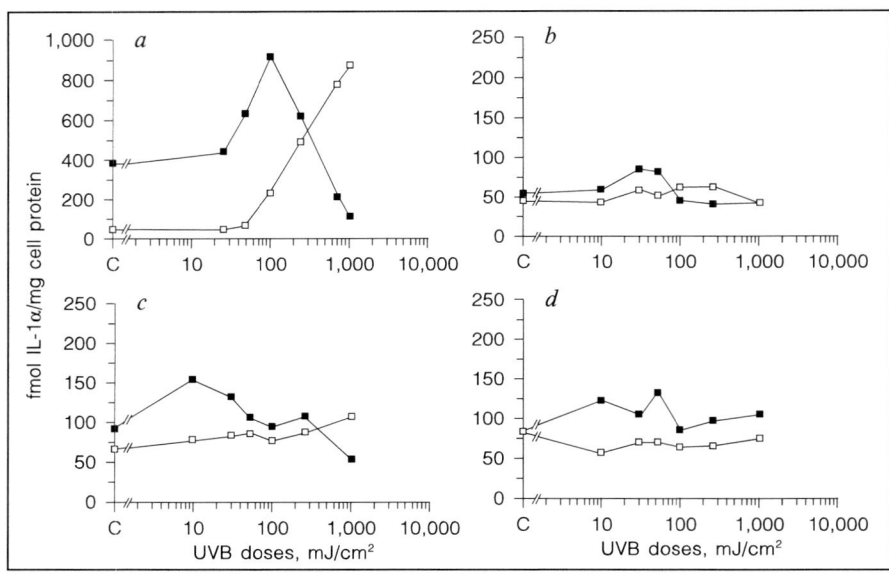

*Fig. 4.* Effect of UVB irradiation on the production of IL-1α by normal and immortalized human keratinocytes. Keratinocytes were irradiated and IL-1α present in the medium (open symbols) or cell lysate (full symbols) was determined as described in 'Materials and Methods'. C = Non-irradiated control. *a* Normal human keratinocytes. *b* HaCaT keratinocytes. *c* SVK 14 keratinocytes. *d* NCTC 2544 keratinocytes.

*Table 1.* Effect of surfactants and UVB irradiation on the release of IL-1α and IL-6 by normal and immortalized human keratinocytes

| Agents | IL-1α release | | | | IL-6 release | | | |
|---|---|---|---|---|---|---|---|---|
| | NHEK | HaCaT | SVK 14 | NCTC 2544 | NHEK | HaCaT | SVK 14 | NCTC 2544 |
| SDS | ++ | – | ++ | – | + | + | ++ | – |
| Miranol | ++ | – | – | – | – | – | + | – |
| CTAB | + | + | ++ | + | – | – | – | – |
| Tween 20 | ++ | – | + | – | – | – | ++ | – |
| UVB | ++ | – | – | – | ++ | ++ | ++ | ++ |

Keratinocytes were treated by surfactancts or UVB irradiated then IL-1α and IL-6 release were measured as described in 'Materials and Methods'.

– = Not significant effect; + = increased between 2- and 5-fold of the control level; ++ = increased between 5- and 10-fold of the control level.

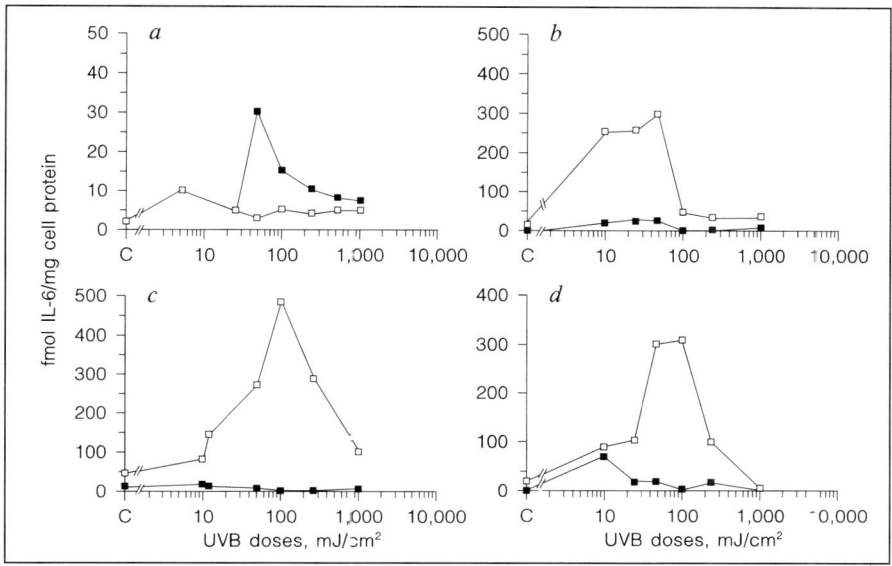

*Fig. 5.* Effect of UVB irradiation on the production of IL-6 by normal and immortalized human keratinocytes. Keratinocytes were irradiated and IL-6 present in the medium (open symbols) or cell lysate (full symbols) was determined as described in 'Materials and Methods'. C = Non-irradiated control. *a* Normal human keratinocytes. *b* HaCaT keratinocytes. *c* SVK 14 keratinocytes. *d* NCTC 2544 keratinocytes.

| *Table 2.* Cytotoxicity of surfactants and UVB on normal immortalized human keratinocyte conventional culture compared to reconstituted epidermis | Agents | IC$_{50}$ NEHK or immortalized keratinocytes monolayer | IC$_{50}$ NEHK reconstituted epidermis |
|---|---|---|---|
| | SDS, mg/ml | 0.08 | 0.96 |
| | Miranol, mg/ml | 0.20 | 7.7 |
| | CTAB, mg/ml | 0.005 | 0.7 |
| | Tween 20, mg/ml | 1.5 | nontoxic |
| | UVB, J/cm$^2$ | 0.20 | 2 |

Cytotoxicity was estimated by the MTT conversion test as described in Material and Methods (IC$_{50}$: concentration of surfactants or UVB dose reducing 50% of the MTT conversion as compared to non-treated control).

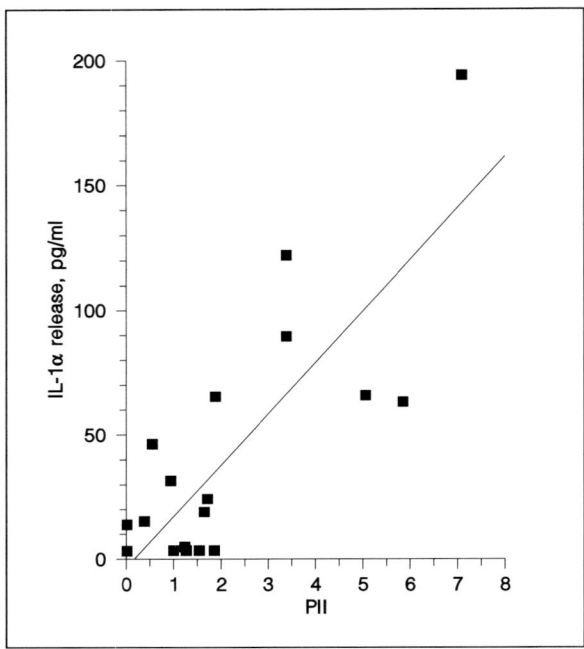

*Fig. 6.* Correlation between the release of IL-1α by Episkin keratinocytes and in vivo cutaneous irritative index of the surfactants. The release of IL-1α from the Episkin keratinocytes was determined as described in 'Materials and Methods'. PII = Primary irritative index (dermal Draize test).

### Surfactant Cytotoxicity and Effect on IL-1α Release by Reconstituted Epidermis: Correlation with Primary Skin Irritancy

Cytotoxicity and effect of different surfactant on the release of IL-1α were assessed on conventional (monolayer) culture of keratinocytes and tridimensional culture of keratinocytes (Episkin model). Cytotoxicity was assessed by the MTT reduction test, and IL-1α release was determined in the underlying medium. As reported in the table 2, the concentration of surfactant inducing the $IC_{50}$ in the two models are very different. The ratio of $IC_{50}$ for monolayer and tridimensional culture of keratinocytes were 12 for SDS, 38 for miranol, 140 for CTAB and superior to 660 for Tween 20.

The effect of the different surfactants on IL-1α release by the Episkin keratinocytes was determined after application of a fixed concentration (1 mg/ml) to the epithelial surface. The results were then compared to historical Draize test data. A good correlation (r = 0.81, n = 20; p < 0.0001) was found (fig. 6).

## Discussion

Keratinocytes are increasingly used in alternative methods to animal experimentation for predicting primary and immunoallergic skin irritancy. Two types of keratinocytes are available: normal cells with the potential of the tissue from which they are derived but which are subject to interindividual variations; and immortalized cell lines, which provide more reproducible results but which first need to be characterized. Besides, commercial systems of epidermis reconstructed from human normal cells show great promise. Given the inflammatory manifestations inherent in skin irritation, studies of soluble mediator synthesis and/or release by these different cell sources may reflect the irritancy of topically applied xenobiotics. Recent studies have shown that interleukins 1 and 6 are involved in the inflammatory response and tissue repair. These two cytokines also have pleiotropic effects on a large variety of cells [24].

Our results confirm previous reports [25–27] of IL-1α production by normal keratinocytes in vitro. In addition, the predominantly intracellular distribution of IL-1α found in this study, in the absence of cytotoxic effects, is explained by the absence of the signal peptide required for its extracellular secretion [28]. It is probable that the IL-1α molecule, as previously reported [12, 29], is pro-IL-1-alpha', which has antigenic properties required for ELISA detection and biological activity similar to that of the mature molecule. IL-1α synthesis has been detected in spontaneously transformed murine (Pam 212) [11, 31] and human keratinocyte lines [11, 30, 31]. Partridge et al. [32] have reported IL-1α secretion by squamous cell carcinoma lines, including SVK 14, the quantity of IL-1α secreted by keratinocytes of this cell line being similar to that secreted by normal keratinocytes. Our results appear to indicate lower IL-1α secretion by immortalized keratinocytes than by normal keratinocytes. The comparison between these two studies is, however, difficult as the term 'normal' applies to keratinocytes of different origins (adult vs. newborn foreskin cells) cultured in different conditions (pure primary culture vs. culture in the presence of 3T3).

IL-6 production by normal and transformed keratinocytes has previously been reported [19, 32–34]. We observed low baseline secretion of this cytokine by both normal and transformed keratinocytes.

Cytokine production by keratinocytes is modulated by a number of intrinsic factors (cell cycle and differentiation) and by cytokines themselves, forming an epidermal 'cytokine network' [35]. In addition, a wide variety of external physical and chemical agents lead to cytokine production and/or secretion. These xenobiotics include physical (UV) and chemical agents. As reported in vivo [11, 36] and in vitro [11, 12, 36] UVB irradiation of normal keratinocytes induces synthesis and release of IL-1α. Similarly, incubation with certain irritants including retinoic acid [37] engenders the same effects [3, 17]. IL-1α production by normal keratinocytes was observed in this study both after UV irradiation and after incubation with potential irritant surfactants. In this latter case, secretion of the cytokine into the extracellular medium only appeared at concentrations close to those at which the surfactants were cytotoxic, in keeping with the results of Gatto et al. [38]. In contrast, it appears that, despite the wide range of UVB doses and surfactant concentrations used, immortalized keratinocytes produced little (SVK 14) or no (HaCaT and NCTC 2544) IL-1$\alpha$. In the case of IL-6, UVB irradiation led to strong release, mainly by transformed keratinocytes; incubation with SDS was far less effective. These results show the different reactivity of normal and established cells. The spontaneous or virus-induced immortalization of these cells, their large number of passages and the culture conditions may explain these differences in the activation of established cell lines by xenobiotics.

Three-dimensional kerationocyte culture models have several advantages over conventional cultures, including the presence of a stratum corneum which acts as a barrier to xenobiotics. The comparison of the IC$_{50}$ values of the four surfactants tested on the two models illustrates this function. The ratios of 10, 100 and 35 for the IC$_{50}$ values of SDS, CTAB and miranol, respectively, between the two systems are in keeping with the different percutaneous absorption of these molecules.

IL-1$\alpha$ release by Episkin keratinocytes correlated with in vivo skin irritancy data after application of a defined concentration (1 mg/ml) of the surfactants. It thus appears that this parameter, which is strongly linked to the mechanisms intervening in skin irritation in vivo, may have predictive value, especially in the case of surfactants and preparations containing these compounds.

### Conclusions

Evaluation of the synthesis and release of inflammatory mediators by cultured human keratinocytes appears to be a promising method for predicting the cutaneous irritancy of cosmetic preparations. However, the use of established cell lines as targets requires initial characterization and validation. Indeed, constitutive and UV- or surfactant-induced synthesis of soluble mediators such as IL-1$\alpha$ and IL-6 appears to differ from that of normal keratinocytes. In three-dimensional culture, normal keratinocytes react to surfactants by releasing IL-1$\alpha$ in amounts that may be predictive of skin irritancy.

# References

1. Mathias CGT: Clinical and experimental aspects of cutaneous irritation; in Maibach HI, Marzulli F (eds): Dermatotoxicity, ed 3. Cambridge, Hemisphere, 1987, pp 173–189.
2. DeLeo A, Harber LC, Kong BM, De Salva J: Surfactant-induced alteration of arachidonic acid metabolism of mammalian cells in culture. Proc Soc Exp Biol Med 1987;184:477–482.
3. Cohen C, Dossou KG, Rougier A, Roguet R: Measurement of proinflammatory mediators produced by human keratinocytes in vitro: A predictive assessment to cutaneous irritation. Toxicol In Vitro 1991;5:407–410.
4. Hawk JLM, Black A, Jaenicke KF, Barr RM, Soter NA, Mallet AI, Gilchrest BA, Hensby CN, Parrish JA, Greaves MW: Increased concentration of arachidonic acid, prostaglandin E2, D2 and 6-oxo-F1, and histamin in human skin following UVA irradiation. J Invest Dermatol 1983;80: 496–499.
5. Larsen CG, Ter T, Larsen FG, Zachariae C, Thestrup-Pedersen K: ETAF/Interleukin-1 and epidermal lymphcyte chemotactic factor in epidermis overlying an irritant patch test. Contact Derm 1989; 20:335–340.
6. Murphy GM, Dowd PM, Hudspith BN, Brostoff J, Greaves MW: Local increase in interleukin-1-like activity following UVB irradiation of human skin in vivo. Photodermatology 1989;6:268–274.
7. Oxholm A, Oxholm P, Staberg B, Bendtzen K: Immunohistological detection of interleukin 1-like molecules and tumor necrosis factor in human epidermis before and after UVB irradiation in vivo. Br J Dermatol 1988;18:369–376.
8. Green FA: Generation of metabolism of lipoxygenase products in normal and membrane-damaged cultured human keratinocytes. J Invest Dermatol 1989;93:486–491.
9. Baker JNWN, Mitra R, Griffiths CEM, Dixit VM, Nicoloff BJ: Keratinocytes an initiator of inflammation. Lancet 1991;337:211–214.
10. Matsue H, Cruz PC, Bergstresser P, Takashima A: Cytokine expression by epidermal cell subpopulations. J Invest Dermatol 1992;99:42s–45s.
11. Ansel JC, Luger TA, Green I: The effect of in vitro and in vivo UV irradiation on the production of ETAF by human and murine keratinocytes. J Invest Dermatol 1983;8:519–523.
12. Kupper TS, Chua AO, Flood P, McGuire J, Gubler U: Interleukin 1 gene expression in cultured keratinocytes is augmented by ultraviolet irradiation. J Clin Invest 1987;80:430–436.
13. Kirnbauer R, Köck A, Krutmann J, Schwarz T, Urbanski A, Luger TA: Different effects of UVA and UVB irradiation on epidermal cell IL-6 expression and release (abstract). J Invest Dermatol 1989;92:459.
14. Urbanski A, Schwarz T, Neuner P, Krutmann J, Kirnbauer R, Köck A, Luger TA: Ultraviolet light induces increased circulating interleukin-6 in humans. J Invest Dermatol 1990;94:808–811.
15. James LC, Moore AM, Wheeler LA, Murphy GM, Dowd PM, Greaves MW: Transforming growth factor-α: In vivo release by normal human skin following UV irradiation and abrasion. Skin Pharmacol 1991;4:61–64.
16. Luger TA, Schwarz T: Epidermal cytokines; in Bos J (ed): Skin Immune System. Boca Raton, CRC Press, 1990, pp 257–292.
17. Corsini E, Marinovich M, Marabini L, Chiesara E, Galli CL: Interleukin-1 production as specific and early event after non-ionic surfactants treatment in murine keratinocytes cell line. Toxicol In Vitro, in press.
18. Gatto H, Richard MH, Viac J, Charveron M, Schmitt D: Effects of retinoic acid on the interleukin 1α and 1β expression by normal human keratinocytes cultured in defined medium. Skin Pharmacol 1993;6:10–19.
19. Gueniche A, Ponec M: Use of human skin cell cultures for the estimation of potential skin irritants. Toxicol In Vitro 1993;7:15–27.
20. Cohen C, Selvi-Bignon C, Barbier A, Rougier A, Lacheretz A, Roguet R: Measurement of proinflammatory mediator production by cultured keratinocytes; a predictive assessment of cutaneous irritancy; in Rougier A, Golberg AM, Maibach HI (eds): In vitro Skin Toxicology. New York, Mary Ann Liebert, 1993, pp 83–96.

21 Lewis RW, McCall JC, Botham PA, Kimber I: Investigation of TNFα release as a measure of skin irritancy. Toxicol In Vitro 1993;7:393–395.

22 Prunieras M, Régnier M, Wodley D: Methods for cultivation of keratinocytes with an air-liquid interface. J Invest Dermatol 1983;83:28S.

23 Mossman T: Rapid colorimetric assay for cellular growth and survival: application to proliferation and cytotoxicity assays. J Immunol Methods 1983;65:55–62.

24 Kishimoto T: B-cell stimulatory factors (BSF's): Molecular structure, biological function, and regulation of expression. J Clin Immunol 1987;7:343–355.

25 Luger TA, Oppenheimer JJ: Characteristics of interleukin 1 and epidermal cell derived thymocyte activating factor (ETAF). Adv Inflam Res 1983;5:1–25.

26 Sauder DN, Carter C, Katz S, Oppenheim JJ: Epidermal cell production of thymocyte activating factor (ETAF). J Invest Dermatol 1982;9:34–39.

27 Oppenheim JJ, Kovacs E, Matsushima K, Durum SK: There is more than one interleukin 1. Immunol Today 1986;7:45–56.

28 Blanton B, Kupper TS, McDougall J, Dower S: Regulation of interleukin 1 and its receptor on human keratinocytes. Proc Natl Acad Sci USA 1989;86:1273–1277.

29 Kupper TS, Horowitz M, Lee F, Coleman D, Flood P: Molecular characterization of keratinocytes cytokines (abstract). J Invest Dermatol 1989;92:501.

30 Bell TV, Harley CB, Stetsko D, Sauder DN: Expression of mRNA homologous to interleukin 1 in human epidermal cells. J Invest Dermatol 1987;88:375–379.

31 Goldminz D, Kupper TS, McGuire J: Keratinocytes membrane-associated epidermal cell-derived thymocyte-activating factor (ETAF). J Invest Dermatol 1987;88:97–100.

32 Partridge M, Chantry D, Turner M, Feldmann M: Production of interleukin-1 and interleukin-6 by human keratinocytes and squamous cell carcinoma cell lines. J Invest Dermatol 1991;96:771–776.

33 Luger TA, Schwarz T, Krutmann J, Kirnbauer R, Neumer P, Köck A, Urbanski A, Borth W, Schauer E: Interleukin-6 is produced by epidermal cells and plays an important role in the activation of human T lymphocytes and natural killer cells. Ann NY Acad Sci 1989;557:405–414.

34 Kirnbauer R, Köck A, Schwarz T, Krutmann J, Borth W, Damm D, Shipley G, Ansel JC, Luger TA: Interferon beta-2, B-cell differentiation factor 2, or hybridoma growth factor (IL6) is expressed and released by human epidermal cells and epidermoid carcinoma cell lines. J Immunol 1989;142:1922–1928.

35 Luger TA, Schwarz T: Evidence for an epidermal cytokine network. J Invest Dermatol 1990;35:100S–104S.

36 Konnokov N, Pincus SH, Dinarello CA: Elevated plasma interleukin-1 levels in humans following ultraviolet light therapy for psoriasis. J Invest Dermatol 1989;92:235–239.

37 Gatto H, Richard MH, Viac J, Charveron M, Schmitt D: Effects of retinoic acid on the interleukin 1α and 1β expression by normal human keratinocytes cultured in defined medium. Skin Pharmacol 1993;6:10–19.

38 Gatto H, Richard MH, Viac J, Charveron M, Schmitt D: Study of immune-associated antigens (IL-1 and ICAM-1) in normal human keratinocytes treated by sodium lauryl sulfate. Arch Dermatol Res 1992;284:186–188.

R. Roguet, L'Oréal Basic Research Center, 1, avenue E.-Schueller,
F–93600 Aulnay-sous-Bois (France)

Elsner P, Maibach HI (eds): Irritant Dermatitis. New Clinical and Experimental Aspects.
Curr Probl Dermatol. Basel, Karger, 1995, vol 23, pp 243–255

..........................

# Transcutaneous Electrical Resistance: Application in Predicting Skin Corrosives

*R.W. Lewis[a], D.A. Basketter[b]*

[a] Zeneca Central Toxicology Laboratory, Alderley Park, Macclesfield, Cheshire, and
[b] Unilever, Environmental Safety Laboratory, Sharnbrook, Bedford, UK

The assessment of skin toxicity is a vital and normally mandatory part of the standard battery of toxicological tests carried out on industrial and other chemicals. Information derived from such assessments provides the basis for avoiding the risk of skin injury through the provision of the appropriate warning labels, packaging requirements for transportation and the use of protective clothing and protective working environments.

The example of skin corrosivity testing can be used to illustrate many of the basic principles in the development of 'alternative' toxicity tests. Once the scientific and ethical needs to develop a replacement test have been established, the underlying mechanism (where known) is examined and a rational hypothesis developed to explain the toxicological endpoint observed in the whole animal. With this information the toxicologist is better able to conceive and develop an alternative test model, ideally avoiding the use of whole animals. The novel model then undergoes in-house evaluation of performance against selected chemicals of known in vivo effect. A period of use often follows where the test is applied to routine safety testing in parallel with existing in vivo methods. Satisfactory performance in this phase (often involving test modifications and optimisation) can lead to comparative assessment of the test in a wider context involving other laboratories in different countries. The development of one particular alternative to in vivo skin corrosivity testing (the transcutaneous electrical resistance test, TER) illustrates these principles from hypothesis formulation to validation (fig. 1).

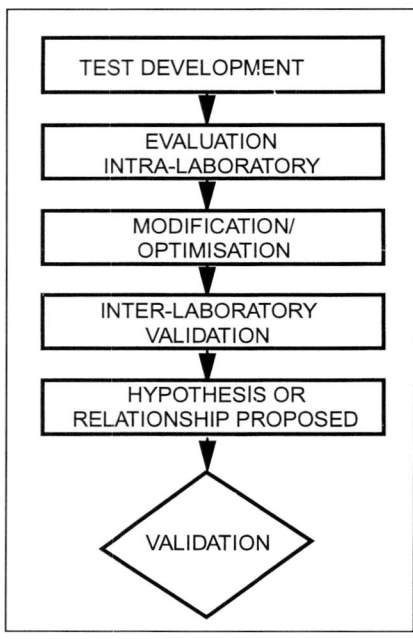

*Fig. 1.* Principles of evolution of alternative tests from development to validation.

## The in vivo Assessment of Skin Irritancy/Corrosion

The potential of chemicals to cause primary skin effects is traditionally determined in animals, usually albino rabbits. The design of the tests has remained relatively unchanged since their introduction almost 50 years ago [1], although the methodology has become standardised, particularly over the last 15 years, through the efforts of organisations such as the Organisation for Economic Co-operation and Development (OECD) [2]. In the standard test, 0.5 ml of liquid or 0.5 g of solid test material is applied to a small area of smooth, shaved skin on the flank of the rabbit and secured in place by means of a semi-occlusive dressing for a period of up to 4 h. After this time the dressing is removed and the application site decontaminated. Subjective assessments of skin irritation (or corrosion) are made in accordance with the Draize scale, at 30–60 min after decontamination and daily thereafter until all signs of irritancy have regressed. The information gained can be used to classify the overall reaction according to local classification needs and procedures.

## Test Development

The development of the transcutaneous electrical resistance test is based on the hypothesis resulting from a retrospective survey by Oliver and co-workers [3, 4] of the onset, duration and nature of the macroscopic skin lesions produced in vivo by a range of irritant and corrosive chemicals.

It was apparent that, for the majority of corrosive chemicals, lesions consistent with a corrosive effect were evident within approximately 4 h of contact and either accompanied or preceded any inflammatory changes (erythema and/or oedema). In a minority of instances, corrosive lesions were noted after the maximal inflammatory response, at approximately 24 h or longer after the initial contact. The irritant chemicals induced mainly transient inflammatory reactions. In addition, some irritant compounds directly, but reversibly, damaged the outermost skin layers (keratin lysis) which imparted a macroscopic 'glossy' appearance to the skin surface.

It was concluded from this analysis that, in general, a direct physico-chemical interaction with skin tissue contributed significantly to the development of a corrosive but not an irritant lesion. It was also concluded that there are occasional exceptions to this hypothesis in that, for a minority of corrosive chemicals, the inflammatory response predominates with the corrosive reaction apparently developing as a consequence of biochemically and physiologically mediated tissue necrosis. Conversely, some irritant chemicals are capable of disrupting the outermost layers of the skin tissue. The relative importance of these physico-chemical and 'biological' processes in skin irritancy and corrosion is illustrated in figure 2.

Consideration of these processes led to the design of the TER test. Both the selection of the tissue and the test measure of effect was based on the property of corrosive chemicals to interact directly with and lyse the cells of the stratum corneum in vivo. The use of epidermal skin discs ensured the inclusion and participation of the normal physiological and anatomical structures that would be exposed in vivo, particularly the stratum corneum. The direct cytolytic action of corrosive chemicals on skin was examined by assessing the functional integrity of the barrier characteristics of the stratum corneum by measurement of changes in electrical resistance [5]. This method was chosen in preference to other measures of barrier function integrity as it was considered the most rapid and simple technique.

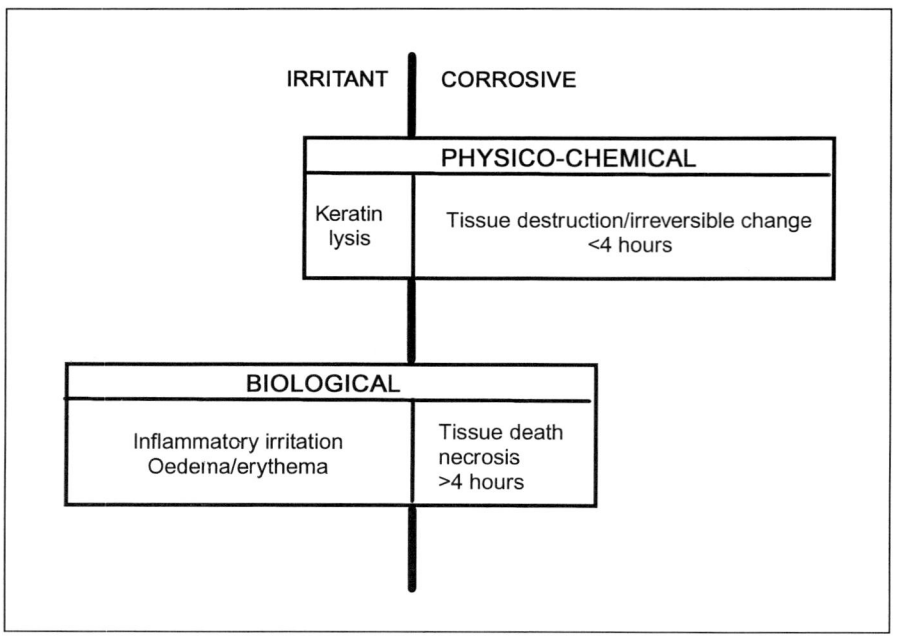

Fig. 2. The relative importance of 'physicochemical' and 'biological' processes in skin irritancy and skin corrosion.

## Main Features of the Original Procedure

The basic apparatus is shown in figure 3.

### Preparation of the Epidermal Skin Discs

The dorsal and flank hair of male albino rats approximately 28 days of age is carefully removed with fine clippers at least 48 h before preparation of the epidermal discs. Epidermal discs are obtained from the shorn dorsal flank of humanely sacrificed animals and excess fat removed. The skin samples are positioned stratum corneum uppermost, over the end of a PTFE tube (diameter 10 mm), and secured in place with a rubber 'O' ring. The whole preparation is suspended in physiological saline and maintained at ambient temperature. A minimum of three discs is used for the assessment of each chemical. Full details of the procedures have been published elsewhere [4].

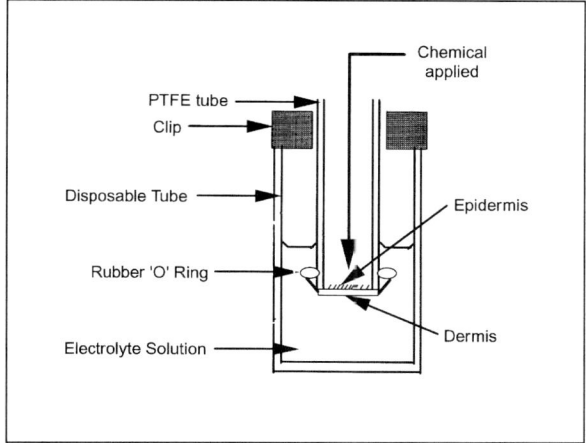

*Fig. 3.* Basic apparatus used in the TER test for the prediction of chemicals corrosive to the skin.

### In vitro Application of Test Material

A standard amount (100 µl of liquid or 100 mg of solid) of each test chemical is applied to the stratum corneum and allowed to remain in contact with the skin for between 1 and 24 h. At the end of the application period the test material is removed with a jet of warm (40–45 °C) water.

### Measurement of Electrical Resistance

After removal of the chemical, the stratum corneum is hydrated by the addition of 70% ethanol (20 µl) and physiological saline (3 ml) added. Using electrodes on either side of the skin discs, resistance is determined by means of a half-bridge apparatus.

### Intra-Laboratory Evaluation

A variety of initial experiments was undertaken [3] to characterise the relationship between chemical contact time with the epidermal skin disc, degree of damage to the stratum corneum and the rate of fall of electrical resistance across the ex vivo preparation. With the knowledge of the in vivo effects of the test chemicals, the optimum skin contact time and the threshold for positive effect in vitro were established. Preliminary experiments were conducted with 63 chemicals (44 corrosive and 19 non-corrosive) selected to represent a range of typical industrial chemicals and formulations of different physico-chemical properties such as pH and physical state. In the majority of cases the pH of the test materials gave no indication or prediction as to the potential for corrosive action. An optimal contact time with the skin of 24 h was determined, together with the definition of 4 kΩ/disc as the threshold below which a corrosive response in vivo could be predicted (skin treated with distilled water under identical conditions gave consistent resistance values of greater than 10 kΩ/disc). Using these criteria and procedures the optimal balance between the sensitivity of the test (91% – the ability to correctly identify corrosive chemicals) and the specificity (74% – the ability to identify non-corrosives) of the test was established. This data set was enlarged [4] to cover a greater number of irritant materials and to encompass non-irritant chemicals. Using the same contact time and threshold of effect, a high predictive ability was again demonstrated (sensitivity 91%, specificity 69% and 95% for irritants and non-irritants, respectively). In subsequent evaluation of the method [4] similar results were obtained for 49 corrosives and 19 non-corrosives.

The consideration of false-positive and false-negative predictions is helpful in understanding and refining the original hypothesis of Oliver and co-workers regarding the mechanism of action of corrosive and irritant chemicals. Further examination of the non-correlates allows the overall performance of the original method to be assessed in more detail. The false-negative predictions (or under-predictions of effect) represent chemicals that can be classified as 'biological corrosives' (fig. 2). In vivo, the corrosive lesion caused by these compounds was relatively slow to develop and was judged to be the sequel to a severe inflammatory response resulting in a cascade of biological events leading to tissue necrosis. The action of these chemicals would not be expected to reduce electrical resistance below the 4-kΩ threshold within the contact time employed.

Examination of the in vivo data for the false-positive predictions (over predictions in the in vitro test) indicates that keratin lysis (removal of the outer layers of the stratum corneum) was a consistent feature and that these chemicals were solvents and surfactants. There was no subsequent development of a corrosive lesion and the fall in electrical resistance in the in vitro test could be ascribed to the surface activity and delipifying properties of these materials.

## Refinement of Original Procedure

The early development and initial validations of the test used a NaCl electrolyte solution coupled with a corrosive threshold of 4 kΩ/disc. As described above, the effective performance of this procedure was compromised by those sub-

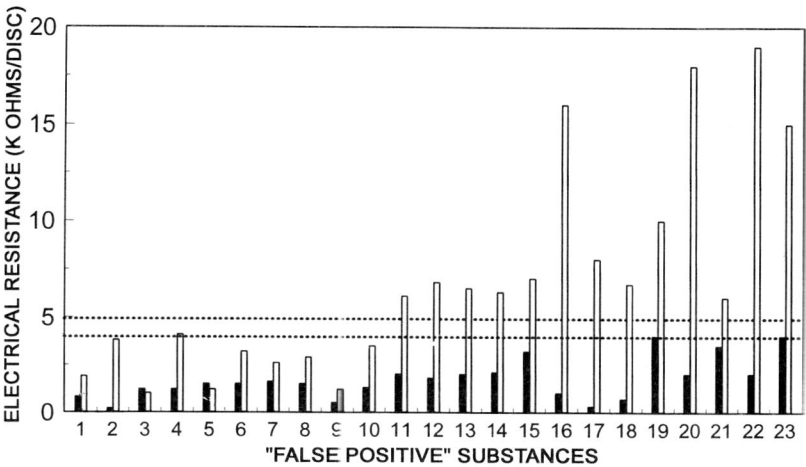

*Fig. 4.* Evaluation of the performance of MgSO$_4$ electrolyte with chemicals known to be 'false positives' when using NaCl as electrolyte. Skin discs were exposed to 23 industrial chemicals known to be non-corrosive in vivo. The electrical resistance values for MgSO$_4$ (open bars) and NaCl (closed bars) are shown. All 23 chemicals are below the 4 k$\Omega$/disc using NaCl electrolyte and are therefore erroneously identified as corrosives. In contrast, with the MgSO$_4$ electrolyte, chemicals 11–23 are correctly identified as being non-corrosive (>5 k$\Omega$/disc).

stances, predominantly organic solvents and surfactants, which reduced skin electrical resistance by enhancing transcutaneous ion flux giving false-positive prediction. It has been demonstrated, for solvents, that these effects are to some extent reversible and do not necessarily involve lysis of subcutaneous tissue. Barlow et al. [6] developed a refinement of the original test based on the proposal that changing the ionic species in the electrolyte solution used for measurement of electrical resistance could modify transcutaneous ion flow and hence could be used to reduce the false-positive rate observed with NaCl as electrolyte. A comparative evaluation of the performance of five different electrolytes indicated that using MgSO$_4$ combined with a revised corrosive threshold of 5 k$\Omega$/disc yielded an optimum lowered false-positive rate with no reduction in sensitivity. An evaluation of 88 industrial chemicals (9 corrosives, 56 non-corrosives and 23 known false positives when tested with NaCl as electrolyte) indicated that all corrosive and non-corrosive chemicals could be correctly identified using MgSO$_4$. Most importantly, the overall false positive rate was reduced from 26 to 9% (fig. 4).

It was concluded, therefore, that the larger ionic radius of the hydrated Mg ion is less likely to penetrate the stratum corneum after exposure to non-corrosive organic solvents and surfactants, thus allowing correct prediction of a material as non-corrosive in the in vitro corrosivity test. However, stratum corneum that had been irreversibly denatured or physically damaged by exposure to true corrosive substances was evidently freely permeable to this ionic species.

### Inter-Laboratory Validation of the in vitro Corrosivity Test

Subsequently, a collaborative study to evaluate the reliability and reproducibility of the in vitro skin corrosivity test was undertaken in three laboratories [7]. Twenty substances were examined in each of the participating laboratories and the results compared with existing data from standard assays in vivo. Magnesium sulphate was used as electrolyte with the corrosive threshold set at 5 k$\Omega$/disc. The conclusions of this inter-laboratory validation included that all three laboratories could correctly identify the six in vivo corrosive substances tested (24-hour mean TER values <5 $\Omega$/skin disc). However, there was no evidence of an ability to rank the irritant materials tested. With a single exception, and in only one of the three laboratories, was a non-corrosive substance identified as a (false) positive.

### Use and Performance of the Modified Test

The in vitro skin corrosivity test is currently conducted to the following specifications:

| | |
|---|---|
| Tissue source | 28 days albino rat |
| Epidermal disc dimensions | 10 mm diameter |
| Amount of test sample applied | 150 µl liquid/100 mg solid |
| Application time | 1, 4 and 24 h |
| Electrolyte | MgSO$_4$ (154 m$M$) |

A recent evaluation of the performance of the in vitro corrosivity test [Lewis 1993; unpubl. data] in predicting the corrosive potential of 338 diverse chemicals showed that this assay correctly predicted 90% (19 out of a total of 21) skin corrosives tested. This high sensitivity was combined with an equally high specificity (94%) ability to correctly identify non-corrosive materials (299 out of a total of 317).

The in vitro corrosivity test is a rapid, technically simple, reproducible and objective technique that can be integrated into assessments of skin toxicity as one part of a testing cascade allowing improved decisions to be made on the subsequent use of live animals. Thus:

(1) If a positive result is obtained then in some circumstances (such as for hazard handling requirements within Zeneca), animal tests are not conducted (replacing the use of animals completely).

(2) If confirmatory animal tests are required (for example, for regulatory purposes) these tests can be designed to reduce the number of animals used (often reducing the number tested from 6 to a single animal) and to refine experiments by shortening the duration of application to the skin (to 1 h or even 3 min compared with the standard 4 h) so that possible pain and distress is kept to a minimum.

### The Use of Human Skin – Towards More Relevant Hazard Identification and Risk Assessment

It is known that rabbit skin can be more sensitive than human skin [8]. This difference between rabbit and human skin may lead to substances being incorrectly classified on the basis of standard in vivo animal skin irritation and corrosivity tests. Oliver and Pemberton [3] reported that one third of chemicals that were 'corrosive' in in vivo rabbit studies (and also in in vitro studies using rat skin) were non-corrosive when human skin was used in the in vitro corrosivity test.

Whittle and Basketter [9, 10] have applied the in vitro corrosivity test to the study of human skin and have defined empirically (i.e. with a battery of simple chemicals) a threshold of 11 k$\Omega$/skin disc, below which test materials are predicted as having the potential to cause corrosive effects on human skin in vivo. These workers anticipated that the more resistant nature of human skin would lead to a number of substances being classified as non-corrosive, contrary to historical in vivo animal data.

Using the 11-k$\Omega$/skin disc threshold, a range of substances with an existing formal EC classification have been evaluated [10]. These include substances tested at specific concentrations such that they fall within predetermined EC classification limits. The total dataset obtained to date is presented in table 1 Of the 39 substances tested, 13 have been classified as corrosive according to EC criteria, whilst 26 were not corrosive. Using human skin in the TER test, 13 of these 39 substances were found to be corrosive, including 1 'false positive'; 26 were found to be non-corrosive, including 1 'false negative'. A number of these substances have been tested on several skin samples obtained from different donors, or with a 3 × greater dose. However, the results have always been consistent [10]. Consequently, it is worth examining the discrepancies between the in vitro result and the EC classification in some detail.

Of the 17 single chemical entities tested in the human skin TER test, 6 are formally classified in the EC [11] as corrosive and all were clearly identified as

*Table 1.* Comparison of in vivo classification with in vitro human skin classification

| Test material | EEC/in vivo classification | Human skin TER, kΩ/disc | In vitro classification |
|---|---|---|---|
| Untreated | – | 44.4 ± 6.0 | – |
| *Common substances* | | | |
| Distilled water | no label required | 41.3 ± 11.4 | non-corrosive |
| 1% sodium lauryl sulphate | no label required | 14.5 ± 3.3 | non-corrosive |
| 9% hydrochloric acid | no label required | 28.5 ± 4.2 | non-corrosive |
| 10% acetic acid | irritant | 16.0 ± 5.7 | non-corrosive |
| 1% sodium hydroxide | irritant | 8.4 ± 6.5 | corrosive |
| 18% hydrochloric acid | irritant | 19.7 ± 9.4 | non-corrosive |
| Sodium perborate monohydrate | irritant | 29.0 ± 5.4 | non-corrosive |
| Sodium perborate tetrahydrate | irritant | 70.2 ± 3.8 | non-corrosive |
| Sodium carbonate | irritant | 25.2 ± 11.2 | non-corrosive |
| Sodium percarbonate | irritant | 23.8 ± 11.2 | non-corrosive |
| Sodium chloride | no label required | 28.7 ± 15.2 | non-corrosive |
| Magnesium sulphate | no label required | 53.4 ± 9.7 | non-corrosive |
| 4% sodium hydroxide | corrosive | 1.1 ± 0.3 | corrosive |
| 30% acetic acid | corrosive | 5.6 ± 1.9 | corrosive |
| 16.3% trichloroacetic acid | corrosive | 8.4 ± 3.3 | corrosive |
| 54% hydrochloric acid | corrosive | 2.4 ± 0.7 | corrosive |
| 8% sodium hydroxide | corrosive | 1.5 ± 0.3 | corrosive |
| *Anionic surfactants* | | | |
| Alkyl sulphate, Na Salt, C12 | irritant | 11.5 ± 2.3 | non-corrosive |
| Alkyl ether sulphate, 2EO, Na salt, C12–C14, 70% | irritant | 21.6 ± 5.3 | non-corrosive |
| Olefinsulphonate, Na salt, C14–C16 | irritant | 24.5 ± 2.2 | non-corrosive |
| Alkyl (4EO) phosphoric acid, Di/Tri ester, C12–C14, 100% | irritant | 25.0 ± 10.6 | non-corrosive |
| *Non-ionic surfactants* | | | |
| Fatty alcohol ethoxylate, 2EO, lauryl, 100% | irritant | 32.4 ± 6.3 | non-corrosive |
| Fatty alcohol ethoxylate, 12EO, lauryl, 100% | irritant | 33.9 ± 8.2 | non-corrosive |
| Fatty acid monoethanolamide, coco | irritant | 25.2 ± 7.7 | non-corrosive |
| *Quaternary surfactants* | | | |
| Cocoalkyl trimethyl ammonium chloride, C8–C18, 35% | irritant | 25.7 ± 5.8 | non-corrosive |
| Behenyl trimethyl ammonium chloride, C20–C22, 80% | irritant | 30.6 ± 4.5 | non-corrosive |
| Cocoalkyl dimethylbenzyl ammonium chloride, C8–C18, 50% | corrosive | 2.1 ± 0.9 | corrosive |

*Table 1* (continued)

| Test material | EEC/in vivo classification | Human skin TER, kΩ/disc | In vitro classification |
|---|---|---|---|
| *Fatty amines and derivatives* | | | |
| Primary amine, hydrogenated tallow, C16–C18 | irritant | $30.7 \pm 3.6$ | non-corrosive |
| Primary amine, tallow, C16–C18 | corrosive | $6.1 \pm 5.0$ | corrosive |
| Secondary amine, dicocoalkyl, C8–C18 | irritant | $29.8 \pm 6.0$ | non-corrosive |
| Tertiary amine, hydrogenated tallow alkyl dimethyl, C16–C18 | corrosive | $8.3 \pm 5.1$ | corrosive |
| Amine, tallow ethoxylated, 5EO | irritant | $27.1 \pm 6.1$ | non-corrosive |
| *Fatty acids* | | | |
| 60% acetic acid | corrosive | $5.7 \pm 9.7$ | corrosive |
| Propionic acid | corrosive | $9.0 \pm 2.2$ | corrosive |
| Butyric acid | corrosive | $2.5 \pm 0.6$ | corrosive |
| Caproic acid | corrosive | $4.7 \pm 3.2$ | corrosive |
| Caprylic acid | corrosive | $25.2 \pm 8.7$ | non-corrosive |
| Capric acid | irritant | $29.9 \pm 5.4$ | non-corrosive |
| C12 fatty acid | no label required | $36.7 \pm 11.6$ | non-corrosive |

Human TER expressed in kΩ/disc $\pm$ SD. A minimum of three skin discs were assessed for each substance.

such by the in vitro assay. However, whilst the 11 other substances are regarded as not corrosive, and 10 were identified as such using human skin in the TER test, in our hands 1% sodium hydroxide generally gives results with human skin in vitro which suggest it is corrosive. Experience with the treatment of the skin of human volunteers tends to support this conclusion [12; Basketter, unpubl. data].

With the 15 surfactant substances tested on human skin in the TER test, the assay correctly identified the three surfactants formally classified as corrosive. The remaining surfactants are classified and labelled as skin irritants and these did not cause any significant reduction in the TER and so were judged not corrosive in the in vitro assay. Notably, this includes the anionic surfactant, sodium lauryl sulphate, which is widely regarded and used as a model skin irritant. Although the data set is small when compared to the rat skin model, it does suggest a lower tendency to false-positive results with this type of chemical.

Seven fatty acids covering the chain lengths C2–C12 were tested on a number of different human skin samples in the TER test. Two of these are not classified as corrosive to skin and hence would not be expected to cause a significant reduction in the TER. The remaining fatty acids have been classified as corrosive on the basis of historic in vivo animal data. However, only the fatty acids of chain length up to C8 caused a significant reduction in the TER of human skin in the TER test and thus would classify as corrosive. It is interesting to note that when tested on rat skin in the TER test, the fatty acids classified as corrosive in the EC were identified as such. However, it has been shown that animal skin can indicate a potential hazard in vivo which may not be reproducible in human skin [8, 13–15]; this has been clearly demonstrated for the C8 fatty acid [15]. It would appear that this difference between animal skin and human skin is also demonstrable in the TER test.

## Conclusion

The in vitro corrosivity test has found use as a true alternative test. Alternatives can be defined as tests which offer the same or better information on hazard, when compared to the standard approach, and which ideally avoid the use of whole animals. As can be seen from the above examples, the TER test (when used as a screen prior to animal tests) can replace the use of animals in testing skin corrosives, can reduce the numbers of animals used in such tests, and can be used to refine protocols such that pain and distress are minimised.

Use of human skin has widened the relevance of the test and is beginning to fulfil the ideal of an alternative test in providing improved hazard information more relevant to the species of interest.

## References

1   Draize JH, Woodward G, Calvery HO: Methods for the study of skin irritation and toxicity of substances applied topically to the skin and mucous membranes. J Pharmacol Exp Ther 1994;82: 377–390.
2   OECD: Acute eye irritation: OECD guidelines for testing of chemicals. Test Guideline 405. Paris, OECD, 1987.
3   Oliver GJA, Pemberton MA: The identification of corrosive agents for human skin in vitro. Food Chem Toxicol 1986;24:513.
4   Oliver GJA, Pemberton MA, Rhodes C: An in vitro model for identifying skin corrosive chemicals. I. Initial validation. Toxicol In Vitro 1988;2:7–17.
5   Blank IH, Finesinger JE: Electrical resistance of the skin. Arch Neurol Psychiatry 1964.
6   Barlow A, Hirst R, Pemberton MA, Rigden A, Hall TJ, Oliver GJA, Botham PA: Refinement of an in vitro test for the identification of skin corrosive chemicals. Toxicol Methods 1991;6:106–115.
7   Botham PA, Hall TJ, Dennett R, McCall JC, Basketter DA, Whittle E, Cheeseman M, Esdaile DJ, Gardner J: The skin corrosivity test in vitro. Results of an inter-laboratory trial. Toxicol In Vitro 1992;6:191–194.
8   Nixon GA, Tyson CA, Wertz WC: Interspecies comparisons of skin irritancy. Toxicol Appl Pharmacol 1975;31:481–490.

9   Whittle E, Basketter DA: The in vitro corrosivity test: Development of the method using human skin. Toxicol In Vitro 1993;7:265–268.

10  Whittle E, Basketter DA: The in vitro corrosivity test: Comparison of in vitro human skin with in vivo data. Toxicol In Vitro 1993;7 269- 274.

11  EEC: Council Directive of 7 June 1988 on the approximation of the laws, regulations and administrative provisions of the Member States relating to the classification, packaging and labelling of dangerous preparations. Off J Eur Commun 1988;L18:14.

12  Dykes PJ, Black DR, York M, Dickens A, Marks R: A stepwise procedure for evaluating irritant or corrosive materials in normal volunteer subjects. Hum Exp Toxicol 1994;in press.

13  Kaestner W: Zur Speziesabhängigkeit der Hautverträglichkeit von Kosmetikgrundstoffen. J Soc Cosmet Chem 1977;28:741.

14  Motoyoshi K, Toyoshima Y, Sato M, Yoshimura M: Comparative studies on the irritancy of oils and synthetic perfumes to the skin of rabbit, guinea pig, rat, miniature swine and man. Cosmet Toilet 1979;84:41–42.

15  Basketter DA, Whittle E, Griffiths HA, York M: The identification and classification of skin irritation hazard potential by human patch test. Food Chem Toxicol 1994;in press.

Dr. R.W. Lewis, Toxicity Section, Zeneca Central Toxicology Laboratory, Alderley Park, Macclesfield, Cheshire, SK10 4TJ (UK)

Elsner P, Maibach HI (eds): Irritant Dermatitis. New Clinical and Experimental Aspects.
Curr Probl Dermatol. Basel, Karger, 1995, vol 23, pp 256–264

..........................

# EEC/COLIPA in vitro Photoirritancy Program: Results of the First Stage of Validation[1]

*H. Spielmann*[a], *M. Liebsch*[a], *W.J.W. Pape*[b], *M. Balls*[c], *J. Dupuis*[d],
*G. Klecak*[e], *W.W. Lovell*[f], *T. Maurer*[g], *O. De Silva*[h], *W. Steiling*[i]

[a]ZEBET Bg VV, Berlin, Germany; [b]Beiersdorf AG, Hamburg, Germany;
[c]FRAME, Nottingham, UK; [d]COLIPA, Brussels, Belgium;
[e]Hoffmann-La Roche, Basel, Switzerland; [f]Unilever Research, Sharnbrook, UK;
[g]Ciba-Geigy Ltd., Basel, Switzerland; [h]L'Oréal, Aulnay-sous-Bois, France;
[i]Henkel KGaA, Düsseldorf, Germany

Phototoxicity is an acute reaction which can be caused by a single treatment with a chemical and UV or visible radiation. In vivo, the reaction can be evoked in all subjects provided that the concentration of chemical and dose of light are appropriate. The current toxicological assays for 'acute dermal phototoxicity' are animal tests using guinea pigs, rabbits, rats or mice. Although a standard protocol for phototoxicity testing in animals has recently been recommended [1], acceptance of an animal test in an OECD guideline for phototoxicity testing could not be achieved. However, a sequential approach using in vitro testing prior to consideration of animal testing was recommended for phototoxicity testing. Consequently, COLIPA and DG XI of the EEC agreed to conduct a joint project on developing validated in vitro phototoxicity tests, which was coordinated by ZEBET.

During phase I of the study (1992/1993), a set of 20 test chemicals (12 phototoxins (PT), 4 non-PT and 4 UV-absorbing non-PT) was carefully selected by a COLIPA task force. Moreover, all laboratories had to perform an in vitro photo-

[1] This study was supported by a grant from DG XI of the EC (Brussels) and by a grant from ECVAM (European Centre for the Validation of Alterative Methods, Ispra, Italy). The authors are indebted to COLIPA (the European Cosmetic, Toiletry and Perfume Association, Brussels) for continuous support.

toxicity assay based on the simple 3T3 cell neutral red uptake (NRU) cytotoxicity assay [2], which was modified for phototoxicity testing at the laboratory of Beiersdorf by exposing 3T3 cells both to test chemicals and UVA irradiation.

Since UVA is most relevant in phototoxic reactions, appropriate filters were used to avoid exposure to UVB, which has strong cytotoxic actions. Standardized exposure to UVA was an important technical aspect of the study, therefore, all laboratories had to use an identical light source and to apply an identical dose of 5 J/cm$^2$ UVA in all of the assays. In the present report, results from 3T3 cell NRU phototoxicity assay (3T3 NRU-PI assay) will be described and data from the other in vitro phototoxicity assays will briefly be summarized.

## Materials and Methods

### UVA Light Source, UVA Meter

In the 3T3 NRU-PI assay all laboratories used an identical UV light source, a doped mercury-metal halide lamp (SOL 500, Dr. Hönle, Martinsried, Germany), which simulates the spectral distribution of the natural sunlight. A spectrum almost devoid of UVB was achieved by filtering with a 50% transmission at a wavelength of 335 nm (filter: H1, Dr. Hönle). Emitted energy was measured with a calibrated UVA meter (Type No. 37, Dr. Hönle). Calibration was controlled with a second UVA meter which had to be kept in the dark. For the red blood cell (RBC) assay the SOL 500 was filtered with a H2 filter (Dr. Hönle, 50% transmission at 320 nm), cut-off wavelength of 305 nm), since RBC are more resistant to UVB. Phototoxicity testing with the dermal model Skin$^2$ was performed under the same UVA irradiation conditions as in the standard 3T3 NRU-PI assay.

### Selection of Test Chemicals

Selection of test chemicals with high quality in vivo data is very difficult, since data from human patients are rare and quite often differ from animal data. Taking this into account, a COLIPA task force selected a list of 20 test chemicals which are supported by high quality human and animal data from the literature, which were assigned to 3 classes (table 1): class I – 12 phototoxins (PT); class II – 4 chemicals which are absorbing UVA light but are not PT, however some are photoallergens; class III – 4 chemicals which are neither UVA-absorbing nor PT.

According to recent studies [3] chemical No. 8 (piroxicam) should not be classified as a phototoxin, since products from UVA photolysis are inducing allergic reactions in patients either after repeated contact or by cross-reaction with contact sensitizers of similar chemical structure, as e.g., merthiolate [4]. Thus, piroxicam is assigned to class II rather than to class I.

### Experimental Design of the 3T3 NRU-PI Assay

The NRU cytotoxicity assay with Balb/c 3T3 fibroblasts was adapted for phototoxicity testing in the following manner [5]: Balb/c 3T3 cells, clone 31 (ICN-Flow) were cultured in 96-well microtiter plates as described earlier [2]. After 24 h DMEM was removed, cells were washed twice in EBSS and eight concentrations of the test chemicals dissolved in EBSS were added. Insoluble chemicals were dissolved in DMSO and added at a maximum of 1% DMSO in EBSS. After 24 h of incubation with test chemicals plates were exposed to UVA (1.67 mW/cm$^2$) for 50 min (= 5 J/cm$^2$). A second set of plates with the same chemicals was kept in the dark. After light exposure, EBSS was replaced by DMEM (without any test chemicals) and NRU was determined 24 h later according to Spielmann et al. [5]. To determine

*Table 1.* Phototoxicity of 20 chemicals tested with the 3T3 NRU phototoxicity assay

| | | In vivo human | In vivo animal | In vitro data | | | n | Result |
|---|---|---|---|---|---|---|---|---|
| | | | | mean IC$_{50}$–UV µg/ml | mean IC$_{50}$+UV µg/ml | mean factor –UV/+UV | | |
| *Class I* | | | | | | | | |
| 1 | Promethazine | + | +/– | 45.9 | 0.8 | 78.5 | 13 | + |
| 2 | Chlorpromazine | ++ | ++ | 24.6 | 0.6 | 46.6 | 13 | + |
| 3 | 6-Methylcoumarine | a | a | * | 32.7 | | 13 | + |
| 4 | TCSA | a | + | 19.8 | 0.4 | 55.6 | 12 | + |
| 5 | Doxycycline | + | + | 1,182 | 6.4 | 255 | 4 | + |
| 6 | 8-MOP | ++ | ++ | * | 14.7 | | 11 | + |
| 7 | Tetracycline | + | + | 1,916 | 16.8 | 374 | 9 | + |
| 9 | Amiodarone | + | + | 24.3 | 4.1 | 6 | 9 | + |
| 10 | Bithionol | + | + | 13.9 | 3.9 | 7 | 13 | + |
| 11 | Neutral red | + | | * | 0.01 | | 14 | + |
| 12 | Rose bengal | +/– | – | 4.2 | 0.2 | 70.2 | 13 | + |
| *Class II* | | | | | | | | |
| 8 | Piroxicam[1] | ? | – | * | * | | 11 | – |
| 13 | Cinnamic aldehyde | a | a | 32.8 | 10.6 | 3.6 | 8 | – |
| 14 | Chlorhexidine | | | 61.5 | 74.4 | 1.5 | 11 | – |
| 15 | Uvinul MS 40 | +/– | | 15,958 | 11,577 | 1.4 | 11 | – |
| 16 | PABA | a | | 10,463 | 9,780 | 1 | 7 | – |
| *Class III* | | | | | | | | |
| 17 | Penicillin G | | | 53,914 | 49,755 | 1.1 | 8 | – |
| 18 | *L*-Histidine[2] | | | * | * | | 12 | – |
| 19 | Thiourea | a | | 17,651 | 16,944 | 1 | 13 | – |
| 20 | Lauryl sulfate | | | 35.6 | 24.2 | 1.5 | 14 | – |

No cytotoxicity could be determined up to the highest test concentration.
n = Number of determinations; means = arithmetic means of n determinations (± SD not shown); ++ = strong phototoxin; + = phototoxic; – = not phototoxic; +/– = inconclusive; a = (photo)allergen.
[1] Highest test concentration: 2,400 µg/ml.
[2] Highest test concentration: 46,400 µg/ml.

phototoxicity cytotoxic concentrations resulting in 50% reduction of viability ($IC_{50}$) were compared in the presence and absence of UVA irradiation, by calculation of the factor between the two $IC_{50}$ values as described previously [5]:

UV factor = $IC_{50}$ (–UV)/$IC_{50}$ (–UV).

The cut-off value of the UV factor for discriminating between PTs and non-PTs was determined by discriminant analysis as outlined in the 'Results' section.

### The Candida albicans Phototoxicity Assay in Yeast

In the simple yeast assay [6] *C. albicans* is seeded on agar plates and zones of growth inhibition around the applied test chemical in duplicate plates with and without UVA exposure are compared 24– 48 h after application. The assay is sensitive to phototoxins acting primarily on DNA [7].

### Histidine Oxidation Test

Due to activation by UVA or UVB some chemicals will oxidize the amino acid histidine via singlet oxygen formation. Reduction of histidine can easily be monitored in a colorimetric assay [7].

### Photohemolysis and Hemoglobin Oxidation with Red Blood Cells

The method is used to study the phototoxic potential of chemicals by their ability to damage the RBC membrane and/or to oxidize hemoglobin when exposed to UV and/or visible light. A new protocol [8] combines the widely used photohemolysis test [7] with the spectral changes characteristic of methemoglobin formation.

### SOLATEX PI Assay

The principle of the commercial SOLATEX PI assay (In Vitro International, IVI, USA) is quantification of an enhanced response in a physicochemical model of dermal irritation upon exposure to UV radiation. Controls not exposed to UV radiation are providing the background irritation response. Two concentrations of a known photoirritant ('positive controls') are included to verify the test performance. The total response is measured spectrophotometrically at 400 nm. Materials inducing an increase >40% in OD after exposure to UV light compared to background, are classified as phototoxic.

### Skin² PI Assay

The commercial Skin² dermal model ZK 1350 (Advanced Tissue Sciences, ATS, USA) is a three-dimensional human skin model consisting of dermal, epidermal and corneal layers. It is produced by seeding neonatal fibroblasts onto an inert nylon mesh and growing them into dermal tissue. Keratinocytes are seeded on top of the fibroblast layer and are differentiating into an epidermis including a multilayered stratum corneum. The tissue is cut into pieces of 9 mm² (ZK 1350) and sterilely sealed test kits are shipped worldwide by airfreight. In the Skin² PI assay, the cytotoxic effects of a chemical are compared to additional exposure to a nontoxic dose of UVA. Chemicals are applied topically and cytotoxicity is determined in the MTT assay in comparison to solvent controls. Cell viability is calculated for each piece of tissue as % of control and the viability of irradiated tissue and dark controls is compared. The cut-off value for classifying a chemical as phototoxic is a reduction in viability of more than 30% due to UVA exposure.

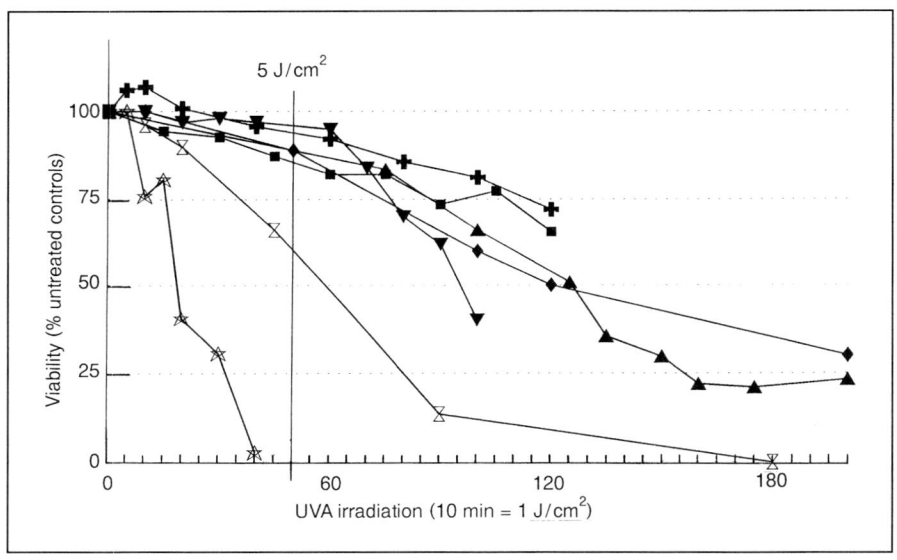

*Fig. 1.* UVA sensitivity of 3T3 mouse fibroblasts in 7 laboratories. 3T3 cells were exposed to UVA (1.67 mW/cm$^2$) in 96-well plates for up to 180 min in order to determine UVA sensitivity. Individual curves are representing data from 7 different laboratories. Curves with open symbols were obtained with older cells. The vertical line at 50 min exposure time indicates the standard UVA dose of 5 J/cm$^2$ which was used for phototoxicity testing in the 3T3 NRU assay.

## Results

### 3T3 NRU Assay

UVA sensitivity of Balb/c 3T3 cells was tested in seven laboratories. Figure 1 shows that in 5 of the laboratories the viability of 3T3 cells with passage numbers between 70 and 80 was not affected within an UVA dose range of about 0–10 J/cm$^2$. However, in the same dose range in the 2 laboratories which were using 3T3 cells with passage numbers of 130–140, a significant reduction in viability was observed in the same dose range. Therefore, it was decided to use 3T3 cells with a low passage (<100) throughout the study. The result described in figure 1 reveals that for 3T3 cells the highest nontoxic UVA dose was 5 J/cm$^2$ (50 min exposure at an intensity of 1.67 mW/cm$^2$). This dose was chosen for further testing in order to detect even weak phototoxins.

A UV factor (IC$_{50}$ –UV/IC$_{50}$ +UV) could only be determined with 15 of the 20 test chemicals, since 5 chemicals (No. 3, 6, 8, 11 and 18) were not cytotoxic to

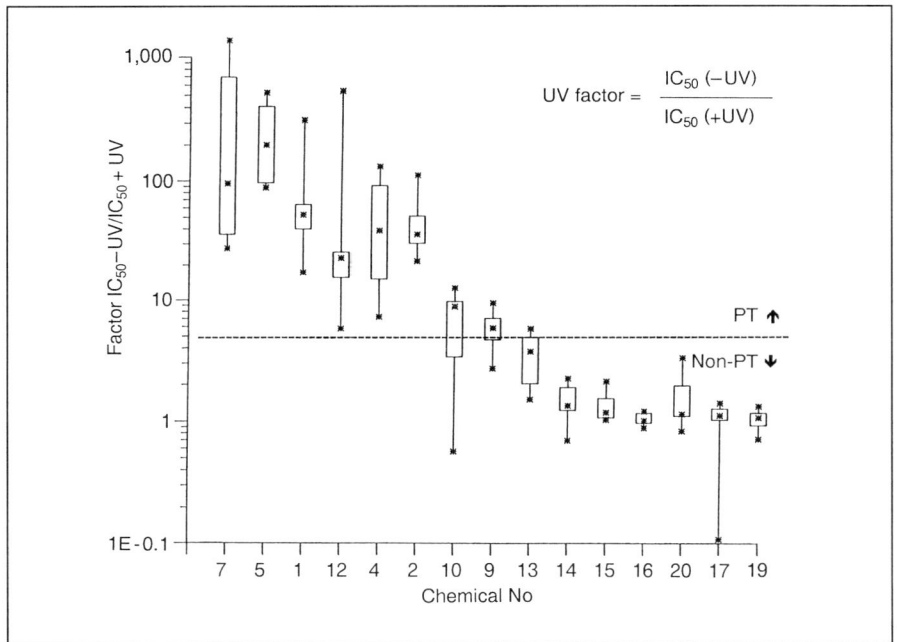

*Fig. 2.* 3T3 NRU phototoxicity assay: determination of UV photoactivation factors for 15 chemicals by discriminant analysis. UVA factors ($IC_{50}-UV/IC_{50}+UV$) were determined with 15 chemicals, which were cytotoxic both, with and without UVA exposure. The numbers of the chemicals correspond to tables 1 and 2. The box plots show medians, 95% confidence limits (boxes) and minima/maxima of factors determined in all of the laboratories. The dashed line indicates the cut-off value determined by discriminant analysis to discriminate between nonphototoxic and phototoxic chemicals.

3T3 cells in the dark even at the highest concentrations tested. Taking into account all 158 UV factors calculated with the 15 cytotoxic chemicals in all of the laboratories, discriminant analysis revealed a UV factor of 5.1 as cut-off value to discriminate between phototoxic and nonphototoxic test chemicals. Figure 2 gives the photoinactivation factors for the 15 chemicals as 'median box plots' of 158 determinations.

Table 1 proves that all of the 11 phototoxins and all of the 9 nonphototoxins were correctly identified in the 3T3 NRU-PI assay. UV factors ranging from 6 (No. 9) to 374 (No. 7) could only be determined with 8 of the 11 phototoxic chemicals (class I). Cytotoxicity could not be determined with chemicals 3, 6 and 11 of class I. However, a cytotoxic effect was measured in the NRU assay after UVA

*Table 2.* Summarized data of assays used in the EC/COLIPA validation study

| | Mechanistic assays | | | Commercial assays | | Growth inhibition assays | | | |
|---|---|---|---|---|---|---|---|---|---|
| | histidine photo-oxida-tion | RBC photo-hemo-lysis | RBC photo-Hb oxi-dation | SOLA-TEX PI | Skin[2] ZK 1300 ZK 1350 | yeast growth inhibi-tion | human lympho-cytes MTT | human keratino-cytes NRU | COMMON standard 3T3-NRU assay |
| *Class I* | | | | | | | | | |
| 1 Promethazine | + | + | + | + | + | + | + | + | + |
| 2 Chlorpromazine | (+) | + | + | + | + | (+) | + | + | + |
| 3 6-MC | + | + | + | + | − | + | + | + | + |
| 4 TCSA | + | + | + | + | + | + | − | + | + |
| 5 Doxycycline | + | − | + | + | + | + | + | + | + |
| 6 8-MOP | + | − | (+) | − | + | + | + | − | + |
| 7 Tetracycline | + | − | + | + | + | − | − | + | + |
| 9 Amiodarone | − | + | − | + | + | + | + | − | + |
| 10 Bithionol | − | + | + | +/− | − | − | − | + | + |
| 11 Neutral red | + | + | − | + | + | − | + | + | + |
| 12 Rose bengal | ++ | + | (+) | + | + | + | n.t. | + | + |
| *Class II* | | | | | | | | | |
| 8 Piroxicam* | − | − | − | − | − | − | − | − | − |
| 13 Cinnamic aldehyde | (+) | − | − | − | − | − | − | − | − |
| 14 Chlorhexidine | − | (+) | − | − | − | − | − | − | − |
| 15 Uvinul MS 40 | − | − | − | + | − | − | − | − | − |
| 16 PABA | − | − | − | − | − | − | − | − | − |
| *Class III* | | | | | | | | | |
| 17 Penicillin G | − | − | − | − | − | − | − | − | − |
| 18 *L*-Histidine | − | − | − | − | − | − | − | − | − |
| 19 Thiourea | − | − | − | n.q. | − | − | − | − | − |
| 20 Lauryl sulfate | − | − | − | − | − | − | − | − | − |

* For classification of piroxicam see 'selection of chemicals'.

n.q. = Test not qualified; n.t. = not tested.

exposure (table 1). The 3 PT were, therefore, correctly classified in the 3T3 NRU-PI assay without using the UV factor.

For the 5 UVA-absorbing chemicals which are not phototoxic in vivo (class II), UV factors between 1 (No. 16) and 3.6 (No. 13) were determined with the exception of piroxicam. UV factors between 1 and 1.5 were observed with the 3 cytotoxic non-UVA absorbing non-PTs (class III). With chemicals 8 (piroxi-cam) and 18 (*L*-histidine) cytotoxicity could not be determined in the absence or presence of UVA exposure, and they were correctly identified as 'non-PT'.

## SOLATEX PI Assay

This new in vitro assay did not allow phototoxicity testing in a reproducible manner mainly due to problems of the software provided by the manufacturer. According to table 2 only 8 of the class I chemicals could be correctly identified in the SOLATEX PI assay, 1 'false-positive' result was obtained with the 4 non-PI chemicals of class II and 1 of the non-PI chemicals of class III could not be tested due to technical problems of the assay.

## Skin$^2$ PI Assay

Table 2 demonstrates that 9 of the 11 PI chemicals of class I were correctly identified in the skin$^2$ ZK 1350 PI assay, No. 3 and No. 10 being the 'false nega-tives'. All of the negative chemicals of classes II and III were correctly identified as 'non-PI'.

## Results Obtained with Additional Assays

Table 2 also gives the data from in vitro phototoxicity testing with assays, which are used to identify specific phototoxic mechanisms, e.g. histidine oxida-tion, RBC photohemolysis and RBC hemoglobin oxidation. Among different lab-oratories, the results were quite reproducible. Since these assays are related to specific mechanisms of phototoxicity, each of the tests gave a positive result only with a few of the phototoxic chemicals of class I. There was some overlap, since the phototoxicity of chemicals is quite often due to more than a single mecha-nism.

In contrast, data obtained with additional in vitro phototoxicity assays, e.g. the human lymphocyte and keratinocyte assays, were not reproducible, since the tests were still under development. As expected, the yeast assay, which is specific for UV-induced damage to DNA, was only positive with some of the PT chemi-cals of class I.

## Discussion and Conclusions

The most important aspect of the EEC/COLIPA validation study of in vitro phototoxicity tests was the use of identical UV exposure conditions in all labora-tories. Another important aspect of quality assurance was the determination of UVA sensitivity of the 3T3 cells before they were used in the study. In addition, it has to be emphasized that the quality of in vivo data of test chemicals was a critical aspect for validating in vitro phototoxicity assays, since human data are difficult to obtain and since data are suffering from differences in species specific-ity. This aspect is illustrated by the misleading animal data of piroxicam which had to be moved from class I to class II due to clinical data in humans.

The 3T3 NRU-PI assay, which was developed in the present study, showed the highest predictivity of all of the in vitro assays evaluated. The predictive value of the 3T3 NRU-PI assay was better than in an earlier study on 3T3 cell phototoxicity [9], in which 8-MOP and doxycycline were classified as 'false negatives' and PABA was classified as 'false positive'. Taking into account the many mechanisms of phototoxicity at the cellular level, it is surprising that a 100% in vitro/in vivo concordance was achieved with the simple 3T3 NRU-PI assay. This assay is, therefore, quite promising for further validation in phase II of the EEC/COLIPA study. Among the more 'general' in vitro phototoxicity tests, which can be used to identify the phototoxic potential of a chemical, the 3T3 cell NRU-PI assay, the RBC assay and the commercial Skin² and SOLATEX PI assays were performing much better in the present study than assays using yeast or human lymphocytes or keratinocytes (table 2). The 3T3 NRU-PI assay, the RBC assay and the two commercial in vitro assays will, therefore, also be candidates for further validation in a blind trial. Although the commercial assays are fairly expensive, they will play an important role in phase II of the validation study (blind trial), since they permit testing of solid and insoluble materials, which cannot be tested in the 3T3 cell NRU-PI assay.

### References

1   Nilsson R, Maurer T, Redmond N: A standard protocol for phototoxicity testing. Results from an interlaboratory study. Contact Derm 1993;28:285–290.
2   Spielmann H, Gerner I, Kalweit S, Moog R, Wirnsberger T, Krauser K, Kreiling R, Kreuzer H, Lüpke N-P, Miltenburger HG, Müller N, Mürmann P, Pape W, Siegemund B, Spengler J, Steiling W, Wiebel FJ: Interlaboratory assessment of alternatives to the Draize eye irritation test in Germany. Toxicol In Vitro 1991;5:539–542.
3   Ljunggren B: The piroxicam enigma. Photodermatology 1989;6:151–154.
4   Hölzle E, Neumann N, Goertz G: Photoallergie: Mechanismus und Diagnostik; in Macher E, Kolde GB, Böcker EB (eds): Jahrbuch Dermatologie: Licht und Haut. Zülpich, Biermann, 1993, pp 135–142.
5   Spielmann H, Balls M, Brandt M, Döring B, Holzhütter HG, Kalweit S, Klecak G, Eplattenier HJ, Liebsch M, Lovell WW, Maurer T, Moldenhauer F, Moore L, Pape WJW, Pfannenbecker U, Potthast J, De Silva O, Steiling W, Wilshaw A: EEC/COLIPA projection on in vitro phototoxicity testing: First results obtained with a Balb/c 3T3 cell phototoxicity assay. Toxicol In Vitro 1994;8:793–796.
6   Daniels F: A simple microbiological method for demonstrating phototoxic compounds. J Invest Dermatol 1965;44:259–263.
7   Johnson BE, Walker EM, Hetherington, AM: In vitro models for cutaneous phototoxicity; in Marks R, Plewig G (eds): Skin Models – Models to Study Function and Dysfunction of Skin. Berlin, Springer, 1986, pp 265–281.
8   Pape WJW, Brandt M, Pfannenbecker U: Combined in vitro assay for photohaemolysis and haemoglobin oxidation as part of a photoirritancy test system assessed with various phototoxic substances. Toxicol In Vitro 1994;8:755–758.
9   Duffy PA, Bennet A, Roberts M, Flint OP: Prediction of phototoxic potential using human A431 cells and mouse 3T3 cells. Molec Toxicol 1987;1:579–587.

Dr. med. Horst Spielmann, Bundesinstitut für gesundheitlichen Verbraucherschutz und Veterinärmedizin, Diedersdorfer Weg 1, D–12277 Berlin (Germany)

Elsner P, Maibach HI (eds): Irritant Dermatitis. New Clinical and Experimental Aspects.
Curr Probl Dermatol. Basel, Karger, 1995, vol 23, pp 265–274

..............................
# Validation of Alternative Tests in the European Union

*Michael Balls*

European Center for the Validation of Alternative Methods (ECVAM),
JRC Environment Institute, Ispra, Italy

Regulatory toxicity testing in animals raises a number of scientific, humanitarian, legislative, practical and economic questions and conflicts. For example, while the use of animal procedures in identifying the potential hazards represented by chemicals and products of many kinds is required by various specific and general laws and regulations, there are other laws of no less importance (e.g. Directive 86/609/EEC in the European Union [1]), which are intended to protect laboratory animals from unnecessary pain, suffering, distress or lasting harm. They stipulate that the use of laboratory animals should only be permitted when it can be justified as necessary on strong scientific grounds. However, the scientific basis of many animal tests themselves and, in particular, of the ways whereby the data they provide are currently applied in human risk assessment, is weak. In addition, the present dependence of hazard prediction on animal tests requires considerable human and economic resources, as a result of which only a very small proportion of the chemicals which might threaten the well-being of humans, and of the environment in general, have been subjected to anything approaching a scientifically satisfactory evaluation.

Nonanimal (i.e. alternative) tests and testing strategies can offer solutions to many of the problems caused by the current over-reliance of regulatory toxicology and risk assessment on the routine, checklist application of animal tests. They can offer a more mechanistic basis for understanding toxic effects at the systemic, organic, cellular and molecular levels. When human cells and tissues are used, or modelling is based on human experience and responses, the problem of species differences, which severely limits the relevance of animal tests, can be avoided.

The change of emphasis from in vivo to in vitro approaches must therefore also embrace the increasing use of properly obtained and safely maintained *human*, rather than *animal*, cells and tissues. Well-designed in vitro studies can also assist with the selection of the most-appropriate laboratory animal species for any subsequent, scientifically necessary, use as a model for man.

Nonanimal methods offer an escape from many of the ethical and legal conflicts inherent in current practices. They can also offer the possibility of more manageable, less time-consuming and less expensive testing strategies, so that the practical and economic limitations which currently prevent the adequate testing of many chemicals and products can be overcome.

## Barriers to Acceptance

Despite the obvious benefits which would accrue from the successful development and use of nonanimal tests and testing strategies, various barriers to their acceptance must be dealt with before they can fully contribute to the transformation of hazard prediction and risk assessment into an effective, modern scientific discipline. Clark [2] has listed these barriers, as follows:

(1) The validation barrier: the method must be adequately validated using a wide range of chemical types.

(2) The scientific barrier: the method must be based on good science and must not make extravagant claims that run counter to common sense.

(3) The legislative barrier: since many toxicological tests are undertaken for legislative purposes, it is pointless to replace them unless the legislation requiring the tests can first be changed.

(4) The development barrier: some toxicologists may not accept methods unless they have been developed in their own laboratories – the 'not invented here' syndrome.

(5) The psychological barrier: some traditional scientists may feel their careers threatened by the introduction of new methodology and techniques.

(6) The fear of litigation barrier: manufacturers are required by specific or general laws to make their products as 'safe' as possible, and, in the case of litigation, the plaintiffs' lawyers would point to current testing practice as the standard to be met.

(7) The regulatory barrier: the attitudes of government regulators reflect their burden of responsibility and accountability.

Clark [2] emphasised that the existence of these barriers must be recognised and understood before they can be overcome. He stressed that validation is the main barrier and is the key to overcoming another important barrier, namely, regulatory acceptance.

## The Validation Barrier

Validation has been defined as *the process by which the reliability and relevance of a procedure are established for a specific purpose* [3]. It is now widely agreed that formal validation studies should be preceded by proper test development (including the production of a protocol or standard operating procedure) and that an interlaboratory trial involving appropriate sets of coded chemicals (of known and different degrees of in vivo toxicity) is an essential part of the process. Only when the outcome of a successful validation study has been subjected to independent assessments should the formal acceptance of a new test, test battery or testing strategy into regulatory practice be proposed and considered [4].

Although there are various forms of validation [5], some of which are more useful than others, the process should primarily be seen as a scientific activity and should be the responsibility of scientists who are committed to new test development and of those toxicologists and regulators who are prepared to assist in the orderly and sensible incorporation of new methodology into regulatory toxicology.

Up to now, the validation process has not been applied with notable success. This is partly because it is much more difficult than we had imagined, and partly because of a misunderstanding of why animal tests are conducted in the first place and therefore, how they should be replaced. Animal testing is performed for two main reasons: (1) in compliance with (or in perceived compliance with) the specific requirements of regulatory authorities for toxicity test data, or (2) in compliance with the (common law-based or product liability-based) need for safety assessments, as interpreted by manufacturing and/or retail companies themselves. Type A testing involves industrial chemicals produced on any significant scale, certain types of cosmetic ingredients, e.g. preservatives, coloring agents, hair dyes, and UV filters, as specified in the EU by the Cosmetics Directive (Directive 76/768/EEC [6]), and certain kinds of products, e.g. medicines, pesticides. Type B testing, not being 'required' by specific legislation, tends to be a matter of decision for companies themselves. That is why many cosmetic companies have declared that they have stopped animal testing. What they have done is to stop doing what they were never specifically required to do, and they may now use in vitro tests instead, often followed by human volunteer studies, as a basis for safety assurance. This has led to a significant, and welcome, reduction in the testing of cosmetic formulations, but has had no impact, of course, on the legally required testing of industrial chemicals, certain cosmetic ingredients, medicines, pesticides, etc. This is an important point of emphasis, because, although it is vital that cosmetic companies produce consumer products (shampoos, etc.) which are as safe as possible, there are no specific regulations, at least in Europe, which make their testing in animals mandatory.

The distinction between regulatory and nonregulatory animal testing has important implications in terms of the way we think about, and conduct, validation studies, which are, as a consequence, of two main types: (1) validation of the relevance and reliability of methods, batteries or testing strategies which could replace, refine or reduce animal testing for *regulatory* purposes, and (2) validation of the relevance and reliability of nonanimal procedures for their general use for *nonregulatory* purposes.

This distinction has two further consequences. Firstly, it leads to *in-house validation*, in which a company evaluates a testing strategy for its own purposes (i.e. for type B testing, perhaps also in the hope of contributing to the development of alternatives for use in type A testing). This often involves the use of chemicals and formulations which are the property of a company, for which it also has (and owns) animal test data, details of which it is not prepared to make public. Many companies, including Beiersdorf, Henkel, L'Oréal, Procter & Gamble, Unilever and Zeneca, have devised in-house procedures for use as screens before a decision is taken about the need for the animal testing of new chemicals or about the comparative acceptability of a new product in terms of consumer safety. Secondly, such in-house validation does not replace (although it can usefully precede) *scientific validation*, in which the names and structures/contents of the chemicals and formulations used must be freely available, as must the animal toxicity test data and the protocols of the tests or testing strategies undergoing validation, and in which the results and conclusions of validation studies must be published in detail in the peer-review literature. Scientific validation is (and should always be) an essential and unavoidable precursor to the acceptance of alternatives for legally required testing (i.e. regulatory testing, type A).

## Obstacles to Successful Validation

Experience gained in validation studies during the last 2 years have revealed a number of serious stumbling blocks, which must be faced up to and eliminated, if the opportunities offered by nonanimal tests and testing strategies are to be grasped [7]. These obstacles include:

(1) Definition of purpose of testing: Especially in the case of cell culture methods, the purpose for which a test is intended is often ill-defined. In addition, where 'endpoint' (e.g. neutral red uptake) is confused with 'test' (defined as the combination of biological system, exposure protocol, endpoint measurement and data analysis) [8], the same basic method can appear to have multiple applications. It seems unlikely that the same in vitro 'test' could be used to predict acute lethal potency, ocular irritancy *and* skin irritancy with any precision.

---

(2) Lack of adequate protocols: While it is essential that the need for continuity in test development is allowed for, a formal validation study should not take place until acceptable protocols or standard operating procedures are available for all the tests to be used. For this reason, a data bank named INVITTOX has been established at Nottingham, UK, by FRAME (Fund for the Replacement of Animals in Medical Experiments) and ERGATT (European Animal Group for Alternatives in Toxicity Testing), with the support of the EU [9].

(3) Reliability and relevance: Reliability and relevance must go hand-in-hand, in that a method which was very reliable, but irrelevant, would be no more useful than one which was highly relevant, but unreliable. However, surveying the in vitro toxicology literature, one gets the distinct impression that reproducibility of test results is easier to establish than their relevance and therefore receives the greater emphasis.

(4) Lack of availability of appropriate test materials: The lack of availability of test materials in sufficient numbers, of sufficient variety, and sufficiently representative of the range of levels of toxicity, remains a major difficulty, despite the efforts of certain individuals and organisations, such as ECETOC [10]. If this applies to toxic manifestations such as ocular irritancy, what prospects are there for finding sufficient model chemicals for validation studies on sensitisation, nephrotoxicity, neurotoxicity, etc.?

(5) Inadequate analysis of data: The quality of most in vivo/in vitro comparisons is not yet good enough for sound conclusions to be drawn. It is very rare for any allowance to be made for the variability or questionable relevance of the animal data, which are thus given a status which they do not deserve. They wrongly become the 'true' values which the nonanimal tests must struggle to reproduce. Also, insufficient allowance is made for the doubt which must be placed on values which fall within the barrier zones on both sides of category cut-off points. This is particularly worrying when Cooper two-way plots are used as a basis for establishing the sensitivity, specificity, predictivity and concordance of in vitro test data. These problems need urgent attention.

(6) Unnecessarily competitive attitudes: The developers of tests often seem to be too devoted to emphasising the advantages of their particular methods in comparison with others. However, it is the strategic use of batteries of tests which provide complementary information or confirmatory support for each other which is likely to provide the basis for a radically different approach to hazard prediction and risk assessment in the future [11].

(7) Advantages and disadvantages: Often linked to the above problem is a failure to take sufficient account of the circumstances in which a test should or should not be used, but to try to show its universal applicability. There is no such thing as a totally 'good' or a totally 'bad' test, as the merits of any test will depend on the precise ways in which it is used. For example, the FRAME neutral red

release method [12], in which cells are exposed to a very high concentration of test material for a very short time, appears to be very useful for assessing the relative potential ocular irritancies of alcohol-based cosmetic formulations, but not surfactant-based formulations. More effort should be put into evaluating the precise circumstances in which new tests and testing strategies should/should not be used. Using a test in circumstances for which it was not intended and for which it is not suitable should be seen for what it is – a waste of effort. It should certainly not be a basis for making adverse comments about its relevance and reliability.

## Validation in Europe: The Role of ECVAM

Validation is now being taken very seriously in Europe – by industrial companies and their trade associations, by academic institutions and professional bodies, by animal welfare organisations, and by governments, including the Commission, Council, Parliament and Member States of the European Union. Above all, validation is seen as a matter for international collaboration, and that is why current and future activities at the European level promise to make substantial and significant contributions to the scientific validation and regulatory acceptance of new test procedures as replacements for the conventional animal tests.

In creating a European Centre for the Validation of Alternative Methods (ECVAM) at their Joint Research Centre (JRC) at Ispra, Italy, the Commission of the EU have made a significant addition to the European commitment to the orderly development, validation and acceptance of replacement alternative methods. In a communication to the Council and the Parliament in October 1991 [13], the Commission pointed to a requirement in Directive 86/609/EEC [1] that:

The Commission and Member States should encourage research into the development and validation of alternative techniques which could provide the same level of information as that obtained in experiments using animals, but which involve fewer animals or which entail less painful procedures, and shall take such other steps as they consider appropriate to encourage research in this field.

The establishment of ECVAM thus reflects the commitment of the EU to the replacement ideal of the Three Rs (reduction, refinement and replacement) concept of Russell and Burch [14], spelled out in Article 7 of Directive 86/609/EEC [1] as follows:

An experiment shall not be performed if another scientifically satisfactory method of obtaining the result sought, not entailing the use of an animal, is reasonably and practicably available.

The Commission's Communication spelled out ECVAM's duties in the following terms. ECVAM was to be set up:

(1) To coordinate the validation of alternative test methods at Community level.
(2) To act as a focal point for the exchange of information on the development of alternative test methods.
(3) To set up, maintain and manage a data base on alternative procedures.
(4) To promote dialogue among legislators, industries, biomedical scientists, consumer organisations and animal welfare groups, with a view to the development, validation and international recognition of alternative test methods.

One of ECVAM's first priorities must be to become well informed about the current state of the art of nonanimal test development in relation to particular types of chemicals, types of products and potential toxic hazards. This will involve consultation, not least with DGXII (the research-funding section) of the Commission, since we see a natural evolution from basic research, through prenormative studies to prevalidation exercises, as a prelude to formal, and rather costly, interlaboratory validation programmes.

With the participation of members of the ECVAM Scientific Advisory Committee and others, we are therefore planning a series of ECVAM Workshops, in order to be able to review the current status of various types of tests and their potential uses and to identify the best ways forward. During 1993–1994, we are inviting small groups of experts to get together to discuss the following topics, and plan to have their reports and recommendations published in ATLA (Alternatives to Laboratory Animals):

(1) Uses of cultured hepatocytes.
(2) In vitro phototoxicity testing.
(3) In vitro corrosivity testing.
(4) The Amden validation principles revisited.
(5) In vitro neurotoxicity testing.
(6) In vitro teratogenicity testing.
(7) Vaccine potency testing.
(8) In vitro tests for acute toxicity and the classification and labelling of chemicals.
(9) In vitro tests for nephrotoxicity.
(10) Alternatives in cosmetics testing.
(11) In vitro tests for dermal penetration.
(12) The quality control of hormones.
(13) In vitro tests for respiratory toxicity.
(14) Alternative strategies in ecotoxicology.
(15) In vitro tests for surfactants and surfactant-based products.

Our strategy also involves the setting-up of ECVAM Task Forces, to focus on the achievement of narrowly defined, specific goals. We see task forces as ways of implementing the recommendations of workshops in the planning of validation studies and of seeking acceptance of the outcomes of successful validation studies. Initially, we also plan to establish task forces on the integrated use of QSAR and in vitro approaches, statistical comparisons involving in vitro and in vivo data, and the conflicting requirements of various EU directives.

ECVAM will have its own research and testing laboratories, which will be used for the development of new test methods, for participation in prevalidation studies and in formal validation programmes, and for training purposes. However, it is clear that ECVAM would never be able to provide expertise in all the different types of tests and areas of pharmacotoxicology for which the validation of alternative tests and test batteries will be necessary in the years to come. Collaboration with academic and industrial alternatives research laboratories in the Member States, and elsewhere, will therefore be essential.

Such collaboration is already under way, and we are building on the investment made in the past by other services of the Commission. Thus, in 1993–1994, in addition to external contracts for information services, ECVAM will be funding prevalidation studies in in vitro phototoxicology and in vitro neurotoxicology, on the relationship between in vitro cytotoxicity and acute lethal potency, and on the further development of the ERGATT/CFN Integrated In Vitro Toxicity Testing Scheme (ECITTS).

### Current Validation of Studies

It is not easy, at least at present, to establish precisely what validation studies are in progress or are being planned. However, at the time of writing, ECVAM itself is involved in six validation and prevalidation studies; namely, an EU/UK international validation study on alternatives to the Draize eye irritancy test, an EU international study on the inhibition of gap junction intercellular communication as a means of identifying tumour promoters, a COLIPA/EU international study on in vitro tests for photoirritancy, a Europe/USA international prevalidation study on in vitro tests for skin corrosivity, an EU/FRAME international study on the use of a battery of in vitro tests for predicting acute lethal potency and for use in the classifying and labelling chemicals, and an EU/FRAME international prevalidation study on in vitro tests for neurotoxicity.

### Alternatives in Skin Pharmacotoxicology

It is very clear from the other sections of this book that very significant progress is being made in the development of in vitro methods for use in skin pharmacotoxicology and safety testing.

---

Balls

In Europe, increasing attention is being focused on the search for nonanimal testing strategies which will be acceptable at the regulatory level, not least because of the Sixth Amendment to the EU Cosmetics Directive [15], which seeks to ban the use of animals in testing cosmetic ingredients and products by 1 January 1998 or as soon as possible thereafter. ECVAM is responding to this challenge, in collaboration with the Consumer Policy Service and other services of the Commission, the Scientific Committee on Cosmetology, the cosmetic industry and its European association, COLIPA, and various laboratories in the Member States and elsewhere. Various prevalidation studies, validation studies and workshops relevant to skin pharmacotoxicology are being planned, initially involving dermal irritancy, dermal penetration and sensitisation.

Much progress toward the replacement of animal tests in this area can be expected in the next few years, partly because of the quality of research being conducted on skin corrosivity, skin irritancy, dermal penetration and phototoxicity, but also because we are on the verge of the breakthroughs necessary to enable nonanimal tests for dermal sensitisation to become a reality. Another vital factor is the role that is already being played by human volunteer studies, which deserve to be given greater recognition as 'alternatives' for use alongside in vitro studies, as a means of reducing animal use and also increasing our understanding of human skin phenomena, on which better standards of human safety and protection must ultimately depend. The development of noninvasive methods for making quantitative measurements of changes to the human skin promises to be particularly important.

### References

1    Anon: Council Directive 86/609/EEC of 24 November 1986 on the Approximation of Laws, Regulations and Administrative Provisions of the Member States Regarding the Protection of Animals used for Experimental and Other Scientific Purposes. Off J Eur Community 1986;L358:1.
2    Clark DG: Barriers to the acceptance of in vitro alternatives. Toxic In Vitro 1994;8:907.
3    Balls M, Blaauboer B, Brusick D, Frazier J, Lamb D, Pemberton M, Reinhardt C, Roberfroid M, Rosenkranz H, Schmid B, Spielmann H, Stammati A-L, Walum E: Report and recommendations of the CAAT/ERGATT workshop on the validation of toxicity test procedures. ATLA 1990;18:313.
4    Balls M, Botham P, Cordier A, Fumero S, Kayser D, Koëter H, Koundakjian P, Gunnar Lindquist N, Meyer O, Pioda L, Reinhardt C, Rozemond H, Smyrniotis T, Spielmann H, van Looy H, van der Venne M-T, Walum E: Report and recommendations of an international workshop on promotion of the regulatory acceptance of validated non-animal toxicity test procedures. ATLA 1990;18:339.
5    Balls M: In vitro test validation: High hurdling but not pole vaulting. ATLA 1992;20:355.
6    Anon: Council Directive 76/768/EEC of 27 July 1976 on the Approximation of the Laws of the Member States Relating to Cosmetic Products. Off J Eur Community 1976;L262:169.
7    Fentem JH, Balls M: Why, when and how in vitro tests should be accepted into regulatory toxicology. Toxic In Vitro 1994;8:923.
8    Balls M, Fentem JH: The use of basal cytotoxicity and target organ toxicity tests in hazard identification and risk assessment. ATLA 1992;20:368.

9   Warren M, Atkinson K, Steer S: Introducing INVITTOX: The ERGATT/FRAME in vitro toxicology data bank. ATLA 1989;16:332.
10  Bagley DM, Botham PA, Gardner JR, Holland G, Kreiling R, Lewis RW, Stringer DA, Walker AP: Eye irritation: Reference chemicals data bank. Toxic In Vitro 1992;6:487.
11  Balls M, Atkinson KA, Gordon VC: Complementation in the development, validation and use of non-animal test batteries, with particular reference to ocular irritancy. ATLA 1991;19:429.
12  Reader SJ, Blackwell V, O'Hara R, Clothier RH, Griffin G, Balls M: A vital dye release method for assessing the short-term cytotoxic effects of chemicals and formulations. ATLA 1989;17:28.
13  Anon: Establishment of a European Centre for the Validation of Alternative Methods (ECVAM): Communication from the Commission to the Council and the European Parliament. Commission of the European Communities. Brussels, CEC, 1991, p 6.
14  Russell WMS, Burch RL: The Principles of Humane Experimental Technique. London, Methuen, 1959, p 238.
15  Anon: Council Directive 93/35/EEC, Amending for the Sixth Time Directive 76/768/EEC on the Approximation of the Laws of the Member States relating to Cosmetic Products. Off J Eur Community 1993;L151:32.

Prof. Michael Balls, European Center for the Validation of Alternative Methods (ECVAM), JRC Environment Institute, I–21020 Ispra (VA) (Italy)

Elsner P, Maibach HI (eds): Irritant Dermatitis. New Clinical and Experimental Aspects.
Curr Probl Dermatol. Basel, Karger, 1995, vol 23, pp 275–287

..........................

# Irritant Reactions on Hairless Micropig Skin: A Model for Testing Barrier Creams?

*B. Gabard*[a], *P. Treffel*[a], *F. Charton-Picard*[b], *R. Eloy*[b]

[a] Biopharmacy Department, Spirig Ltd., Egerkingen, Switzerland;
[b] Biomatech, Chasse-sur-Rhône, France

Occupational dermatoses are most numerous among recognized occupational diseases. Their frequency is increasing. Skin barrier creams are designed to prevent or reduce the irritancy of hazardous materials in the working and/or home environment. Used repeatedly, detergents, organic solvents or other compounds such as cutting oils for example are presumed to be responsible for the development of numerous chronic irritant dermatitis [1, 2]. Many methods have been used to identify the potential protective or skin regeneration efficacy of barrier creams, but there is no widely accepted model. Main difficulties reside in the wide range of possible irritants and in the obvious need to reproduce the frequent repetition of a low-grade exposure [3, 4].

We looked for an animal model that would present the following characteristics: (1) pharmacological reactions similar to the ones of human skin, allowing a meaningful comparison of the irritant reactions to be made; (2) possibility of easily conducting an irritation with various concentrations of the irritants; (3) possibility of quantifying the irritation with noninvasive skin measurement techniques.

For these purposes, we chose the Yucatan hairless micropig (YHM), the skin of which is known to be very close to human skin, at least morphologically [5].

## Methods

*Animals*

Five female and 2 male YHM (Micropig®; Charles River France, Saint-Aubin-les-Elbeuf, France; age 9–12 months, weight 9–14 kg) were trained to stand or lay quietly on a table during the measurements. This was attained after a few days of training for all animals which were neither anesthetized nor restrained.

*Noninvasive Measurements*

All measurements were done in a room where the environmental conditions were kept as far as possible constant for the duration of the experiments, close to the usual environment of the animals.

*Skin Color (SkCo).* SkCo was measured in the CIELAB colour system [6, 7] by the Chroma Meter CR 200 (Minolta, Osaka, Japan) calibrated to the standard white tile provided by the manufacturer. The method has already been described in numerous publications for human skin [7–12], but, to our knowledge, there is only one report of using a Chroma Meter on porcine skin [13].

*Skin Hydration (SkHy).* SkHy was assessed indirectly through measurement of the electrical capacitance of the outer epidermis [14] with the Corneometer CM820PC (Courage & Khazaka, Cologne, Germany). The depth of measurement in human skin, about 60–100 µm [14–17], is not known in porcine skin. It was assumed that the measuring principle was the same as in human skin.

*Transepidermal Water Loss (TEWL).* TEWL was measured with the Evaporimeter EP1 (Servomed, Stockholm, Sweden). The protective shield provided by the manufacturer was used throughout. There are already reports on the use of the evaporimeter on animal skin [3, 18] and particularly on porcine skin [19]. It was not possible to conform to the recently published guidelines [20] regarding the use of a protective box due to the nonrestraining of the animals.

*Sites of Measurements*

Physiological measurements were taken on the back (front and rear), the ear, the flank (front and rear), the middle belly and both legs on each side (right and left). Pharmacologic and irritative tests were conducted on the back and upper flank.

*Pharmacologic and Irritative Tests*

*Histamine Prick Test.* The histamine prick test was performed as usual with a commercial solution of 1.7 mg/ml of histamine-HCl in physiological saline (Allergopharma J. Ganzer KG, Reinbeck, FRG). SkCo and diameter of the weal were measured at regular intervals during 90 min.

*Nicotinic Acid Ester-Erythema.* The site was first occluded during 2 h with a large Finn chamber. Then, a piece of filter paper soaked in a solution of nicotinic acid methyl ester (methylnicotinate; in $H_2O$) or in a solution of nicotinic acid hexyl ester (hexylnicotinate; in propylene glycol/isopropanol 60:40; both Fluka, Buchs, Switzerland) was laid down under the Finn chamber for further 15 min. SkCo was measured up to 80 min afterwards.

*Sodium Hydroxide Irritation.* Different concentrations of NaOH were applied under occlusion as described above for 4 min. SkCo was measured after the removal of the occlusion and in one animal up to 40 min thereafter.

*Toluene Irritation.* A piece of filter paper was soaked in toluene and applied on the skin under a large Finn chamber for 5 min. SkCo was measured up to 60 min after removal of the occlusion.

*Sodium Lauryl Sulfate Irritation.* Pieces of filter paper soaked in the aqueous solutions were applied as before under large Finn chambers for 24 h. SkCo, TEWL and SkHy were measured until day 7.

*Table 1.* Results of the ANOVA: p values

| Parameter | TEWL g/m²·h | Capacitance arbitrary units | Skin color, arbitrary units | | |
|---|---|---|---|---|---|
| | | | L* | a* | b* |
| Overall mean ± SD | 7.56±2.90 | 53.9±16.3 | 56.8±7.7 | 5.83±1.15 | 6.38±1.35 |
| Source of variation | | | | | |
| Animal | <0.001 | <0.001 | NS | 0.023 | 0.020 |
| Side | NS | NS | NS | 0.020 | NS |
| Site | NS | 0.046 | <0.001 | <0.001 | <0.001 |
| Variation coefficient, % | 32.4 | 18.3 | 6.4 | 15.9 | 16.3 |

NS = No statistical significance ($p > 0.05$).

*Cutting Oils.* Ten different cutting oils generously provided by the companies were applied under occlusion using the similar system piece of filter paper/large Finn chamber. In a first step, the application lasted 6 h and SkCo was measured immediately after removal of the patch and at 24 and 48 h thereafter. In a second step, 2 applications of 6 h duration were conducted 24 h apart. SkCo was measured between the two applications and afterwards up to 7 days.

### Calculations and Statistics

Means and standard deviations were calculated in the usual manner. The data from the physiological measurements were submitted to an analysis of variance (ANOVA). If statistical significance was attained ($p < 0.05$), a multiple-range test was performed (Duncan's test). As to the pharmacological measurements, maximal changes were compared with start values with the Student's t test. SPSS-PC⁺ 4.0 was used for statistical calculations.

## Results and Discussion

### Physiological Measurements

The results of the ANOVA are shown in table 1. Data from different animals differed from each other (exception: SkCo-L*), as well as data from different sites of measurement (exception: TEWL). No statistical differences were found between both sides of the animals (exception: SkCo-a*; this is considered to be fortuitous).

The coefficients of variation, calculated from the ANOVA residual mean square, were highest for TEWL and lowest for the SkCo-L*.

*Table 2.* Means (SD) of the different parameters for each animal

| SkCo-L* | 53.4 (8.6) | 53.7 (5.1) | 55.6 (6.9) | 58.2 (8.1) | 58.4 (7.6) | 59.1 (8.0) | 59.2 (7.7) |
|---|---|---|---|---|---|---|---|
| animal | (No. 2) | (No. 4) | (No. 7) | (No. 5) | (No. 6) | (No. 3) | (No. 1) |
| SkCo-a* | 5.25 (1.01) | 5.36 (1.25) | 5.63 (1.36) | 5.77 (0.93) | 5.95 (0.91) | 6.25 (1.09) | 6.46 (1.08) |
| animal | (No. 1) | (No. 5) | (No. 7) | (No. 2) | (No. 3) | (No. 6) | (No. 4) |
| SkCo-b* | 5.77 (1.32) | 5.95 (1.45) | 6.07 (1.11) | 6.19 (0.84) | 6.6 (1.65) | 6.85 (1.09) | 7.22 (1.42) |
| animal | (No. 1) | (No. 3) | (No. 6) | (No. 4) | (No. 7) | (No. 5) | (No. 2) |
| SkHy | 32.5 (15.1) | 45.0 (9.3) | 45.8 (8.4) | 59.0 (8.8) | 61.6 (12.2) | 63.1 (11.3) | 69.9 (11.8) |
| animal | (No. 7) | (No. 1) | (No. 3) | (No. 6) | (No. 5) | (No. 2) | (No. 4) |
| TEWL | 5.12 (1.82) | 6.50 (2.03) | 7.0 (3.18) | 7.25 (2.84) | 8.0 (2.68) | 8.12 (2.25) | 10.94 (1.91) |
| animal | (No. 1) | (No. 6) | (No. 2) | (No. 5) | (No. 3) | (No. 7) | (No. 4) |

SkCo = Skin color (L*, a*, b*); SkHy = skin capacitance.
Underlined groups show animals not statistically different from each other.

Animal Data (table 2)

*Skin Color.* Contrary to the other parameters measured, SkCo appeared homogeneous. No important variations were noticed between extreme values. Consequently, statistical significance for possible differences between the animals was low or even absent. SkCo-L* values differed by no more than 10%, -a* and -b* values by about 20%.

*Skin Hydration.* 'Dry' and 'hydrated' skin animals could be distinguished. Extreme values differed by a factor 2 (for example animal 7: 32.5 units, and animal 4: 69.9 units).

*Transepidermal Water Loss.* Animals with high TEWL (e.g. No. 4) could be distinguished from animals with low TEWL (e.g. No. 1). The majority of the data was comprised between 6 and 8 g/m$^2$ × h. The values agree with published data from porcine skin [19].

Site of Measurement (data not shown)

Clear intraindividual anatomical variations were noticed for SkCo. The ear showed the lowest SkCo-L* and -b*. The highest SkCo-L* was measured on the

Table 3. Histamine prick: mean maximal changes from start values ± SD (data from 4 animals)

| | Skin color, arbitrary units | | | Weal diameter mm |
|---|---|---|---|---|
| | SkCo-L* | SkCo-a* | SkCo-b* | |
| Means ± SD | −4.75±4.25 | 4.33±1.71 | 0.07±2.0 | 6.6±1.2 |
| Student's t test | 2.2287 (NS) | 5.0477 (<0.05) | | |

belly and the flank, the highest SkCo-a* and -b* on the legs. Similarly, high SkHy values were measured on the ears, low ones on the back. On the contrary, no such anatomical variations could be demonstrated for TEWL as no statistical differences were found between different sites of measurements. This probably does not reflect the reality, as anatomical variations of TEWL are well known from human experience [20]. A possible explanation may be the lack of good protection against air turbulences during the measurements, leading to variations masking anatomical differences of smaller magnitude.

*Pharmacological Measurements*
Histamine Prick Test (fig. 1; table 3)
An erythema appeared early after the prick and was maximal 5–10 min thereafter (decrease of SkCo-L*, increase of SkCo-a*, no change of SkCo-b*). These changes were observed in all tested animals and were statistically significant for SkCo-a*. A weal developed soon and attained a maximum of about 7 mm in diameter between 20 and 30 min after the prick, that means clearly delayed when compared with the time course of the erythema. It almost disappeared about 90 min after the prick. Thus, this reaction seemed quite similar to that known from human skin [21]. The evaluation of a flare was not attempted, but it is known from the literature that the flare phenomenon does occur in porcine skin [22].

Nicotinic Acid Esters Erythema (data not shown)
Use of application time and of methylnicotinate concentrations as known from human experimentation [12, 23] did not provoke any measurable reaction. In order to change the conditions of penetration, we tried to modify the time of application, to occlude the skin for some time before application of the chemical agent and to increase the concentration in the applied solution. A measurable reaction, characterised by a slight decrease of SkCo-L* and -b* and an increase of SkCo-a*, was obtained after application for 15 min of a 10 M solution on a site previously occluded for 2 h with a large Finn chamber. These changes were simi-

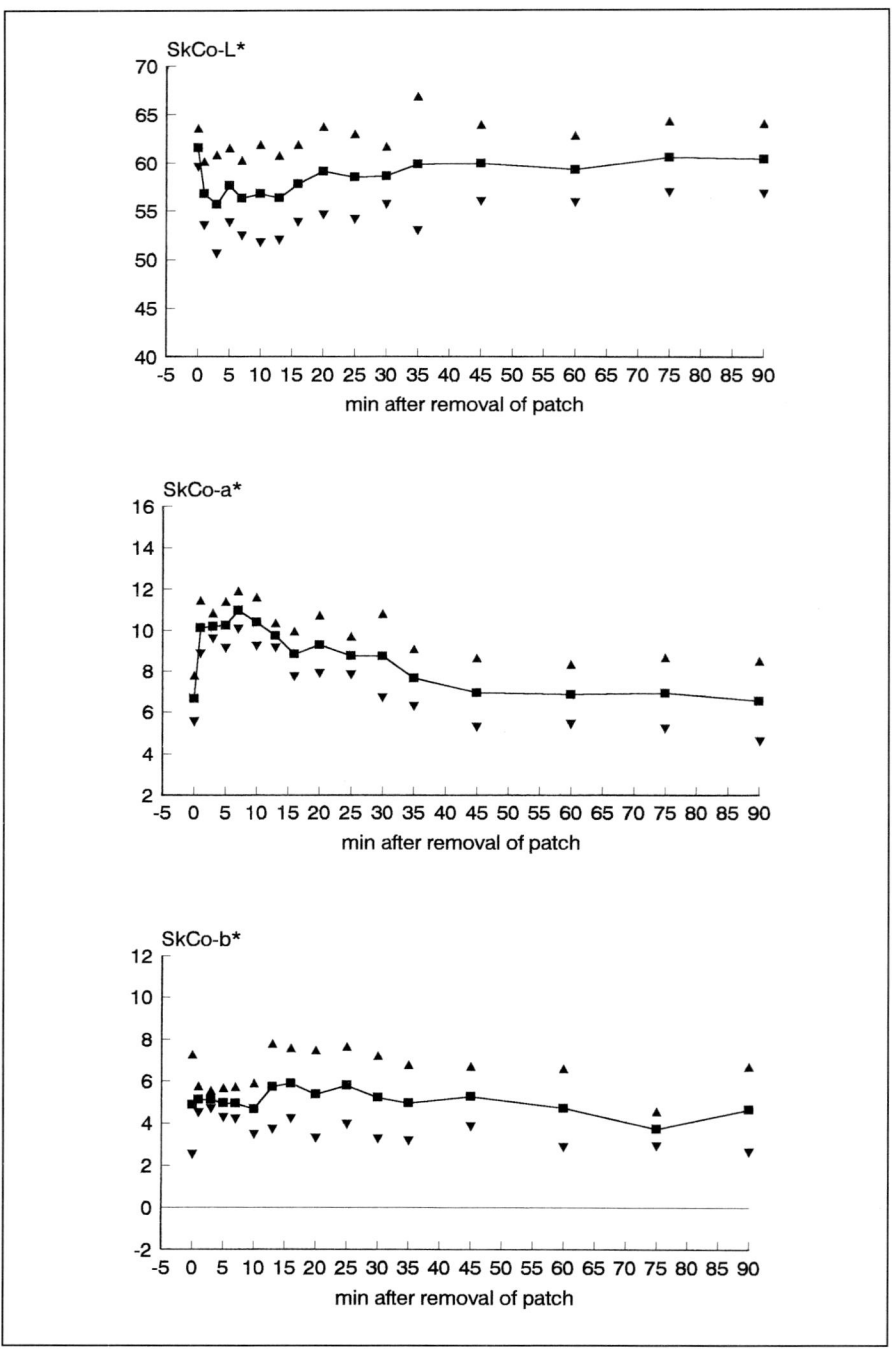

lar to those known from human skin [12, 23]. In order to further characterize the parameters of the penetration of nicotinic acid esters on the micropig skin, the more lipophilic hexylnicotinate was used. Here too, high concentrations ($3\,M$) and long application times (15 min) were needed to detect a change in skin color. Thus, in our experimental conditions, the well-known nonimmunologic contact urticarial reaction of human skin could only be reproduced with concentrations and application times without any relationship to those known from human experimentation [12, 23–25]. Further investigations are necessary to detect the causes of these discrepancies, but our tests already point to lack of penetration as a possible factor.

### Application of Irritative Agents
### Sodium Hydroxide (table 4)

A short (4 min) occlusive application of NaOH in different concentrations (3–5%, i.e. 0.75–1.25 N) provoked an erythema that appeared soon and was maximal 5 min after the removal of the patch. The animals reacted very differently, between nil and very strong for the same concentration of NaOH, with a decrease of SkCo-L* and an increase of both SkCo-a* and -b*. There was a tendency toward stronger reactions after application of higher concentrations of NaOH. No change attained statistical significance due to the high variability and large standard deviations. In an animal which was a good responder, the time course of the irritation was measured. Forty min after the removal of the patch, SkCo-L* and -a* were not back to start values.

Compared to human skin, porcine skin appeared to be more sensitive to NaOH, as only very short application times of usual concentrations were needed to provoke a clear irritation. On the other hand, as in humans, the reactions were heterogenous, and 'bad' or 'good' responders could be found [26].

### Toluene (table 5)

The erythema observed after the occlusive application of undiluted toluene for 5 min was characterized by a decrease of SkCo-L* and an increase of both SkCo-a* and -b*. The animals reacted homogeneously and the changes were statistically significant. The parameters returned to the start values 30–45 min after the removal of the patch. Here too, compared with human skin, porcine skin appeared sensitive to toluene as an irritative agent.

*Fig. 1.* Measurement of skin color before (t = 0) and after histamine prick test in YHM. SkCo-L* = Clarity; SkCo-a* = green-red chromaticity; SkCo-b* = blue-yellow chromaticity; all arbitrary units. Means and SDs of 4 animals.

*Table 4.* NaOH-patch: mean maximal changes from start values ± SD (data from 4 animals)

| NaOH concentration | Skin color, arbitrary units | | |
|---|---|---|---|
| | SkCo-L* | SkCo-a* | SkCo-b* |
| 5.0% | −12.5 ± 1.0 | 2.4 ± 0.3 | 3.2 ± 0.5 |
| 4.0% | −5.6 ± 5.8 | 3.7 ± 3.3 | 1.8 ± 1.4 |
| 3.5% | −7.0 ± 8.7 | 3.3 ± 4.1 | 1.7 ± 2.1 |
| 3.0% | −0.8 ± 2.5 | 2.0 ± 2.9 | 0.7 ± 2.1 |

*Table 5.* Toluene patch: mean maximal changes from start values ± SD (data from 7 animals)

| | Skin color, arbitrary units | | |
|---|---|---|---|
| | SkCo-L* | SkCo-a* | SkCo-b* |
| Means ± SD | −2.57 ± 1.64 | 4.19 ± 1.81 | 1.0 ± 0.31 |
| Student's t test | 3.840 | 5.6746 | 7.9097 |
| | (<0.05) | (<0.05) | (<0.05) |

Sodium Lauryl Sulfate (SLS)

SLS, applied at 2.5, 5 and 10% for 24 h under occlusion, led to changes in TEWL, skin color and skin capacitance as well.

*TEWL.* TEWL showed a strong increase after patch removal (fig. 2), without any dose dependency. These increased values returned rapidly to normal level which was attained between days 3 and 4 for 2.5 and 5% SLS. There was a tendency for TEWL to remain at a higher level for a longer time after SLS 10%.

*Skin Color.* A strong erythema (decrease of SkCo-L*, an increase of SkCo-a* and no changes of -b*) was measured after the removal of the patch (fig. 3). Greater variations were noticed with increased SLS concentrations. These changes were maximal on day 2 (24 h after patch removal), with exception of the variations measured after application of the highest SLS concentration. In this last case, skin color values remained on an altered level for a longer time. Start values were attained between days 4 and 7, depending on the concentration.

*Skin Hydration.* SkHy was significantly increased after patch removal, without showing dose dependency (fig. 4). Thereafter, if fell under the start level, indi-

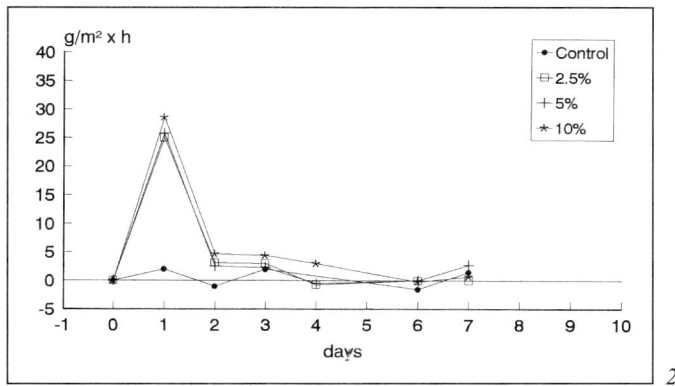

2

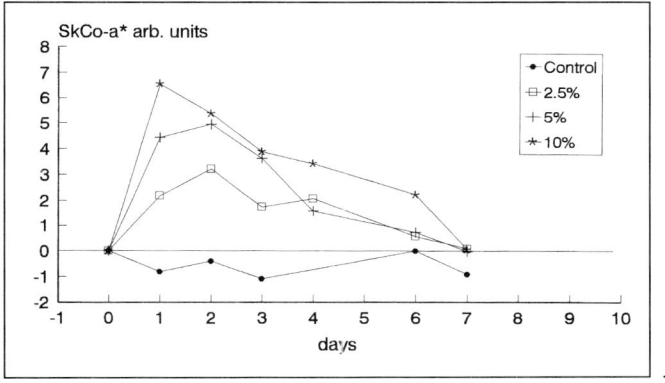

3

*Fig. 2.* TEWL changes from start values (g/m² × h) of YHM skin due to a 24-hour occlusive application of different SLS solutions. Means (for the sake of clarity, SDs are omitted). Start values: control, 7.0 ± 1.4 g/m² × h; SLS 2.5%, 7.8 ± 3.6 g/m² × h; SLS 5.0%, 8.4 ± 1.3 g/m² × h; SLS 10.0%, 7.6 ± 1.1 g/m² × h.

*Fig. 3.* Skin color changes from start values (illustrated by the green-red chromaticity a*) of YHM skin due to a 24-hour occlusive application of different SLS solutions. Means (for the sake of clarity, SDs are omitted). Start values: control, 5.52 ± 0.16; SLS 2.5%, 5.47 ± 1.53; SLS 5.0%, 5.57 ± 1.27; SLS 10.0%, 5.81 ± 1.40.

cating skin dryness. The lowest values were attained between days 4 and 7, depending on the concentration of SLS. The variations were smallest with 2.5% SLS.

These results point to some differences with the well-known reaction of human skin to application of detergents such as SLS [23, 27]. The increase of TEWL and SkHy after removal of the patch, without dose dependency, leads

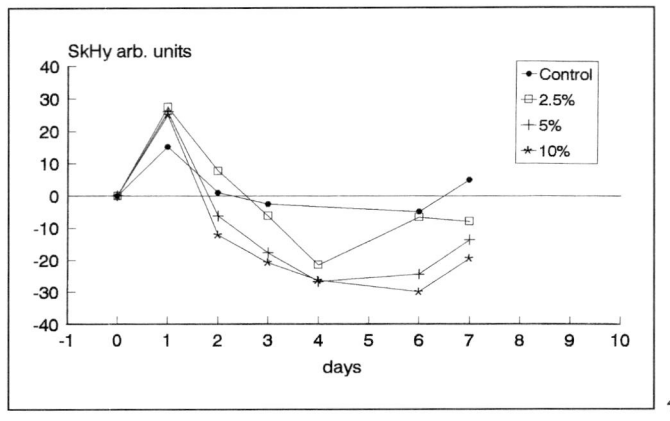

*4*

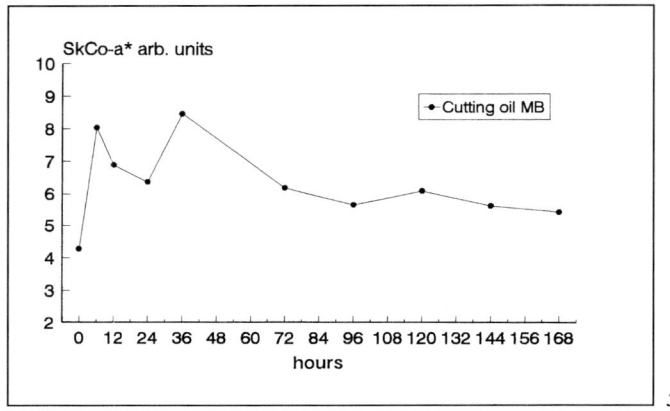

*5*

*Fig. 4.* Skin hydration (capacitance) changes from start values of YHM skin due to a 24-hour occlusive application of different SLS solutions. Means (for the sake of clarity, SDs are omitted). Start values: control, 43.6 ± 19.3; SLS 2.5%, 49.6 ± 14.3; SLS 5.0%, 56.1 ± 13.0; SLS 10.0%, 61.3 ± 13.0.

*Fig. 5.* Skin color (illustrated by the green-red chromaticity a*) in 1 YHM after 2 × 6 h (0–6 and 24–30 h) occlusive application of a cutting oil.

to the conclusion that the evaporation of water was measured that accumulated under the Finn chamber during the occlusion, and not that the skin barrier was disturbed. Compared to human skin, YHM skin seemed to react rather with a clear erythema, which could be precisely quantified. As in human skin, the dryness appeared with a few days delay with respect to the other parameters [23].

Cutting Oils (fig. 5)

In our experimental conditions, most of the cutting oils, which were applied undiluted, did not provoke any measurable reaction after 6 h or 2 × 6 h application. However, an erythema could be measured after application of some particular cutting oils. In these cases, repetition of the application led to an increase of the reaction, characterized by a decrease of SkCo-L*, an increase of SkCo-a* and variable changes of SkCo-b*.

## Conclusion

These experiments were conducted to look at the possibility of defining an animal model that could be used to test the efficacy of barrier creams or creams designed to protect or regenerate a skin challenged by a working environment. For this purpose, it was necessary to first characterise the reaction of the chosen animal skin to 'standard' irritants or at least irritants well known from human experience.

We chose to look at the hairless micropig for several reasons:

(1) Similarity of porcine skin with human skin: Morphological similarity has been thoroughly investigated [5] and it has been known for some time that porcine skin has the closest permeability characteristics to that of human skin for foreign compounds [28]. We could show similarity with human skin, concerning individual values of some physiological parameters, anatomical variations, no difference between both sides of the body, thus allowing contralateral comparisons. Although a comparison of our results with values from the literature is difficult due to the paucity of published data, at least TEWL was similar to that already reported [19]. SkHy values seemed lower than those of human skin [14–17, 23, 27] but further experience is needed as porcine skin is known to be sensitive to changes in ambient humidity [29].

(2) Convenience of the animal due to its size.

(3) Possibility to use noninvasive skin measuring techniques to precisely quantify an irritation. The results showed that this last goal has been attained. We could show that differences may exist between the irritant reactions as known from human experience and the reactions that were measured under our experimental conditions. Besides penetration factors, also a different pharmacology of the YHM skin may play a role in these differences. Nevertheless, agents such as cutting oils, known to be injurious to human skin in a working environment and under certain conditions, also provoked a reaction of the YHM skin.

We conclude that these results, although preliminary, encourage further development of the model and will allow the design and evaluation of barrier or regenerating creams under conditions near to those of their future use.

## References

1    Malten KE: Thoughts on irritant contact dermatitis. Contact Derm 1987;7:238–247.
2    Lachapelle JM, Frimat P, Tennstedt D, Ducombs G: Dermatologie Professionnelle et de l'Environ-nement. Paris, Masson, 1992.
3    Frosch PJ, Schulze-Dirks A, Hoffman M, Axthelm I, Kurte A: Efficacy of skin barrier creams (I). The repetitive irritation test (RIT) in the guinea pig. Contact Derm 1993;28:94–100.
4    Frosch PJ, Kurte A, Pilz B: Efficacy of skin barrier creams (III). The repetitive irritation test (RIT) in humans. Contact Derm 1993;29:113–118.
5    Lavker RM, Dong G, Zheng P, Murphy GF: Hairless micropig skin. A novel model for studies of cutaneous biology. Am J Pathol 1991;138:687–697.
6    Weatherall IL, Coombs BD: Skin color measurements in terms of CIELAB color space values. J Invest Dermatol 1992;99:468–473.
7    Takiwaki H, Overgaard L, Serup J: Comparison of narrow-band reflectance spectrophotometric and tristimulus colorimetric measurements of skin color. Skin Pharmacol 1994;7:217–225.
8    Wilhelm KP, Maibach HI: Skin color reflectance measurements for objective quantification of erythema in human beings. J Am Acad Dermatol 1989;21:1306–1308.
9    Serup J, Agner T: Colorimetric quantification of erythema: A comparison of two colorimeters (Lange Micro Color and Minolta Chroma Meter CR-200) with a clinical scoring scheme and laser-Doppler flowmetry. Clin Exp Dermatol 1990;15:267–272.
10   Queille-Roussel C, Poncet M, Schaefer H: Quantification of skin-colour changes induced by topical corticosteroid preparations using the Minolta Chroma Meter. Br J Dermatol 1991;124:264–270.
11   Waring MJ, Monger L, Hollingsbee DA, Martin GP, Marriott C: Assessment of corticosteroid-induced skin blanching: Evaluation of the Minolta Chromameter CR 200. Int J Pharm 1993;94: 211–222.
12   Chan SY, Li Wan Po A: Quantitative evaluation of drug-induced erythema by using a tristimulus colour analyser: Experimental design and data analysis. Skin Pharmacol 1993;6:298–312.
13   Sambuco CP, Cole CA, Forbes PD, Urbach F: Photobiologic effects of cosmetic tanning enhancers in miniature pigs; in Fitzpatrick TB, Forlot P, Pathak MA, Urbach U (eds): Psoralens: Past, Present and Future of Photochemoprotection and Other Biological Activities. Paris, Libbey Eurotext, 1989, pp 399–406.
14   Courage W: Hardware and measuring principle: The corneometer; in Elsner P, Berardesca E, Maibach HI (eds): Bioengineering of the Skin: Water and the Stratum corneum. Boca Raton, CRC Press, 1994, chap 14.
15   Barel AO, Clarys P, Wessels B, de Romsee A: Non-invasive electrical measurements for evaluating the water content of the horny layer: Comparison between capacitance and conductance measure-ments; in Scott RC, Guy RH, Hadgraft J, Boddé HE (eds): Prediction of Percutaneous Penetration: Methods, Measurements, Modelling. London, IBC International Services Ltd, 1991, vol 2, pp 238–247.
16   Blichman CW, Serup J: Assessment of skin moisture: Measurement of electrical conductance, capa-citance and transepidermal water loss. Acta Derm Venereol (Stockh) 1988;68:284–290.
17   Gabard B, Treffel P: Hardware and measuring principle: The NOVA™ DPM 9003; in Elsner P, Berardesca E, Maibach HI (eds): Bioengineering of the Skin: Water and the Stratum corneum. Boca Raton, CRC Press, 1994, chap 15.
18   Jackson SM, Wood LC, Lauer S, Taylor JM, Cooper AD, Elias PM, Feingold KR: Effect of cuta-neous permeability barrier disruption on HMG-CoA reductase, LDL receptor, and apolipoprotein E mRNA levels in the epidermis of hairless mice. J Lipid Res 1992;33:1307–1314.
19   Fourtanier A, Berrebi C: Miniature pig as an animal model to study photoaging. Photochem Photo-biol 1989;50:771–784.
20   Pinnagoda J, Tupker RA, Agner T, Serup J: Guidelines for transepidermal water loss (TEWL) measurement. Contact Derm 1990;22:164–178.
21   Carey C, Clark M, Lopez-Gil JA, Vere DW: The basis of histamine assay in human skin. II. Weal formation. Br J Clin Pharmacol 1993;36:39–43.
22   Jancsó G, Pierau FK, Sann H: Mustard oil-induced cutaneous inflammation in the pig. Agents Action 1993;39:31–34.

23   Gabard B: Einsatz von nichtinvasiven Messmethoden zu pharmakologischen Untersuchungen an der menschlichen Haut; in Lange L, Jaeger H, Seifert W, Klingmann I (eds): Pharmakodynamische Modelle für die Arzneimittelentwicklung. Berlin, Springer, 1993, pp 80–95.

24   Oestmann E, Lavrijsen APM, Hermans J, Ponec M: Skin barrier function in healthy volunteers as assessed by transepidermal water loss and vascular response to hexyl nicotinate: Intra- and inter-individual variability. Br J Dermatol 1993;128:130–136.

25   Guy RH, Tur E, Bugatto B, Gaebel C, Sheiner LB, Maibach HI: Pharmacodynamic measurements of methyl nicotinate percutaneous absorption. Pharm Res 1984;1:76–81.

26   Wilhelm KP, Pasche F, Surber C, Maibach HI: Sodium hydroxide-induced subclinical irritation. Acta Derm Venereol (Stockh) 1990;70:463–467.

27   Agner T, Serup J: Sodium lauryl sulfate for irritant patch testing: A dose response study using bioengineering methods for determination of skin irritation. J Invest Dermatol 1990;95:543–547.

28   Dick IP, Scott RC: Pig ear skin as an in-vitro model for human skin permeability. J Pharm Pharmacol 1992;44:640–645.

29   Bissett DL, McBride JF: The use of the domestic pig as an animal model of human dry skin and for comparison of dry and normal skin properties. J Soc Cosmet Chem 1983;34:317–326.

Dr. B. Gabard, Biopharmacy Department, Spirig Ltd., CH–4622 Egerkingen (Switzerland)

Elsner P, Maibach HI (eds): Irritant Dermatitis. New Clinical and Experimental Aspects.
Curr Probl Dermatol. Basel, Karger, 1995, vol 23, pp 288–295

..........................

# Objective and Reproducible Assessment of Irritants in vivo

## A Reappraisal of the $IT_{50}$ in Honour of *Kligman* and *Wooding*

*Friedrich A. Bahmer*[a], *Uwe Feldmann*[b]

[a] Department of Dermatology, and
[b] Department of Biostatistics and Informatics, Medical Faculty, University of the
Saarland, Homburg/Saar, Germany

Evaluation of the irritating potential of many substances, especially that of fluids, is of paramount medical, toxicological, and socioeconomic importance. Therefore, several in vitro and in vivo methods have been developed during the past decades with the aim to quantify the irritant potential of at least a number of often-used fluids such as cutting emulsions [1]. For in vivo testing, the soap chamber test of Kligman and Frosch, or a modification thereof, is widely used [2].

Because of the rather poor reproducibility of the methods employed for in vivo testing, attempts are made to complement the clinical methods with bioengineering methods such as the measurement of the transepidermal water loss (TEWL), the measurement of the blood flow in the dermal plexus with laser doppler flow meter [3, 4], electrical impedance measurements [5], and visual evaluation based on skin replica methods [6], to name just a few. In addition, in vitro models have been developed to assess the irritancy potential of various substances [7, 8].

From the data published it is obvious that the instrument-based methods as well as the clinical evaluation methods show a rather high variability. Thus, it seems that methods, both accurate and reproducible, are still lacking.

As early as 1967, Kligman and Wooding [9] convincingly demonstrated that the weak point in all clinical evaluation systems is the assessment of the intensity of the irritant reaction. They pointed out that different substances might show an extremely variable reaction, which is neither well correlated to the concentration of the substance tested, nor to the different skin characteristics of the individuals. They came to the conclusion that all clinical scoring systems are both difficult to employ and rather unreliable [9].

Since the assessment of the intensity of an irritant reaction shows such a large interobserver variability, they proposed to repeat the application of the substance until a reaction ensues, because a good agreement exists between different observers whether a reaction is present or not. Then, the question is not *how strongly* a substance is irritating, but *at which concentration* or *after what time interval* if applied repeatedly. However, the S-shaped curve of the proportion of reactors versus different concentrations or time elapsed requires special statistical methods. Therefore, they adopted the Probit analysis, developed some years before by Finney [10]. Since at that time powerful computers and statistical programs were barely available, they used a graphic analysis method based on the work of Litchfield [11].

For strong irritants, they proposed to determine the 'irritant dose' ($IT_{50}$), which is the dose eliciting a positive reaction in 50% of the individuals tested. Accordingly, for weak irritants the 'irritative time' ($IT_{50}$) is the appropriate measure. This is the time (usually measured in days) when half of the probands tested shows a reaction. Since we are concerned mainly with the determination of weak irritants, we report on the use of the $IT_{50}$ for the assessment of the irritancy of four different concentrations of a cutting fluid (CF) in comparison to 3 different concentrations of the known irritant sodium lauryl sulfate (SLS).

## Probands

The study was carried out on 22 volunteers free of skin disease at the time of testing. There were 13 women (age 22–48 years, mean 27.5 years), and 9 men (age 24–56 years, mean 29.3 years). Amongst them, 4 women and 2 men had a history of atopy, manifesting as rhinoconjunctivitis, but showing no symptoms at the time of testing. All had given their informed consent.

## Test Method

Sodium lauryl sulfate (SLS), minimum 99.5% purity (Sigma Chemicals, Munich) at 0.05, 0.25 and 0.1% dissolved in water (weight/volume) was used. As cutting fluid (CF), a commercially available product from a Swiss manufacturer was used at dilutions of 0.5, 1, 2, 4 and 8% in distilled water. Of each test solution, 0.05 ml was applied in a large Finn Chamber (Hermal, Reinbek, Germany), containing a cotton disc (Webril, Kendal), on the volar side of the forearm of the test persons. The chambers were fixed with a 5-cm wide adhesive tape (Micropore, 3M, St. Paul, Minn., USA). The chambers were left in situ for 24 h. If no erythema was present, the test procedure was repeated on exactly the same site, easily identifiable by the impression of the rim of the chamber. This procedure was carried on for a maximum of 12 days if no erythema was observed at the test site. If an unequivocal erythema was present, the test was discontinued.

*Table 1.* Median irritative time and skewness (with standard errors) of CF dilutions

|  | Concentration | | | | |
|---|---|---|---|---|---|
|  | 8% | 4% | 2% | 1% | 0.5% |
| $IT_{50}$, days, $\tau$ | 4.02 (0.04) | 7.04 (0.26) | 7.38 (0.26) | 7.87 (0.15) | 8.69 (0.15) |
| Skewness, $\lambda$ | 1.75 (0.04) | 1.76 (0.15) | 1.92 (0.18) | 1.95 (0.11) | 2.36 (0.12) |

*Table 2.* Median irritative time and skewness (with standard errors) of SLS dilutions

|  | Concentration | | |
|---|---|---|---|
|  | 0.25% | 0.1% | 0.05% |
| $IT_{50}$, days, $\tau$ | 3.63 (0.14) | 5.16 (0.07) | 6.92 (0.13) |
| Skewness, $\lambda$ | 4.11 (0.64) | 4.57 (0.26) | 3.30 (0.22) |

## Statistical Analysis and Results

For purposes of statistical analysis, an appropriate model was sought to describe the relationship among skin irritations and the delay time after application of the solutions. Several well-known dose-response and time-response models were investigated, such as logit, probit, Weibull and Gompertz. The log-logistic model turned out to assure the best maximum likelihood fit to the data:

$$p(t) = \frac{1}{1 + \left(\frac{\tau}{t}\right)^{\lambda}} \tag{1}$$

In this time-response relationship p(t) is the fraction of irritations occurring up to the delay time t after application. The parameter $\tau$ indicates the median irritative time $IT_{50}$ and $\lambda$ is a skewness parameter indicating the slope of the time-response curve at the $IT_{50}$.

Parameter estimation was carried out by the maximum likelihood method, available in the procedure NLIN of the statistical program package SAS. The parameter estimates and their standard errors read as shown in tables 1 and 2.

In order to demonstrate the vividness of the statistical time-response model (eq. 1) used in this paper, we present the data and their curve estimations. Figures 1a–e show the curves for CF dilutions and figures 2a–c the curves of the SLS dilutions.

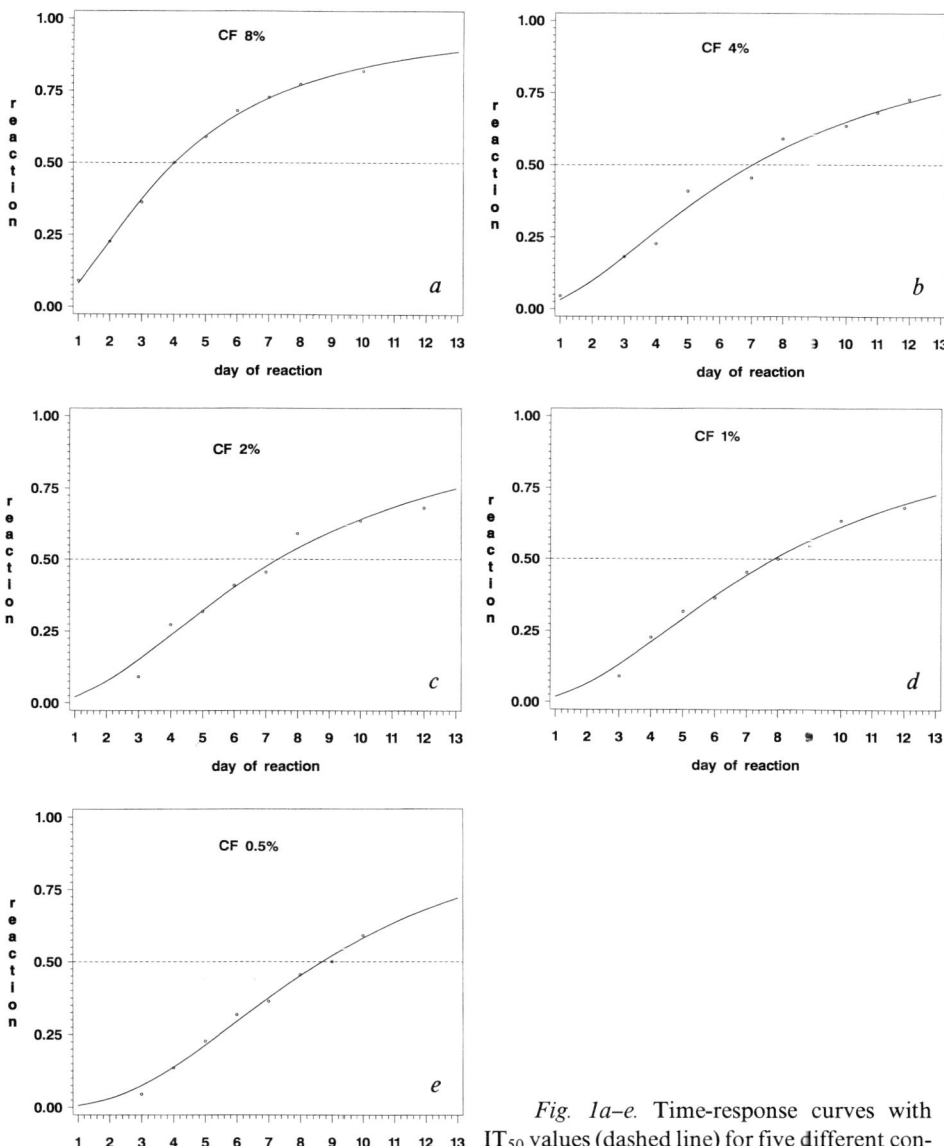

*Fig. 1a–e.* Time-response curves with $IT_{50}$ values (dashed line) for five different concentrations of a cutting fluid (CF.

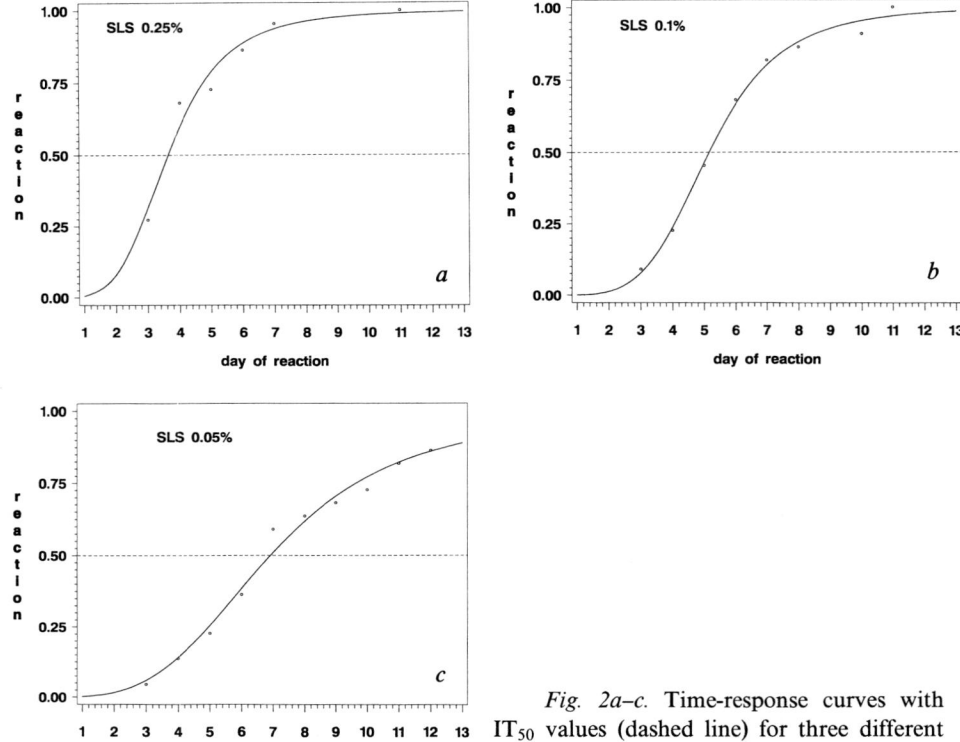

*Fig. 2a–c.* Time-response curves with $IT_{50}$ values (dashed line) for three different concentrations of sodium lauryl sulfate (SLS).

Although with the public availability of microcomputers there is no need to linearize time-response curves, in honour of Kligman and Wooding we will show that their model is equivalent to the model used in this paper.

Take the log odds of the responses, i.e.

$$y = \log\left(\frac{p(t)}{1 - p(t)}\right)$$

and the (natural) logarithms of the delay times, i.e.

$$x = \log(t).$$

Then one obtains the linear relation

$$y = \alpha + \beta x \tag{2}$$

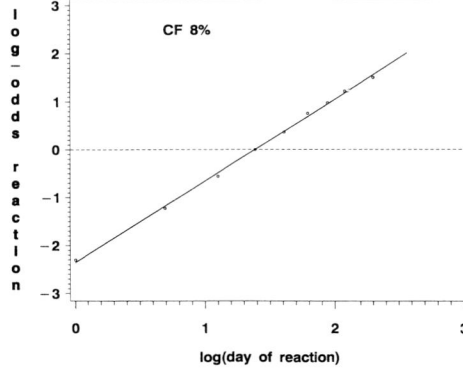

*Fig. 3.* Fit of a straight line after linearization of the data points for the 8% concentration of the cutting fluid (CF).

with intercept $\alpha = -\lambda\log(\tau)$ and slope $\beta = \lambda$. In this model the median irritative time can be computed by

$$\log(\tau) = -\frac{\alpha}{\beta} \text{ or } \tau = e^{\frac{-\alpha}{\beta}}. \tag{3}$$

As a practical demonstration of this linearization we use the 8% CF dilution (fig. 3).

Applying ordinary least-squares regression, the intercept estimate is $\alpha = -2.35$ (0.42) and the slope estimate $\beta = 1.70$ (0.03). The median irritative time estimate therefore is $\tau = 3.98$ days, according to equation (3). This result is very close to the above obtained result $\tau = 4.02$ days.

The advantage of the time-response model (eq. 1), however, is its vividness in presenting the original time-response relationship and the direct estimation of the parameters of interest, inclusive of their standard errors.

## Discussion

To determine the irritating potential of weak irritants or fluids in vivo, we propose to use the slightly modified $IT_{50}$ method of Kligman and Wooding, because it avoids the weakest point of the widely used assessment methods: the purely subjective scoring of the intensity of an irritant reaction. With the $IT_{50}$ procedure, the investigator has to determine only whether there is a reaction – erythema – or not. Little disagreement exists between different observers on this point [9].

In addition, the application of different concentrations of irritants to the human skin in order to assess the range of irritating concentrations subjects the patient to the risk of severe, sometimes even caustic, reactions [12, 13]. This serious hazard is almost completely eliminated with the $IT_{50}$ procedure. Since this test is discarded at an early point, i.e. as early as an erythema ensues, there is very little risk eliciting a severe reaction.

Data for statistical analysis of this test can easily be obtained by daily testing an appropriate number of probands over a certain time period. Usually, 12–15 days are sufficient [9]. The analysis of the dose-response or time-response, S-shaped curves requires special methods, however [14]. Our investigation clearly shows that an excellent curve fitting to the data points can be obtained by log-logistic models. The median irritative time $IT_{50}$ can thus be calculated with high accuracy. Furthermore, the slope of the time-response curve at the $IT_{50}$ yields additional information. For the various concentrations of the cutting fluids tested, the skewness $\lambda$ was between 1.75 and 2.36, whereas the different concentrations of the known irritant SLS showed $\lambda$ values between 3.30 and 4.11. The irritating potential of a substance might thus be characterized by just two values: $IT_{50}$ and skewness of the slope.

The graphical presentation of the curve fit is very easy to interpret. Thus, there is no need to linearize the data points by transforming them. To demonstrate the value of the model of Kligman and Wooding, however, the data of the 8% concentration of CF were linearized and the $IT_{50}$ calculated, yielding only a negligible difference to the value obtained by log-logistic maximum likelihood fit.

In our opinion, the rather old method of Kligman and Wooding, updated by the use of chambers instead of patches, and a computer-based curve fitting program, is well suited for the in vivo assessment of the irritating potential of substances which might come into contact with the skin of humans and animals. The $IT_{50}$ method might also provide the basis for the important comparison of clinical data with those from in vitro irritancy assays, e.g. from studies of keratinocyte cultures [7, 8], with the aim to replace the studies in volunteers by experimental methods.

## References

1  Frosch PJ, Kurte A, Pilz B: Biophysical techniques for the evaluation of skin protective creams; in Frosch PJ, Kligman EM (eds): Non-Invasive Methods for the Quantification of Skin Functions: An Update on Methodology and Clinical Applications. Heidelberg, Springer, 1993.
2  Frosch PJ, Kligman AM: The chamber scarification test: A new method for assessing the irritancy of soap. Contact Derm 1976;2:314–324.
3  Freeman S, Maibach H: Study of irritant contact dermatitis produced by repeat patch test with sodium lauryl sulfate and assessed by visual methods, transepidermal water loss, and laser doppler velocimetry. J Am Acad Dermatol 1988;19:496–502.
4  Van Neste D, De Brouwer B: Monitoring of skin response to sodium lauryl sulphate: Clinical scores versus bioengineering methods. Contact Derm 1992;27:151–156.

5    Ollmar S, Nyrén M, Nicander I, Emtestam L: Electrical impedance compared with other non-invasive bioengineering techniques and visual scoring for detection of irritation in human skin. Br J Dermatol 1994;130:29–36.

6    Kawai K, Nakagawa M, Kawai J, Kawai K: Evaluation of skin irritancy of sodium lauryl sulphate: a comparative study between the replica method and visual evaluation. Contact Derm 1992;27:174–181.

7    Wilke B, Hoth I, Bandemir B: Die Erfassung der irritativen Potenz von Umweltnoxen in einer Epidermiszellsuspension in vitro. Dermatosen 1988;36:147–152.

8    Wilhelm K-P, Samblebe MS, Siegers C-P: Quantitative in vitro assessment of N-alkyl sulphate induced cytotoxicity in human keratinocytes (HaCaT). Comparison with in vivo human irritation tests. Br J Dermatol 1994;130:18–23.

9    Kligman AM, Wooding WM: A method for the measurement and evaluation of irritants on human skin. J Invest Dermatol 1967;49:78–94.

10   Finney DJ: Probit Analysis, ed 2. London, Cambridge University Press, 1962.

11   Litchfield JT: A method for rapid graphic solution of time-per cent effect curves. J Pharmacol Exp Ther 1949;97:399–408.

12   Frosch PJ: Hautirritation und empfindliche Haut. Grosse Scripta 7. Berlin, Grosse, 1985.

13   Willis CM, Stephens CJM, Wilkinson JD: Experimentally-induced irritant contact dermatitis. Determination of optimum irritant concentrations. Contact Derm 1988;18:20–24.

14   Armitage P, Berry G: Statistical Methods in Medical Research, ed 2. London, Blackwell, 1987.

Prof. Dr. med. Friedrich A. Bahmer, Direktor der Hautklinik, Zentralkrankenhaus,
St. Jürgen-Strasse, D–28205 Bremen (Germany)

# Subject Index

Mometasone furoate, efficacy (cont.)
    atopic dermatitis 213
        pediatric dermatoses 213, 215
    inhibition of cytokine synthesis 208,
        209, 221
    receptor binding affinity 209
    safety profile 208, 215, 221
        contact sensitization 219–222
        hypothalamic-pituitary-adrenal axis
            effects 217–219
        skin thinning 216, 217, 221, 222
    structure 208

Nonanoic acid, evaluation of irritation by
    echography 173–175

Occlusion, *see also* Gloves
    dressing types 181
    effects
        barrier repair 181–184
        transepidermal water loss 185, 194
    side effects 185, 186
Occupational skin disease
    automobile workers, *see* Automobile
        workers
    Finland
        data collection agencies 28
        effect of patient sex 30
        frequency 29–31
        irritants 31–33
        occupational distribution 36, 37
    hairdressers, *see* Hairdressing
        apprentices
    health care workers, *see* Health care
        workers
    incidence rate 20, 21
    metalworkers, *see* Metalworking fluid
        dermatitis
Octanol/water partition coefficient
    correlation to cytotoxicity of
        compounds 226–229
    determination 225
    effect on skin penetration 229
OKM5
    expression in contact dermatitis 111,
        112
    immunohistochemical staining 109

Organic solvents, irritant in occupational
    skin disease 31
Ornithine decarboxylase, role in
    inflammatory response 4
Osmium tetroxide, skin fixative 96, 97,
    100

Papular and follicular contact dermatitis
    histology 12
    lesion characteristics 12
    outbreak in Switzerland 9, 10
    patch testing 12, 13
    use testing 15
    vitamin E linoleate as cause 15, 16
Patch test
    cleaning products 42, 43, 45, 46
    cosmetic products 12, 13
    hygiene products 42, 43, 45, 46
    phases of irritant response 139, 140
Phototoxicity assay
    3T3 cell neutral red uptake cytotoxicity
        assay, *see* Fibroblast
    animal testing 256
    *Candida albicans* 259
    histidine oxidation test 259
    red blood cell assay 259, 263
    Skin$^2$ PI assay 259, 263
    SOLATEX PI assay 259, 263

Quantitative structure-activity relationship
    fibroblast cytotoxicity
        assessment 224
        hydrophobicity correlation 226–229
        phenylalcohols 225, 226, 229
        therapeutic compounds 226, 227, 229
    potency ranking of compounds 224

Retinoic acid, mechanism of skin damage,
    skin irritation 133–135
Rubber chemicals, irritants in allergic
    contact dermatitis 33, 36
Ruthenium tetroxide, skin fixative 96

Sebumetry, evaluation of barrier creams
    188
Skin accommodation
    blunting of inflammation 105

Tumor necrosis factor-α
  induction by ultraviolet B 198
  role in inflammatory response 3, 5, 198

Ultrasonography, *see* Echography
Ultraviolet A, *see also* Phototoxicity assay
  filters 257
  light sources 257
Ultraviolet B
  effect on keratinocyte cytokine expression
    235, 240
  induction of tumor necrosis factor 198

Validation, *see* Toxicity testing
Vegetables, allergic contact dermatitis 46
Vitamin E linoleate
  contact dermatitis from metabolites
    course of disease 15
    histology 12
    lesion characteristics 12
    patch testing 12, 13
    use testing 15
  oxidation products 16

Wet work, risk factor in occupational skin
  disease 31, 48, 49, 54

Yucatan hairless micropig
  histamine prick test 276, 279
  nicotinic acid ester erythema 276, 279,
    281
  patch testing 276
    cutting oils 277, 285
    sodium hydroxide 281
    sodium lauryl sulfate 282–284
    toluene 281
  sites for skin measurement 276, 278, 279
  skin similarity to humans 275, 285
  variability in skin tests 277, 278